Porcine Models of Neurotrauma and Neurological Disorders

Porcine Models of Neurotrauma and Neurological Disorders

Editors

John O'Donnell
Dmitriy Petrov

Basel • Beijing • Wuhan • Barcelona • Belgrade • Novi Sad • Cluj • Manchester

Editors
John O'Donnell
University of Pennsylvania
Philadelphia, PA
USA

Dmitriy Petrov
University of Pennsylvania
Philadelphia, PA
USA

Editorial Office
MDPI
St. Alban-Anlage 66
4052 Basel, Switzerland

This is a reprint of articles from the Special Issue published online in the open access journal *Biomedicines* (ISSN 2227-9059) (available at: https://www.mdpi.com/journal/biomedicines/special_issues/Neurotrauma_Neurological_Disorders).

For citation purposes, cite each article independently as indicated on the article page online and as indicated below:

Lastname, A.A.; Lastname, B.B. Article Title. *Journal Name* **Year**, *Volume Number*, Page Range.

ISBN 978-3-7258-0505-1 (Hbk)
ISBN 978-3-7258-0506-8 (PDF)
doi.org/10.3390/books978-3-7258-0506-8

© 2024 by the authors. Articles in this book are Open Access and distributed under the Creative Commons Attribution (CC BY) license. The book as a whole is distributed by MDPI under the terms and conditions of the Creative Commons Attribution-NonCommercial-NoDerivs (CC BY-NC-ND) license.

Contents

About the Editors vii

Preface ix

John C. O'Donnell and Dmitriy Petrov
Porcine Models of Neurotrauma and Neurological Disorders
Reprinted from: Biomedicines 2024, 12, 245, doi:10.3390/biomedicines12010245 1

Michael R. Grovola, Alan Jinich, Nicholas Paleologos, Edgardo J. Arroyo, Kevin D. Browne, Randel L. Swanson, et al.
Persistence of Hyper-Ramified Microglia in Porcine Cortical Gray Matter after Mild Traumatic Brain Injury
Reprinted from: Biomedicines 2023, 11, 1960, doi:10.3390/biomedicines11071960 7

Anna Oeur, Mackenzie Mull, Giancarlo Riccobono, Kristy B. Arbogast, Kenneth J. Ciuffreda, Nabin Joshi, et al.
Pupillary Light Response Deficits in 4-Week-Old Piglets and Adolescent Children after Low-Velocity Head Rotations and Sports-Related Concussions
Reprinted from: Biomedicines 2023, 11, 587, doi:10.3390/biomedicines11020587 25

Mackenzie Mull, Oluwagbemisola Aderibigbe, Marzieh Hajiaghamemar, R. Anna Oeur and Susan S Margulies
Multiple Head Rotations Result in Persistent Gait Alterations in Piglets
Reprinted from: Biomedicines 2022, 10, 2976, doi:10.3390/biomedicines10112976 41

Anna Oeur, William H. Torp, Kristy B. Arbogast, Christina L. Master and Susan S. Margulies
Altered Auditory and Visual Evoked Potentials following Single and Repeated Low-Velocity Head Rotations in 4-Week-Old Swine
Reprinted from: Biomedicines 2023, 11, 1816, doi:10.3390/biomedicines11071816 59

John C. O'Donnell, Kevin D. Browne, Svetlana Kvint, Leah Makaron, Michael R. Grovola, Saarang Karandikar, et al.
Multimodal Neuromonitoring and Neurocritical Care in Swine to Enhance Translational Relevance in Brain Trauma Research
Reprinted from: Biomedicines 2023, 11, 1336, doi:10.3390/biomedicines11051336 77

Samuel S. Shin, Vanessa M. Mazandi, Andrea L. C. Schneider, Sarah Morton, Jonathan P. Starr, M. Katie Weeks, et al.
Exploring the Therapeutic Potential of Phosphorylated *Cis*-Tau Antibody in a Pig Model of Traumatic Brain Injury
Reprinted from: Biomedicines 2023, 11, 1807, doi:10.3390/biomedicines11071807 108

Mark Pavlichenko and Audrey D. Lafrenaye
The Central Fluid Percussion Brain Injury in a Gyrencephalic Pig Brain: Scalable Diffuse Injury and Tissue Viability for Glial Cell Immunolabeling following Long-Term Refrigerated Storage
Reprinted from: Biomedicines 2023, 11, 1682, doi:10.3390/biomedicines11061682 118

Chase A. Knibbe, Rakib Uddin Ahmed, Felicia Wilkins, Mayur Sharma, Jay Ethridge, Monique Morgan, et al.
SmartPill™ Administration to Assess Gastrointestinal Function after Spinal Cord Injury in a Porcine Model—A Preliminary Study
Reprinted from: Biomedicines 2023, 11, 1660, doi:10.3390/biomedicines11061660 135

Will Ao, Megan Grace, Candace L. Floyd and Cole Vonder Haar
A Touchscreen Device for Behavioral Testing in Pigs
Reprinted from: *Biomedicines* **2022**, *10*, 2612, doi:10.3390/biomedicines10102612 . . . . **148**

Alesa H. Netzley and Galit Pelled
The Pig as a Translational Animal Model for Biobehavioral and Neurotrauma Research
Reprinted from: *Biomedicines* **2023**, *11*, 2165, doi:10.3390/biomedicines11082165 . . . . **164**

Erin M. Purvis, Natalia Fedorczak, Annette Prah, Daniel Han and John C. O'Donnell
Porcine Astrocytes and Their Relevance for Translational Neurotrauma Research
Reprinted from: *Biomedicines* **2023**, *11*, 2388, doi:10.3390/biomedicines11092388 . . . . **179**

Connor A. Wathen, Yohannes G. Ghenbot, Ali K. Ozturk, D. Kacy Cullen, John C. O'Donnell and Dmitriy Petrov
Porcine Models of Spinal Cord Injury
Reprinted from: *Biomedicines* **2023**, *11*, 2202, doi:10.3390/biomedicines11082202 . . . . **209**

About the Editors

John O'Donnell

Dr. O'Donnell is a Research Assistant Professor of Neurosurgery at the University of Pennsylvania and the Director of Operations at the Center for Neurotrauma, Neurodegeneration, and Restoration at the Corporal Michael J. Crescenz VA Medical Center (CMC-VAMC). He has collaborated on several neural tissue engineering projects to facilitate regenerative rehabilitation in the brain and has led development of the tissue-engineered rostral migratory stream. However, his lab is primarily focused on the preclinical study of moderate-to-severe traumatic brain injury (TBI) with coma. The loss of consciousness from TBI requires rotational acceleration, acting upon a sufficient enough brain mass to generate injurious forces, affecting the reticular activating system. The swine model of rotational acceleration at Penn satisfies these needs, but coma had historically only been studied up to 8 hours post-injury, due to critical care challenges. To overcome these challenges and carve out this research niche, Dr. O'Donnell worked closely with neurosurgeon Dmitriy Petrov, MD, to develop a neurointensive care unit for pigs that employs clinical equipment and critical care techniques. The integration of neuromonitoring and critical care into a model that uniquely recreates the forces of human TBI to create true moderate and severe TBI with coma has further increased its translational relevance. This model allows Drs. O'Donnell and Petrov to engage in preclinical study of the unique neurocritical care environment, while also extending the study period for moderate-to-severe TBI with coma and offering a path to preclinical study of traumatic disorders of consciousness.

Dmitriy Petrov

Dr. Petrov is an Assistant Professor of Neurosurgery at the Perelman School of Medicine, University of Pennsylvania. His clinical enterprise focuses on the treatment of complex neurotrauma. In addition to his clinical research in neurocritical care and related fields, his preclinical research program with Dr. O'Donnell focuses on the optimization and utilization of a translational large animal model of central and peripheral nervous system trauma. Through a multi-disciplinary approach, in collaboration with experts in neurocritical care and bioengineering, his lab emphasizes the development of novel and innovative therapeutic options for complex neurotrauma pathologies.

Preface

Preclinical treatments found to be effective in rodent models of neurotrauma and neurological disorders are prone to translational failure. While the heterogeneity of these injuries and disorders is greater in the wider human population, this is true to some degree for all models of human maladies, and thus the uniquely high translational failure rate for neurotherapeutics cannot be attributed to clinical heterogeneity alone. Models reduce heterogeneity by design to allow for the controlled testing of select variables. As such, when we seek to improve our modeling of a condition, it would be counterproductive to increase the heterogeneity (within-group variability) in our models. Therefore, to achieve translational success, we must focus on improving the fidelity with which we replicate the mechanisms and manifestations of the human condition being studied. Our rodent models are vital for making foundational progress in our basic understanding of neurotrauma and neurological disorders, with their low cost and ease-of-use allowing for widespread utilization, high sample sizes, and frequent replication for studies identifying candidate therapeutics and targets. However, pathological mechanisms and manifestations in these vital rodent models are often quite different from the human maladies they are modeling. In many cases, greater fidelity requires utilizing a more costly large animal model to replicate clinical mechanisms and manifestations, and pigs have emerged as ideal models for replicating human neurotrauma and neurological disorders. While more costly than rodents, pig studies are significantly cheaper than clinical trials, and therefore provide a relatively low-cost opportunity to test candidate therapeutics in the context of clinically relevant mechanisms and manifestations. By using and promoting the use of these pig models (e.g., for rotational acceleration brain injury), we can improve our ability to fail early when developing neurotherapeutics, which is an essential element of any functioning translational pipeline. In this Special Issue, we provide examples of such pig models in use from leaders in the field, as well as scholarly reviews and reports of new tools for use in pigs to study neurotrauma and neurological disorders.

John O'Donnell and Dmitriy Petrov
Editors

Editorial

Porcine Models of Neurotrauma and Neurological Disorders

John C. O'Donnell [1,2,*] and Dmitriy Petrov [2,*]

[1] Center for Neurotrauma, Neurodegeneration & Restoration, Corporal Michael J. Crescenz Veterans Affairs Medical Center, Philadelphia, PA 19104, USA
[2] Center for Brain Injury & Repair, Department of Neurosurgery, Perelman School of Medicine, University of Pennsylvania, Philadelphia, PA 19104, USA
* Correspondence: odj@pennmedicine.upenn.edu (J.C.O.); dmitriy.petrov@pennmedicine.upenn.edu (D.P.)

The translation of therapeutics from lab to clinic has a dismal record in the fields of neurotrauma and neurological disorders. This is due in part to the challenging heterogeneity of the clinical population common to all translational research, but also due to the unique challenges of recreating the mechanisms and manifestations of human neurological injury/disorders in small animals. Large animal models are an essential component of successful pipelines for moving discoveries from bench to bedside in other fields when exploring device or therapeutic scale-up and/or investigational new drug/investigational device exemption (IND/IDE)-enabling studies, and neuroscience has made significant progress toward establishing such pipelines in its many unique subfields. Due to their size, neuroanatomy, and other factors, pigs have proven to be ideal for providing high-fidelity, clinically relevant studies to bridge the gap between small animals and humans.

This Special Issue collects a dozen papers from leaders in the fields of neurotrauma and neurological disorders detailing clinically relevant studies and sophisticated swine model systems that demonstrate their potential to empower translational research. There are five primary research papers utilizing the rotational acceleration injury model of traumatic brain injury (TBI), which requires a brain of sufficient mass (like the pig) to generate injurious forces. Dr. Grovola and colleagues in the Cullen laboratory reported hyper-ramified microglia in the gray matter after mild TBI, with fibrinogen indicative of blood–brain barrier disruption also predominantly in the gray matter, adding to the body of evidence that rotational acceleration pathology is not limited to white matter [1]. The Margulies group contributed three primary research articles utilizing rotational acceleration in a pediatric pig model. Oeur et al. applied single and repeated low-velocity head rotations in piglets and also gathered data from adolescent humans presenting with concussion, and found that deficits in the pupillary light reflex were altered after injury compared to reference ranges, suggesting that pupillometry could be a valuable tool for neurofunctional assessment [2]. Mull et al. reported significant gait alterations in piglets following rotational acceleration that were more severe and longer lasting when multiple rotations were applied, and they also validated reference ranges for assessing gait alterations in piglets [3]. Dr. Oeur and colleagues also investigated auditory and visually evoked potentials following rotational acceleration in piglets, and found that single and repeated injury groups exhibited different alterations to their evoked potential responses [4]. Finally, we reported the development of neurocritical care techniques and a neurointensive care unit stocked with clinical equipment for use in pigs, which allowed us to extend the study period for TBI with coma in pigs beyond 8 h for the first time while gathering extensive, clinically relevant neuromonitoring data [5].

Beyond the papers featuring rotational acceleration injury, Dr. Shin and colleagues in the Kilbaugh laboratory reported on the efficacy of a promising biological approach targeting phosphorylated tau and reducing pathology after a controlled cortical impact (CCI) in pigs [6]. Additionally, Pavlichenko and Lafrenaye employed central fluid percussion injury (CFPI) in pigs and found that while cellular signs of pathology and inflammation correlated

Citation: O'Donnell, J.C.; Petrov, D. Porcine Models of Neurotrauma and Neurological Disorders. *Biomedicines* **2024**, *12*, 245. https://doi.org/10.3390/biomedicines12010245

Received: 28 November 2023
Accepted: 9 January 2024
Published: 22 January 2024

Copyright: © 2024 by the authors. Licensee MDPI, Basel, Switzerland. This article is an open access article distributed under the terms and conditions of the Creative Commons Attribution (CC BY) license (https://creativecommons.org/licenses/by/4.0/).

with the pressure pulse, blood gasses and mean arterial pressure did not; furthermore, they reported that preserved, refrigerated pig brain tissue was histologically viable after 10 years [7]. We also received papers detailing new tools for use in pig neurotrauma studies. Knibbe and colleagues from the Boakye laboratory detailed the use of the SmartPill™ device for data collection from the gastrointestinal tract following spinal cord injury [8]. Ao and colleagues from the Vonder Haar laboratory also described a new reinforced touchscreen system for administering behavioral assessments in pigs [9]. This behavioral touchscreen work ties into a review in this collection from Alesa Hughson Netzley that provides a wider overview of behavioral testing techniques for use in pig models of neurotrauma, following up on their previous behavioral work in Yucatan minipigs [10,11]. We also provided a review, led by Erin Purvis, of astrocytic studies in pigs, including tables of antibodies that were utilized in various studies, as we are sympathetic to the difficulties associated with finding antibodies and reagents compatible with pigs [12]. Finally, Dr. Wathen and colleagues provided an extensive review of spinal cord injury studies in pig models and the significant translational advantages that pigs present [13].

Recognizing the advantages (e.g., translational, logistical) as well as the limitations of the models that we utilize is vital for accurate reporting and establishing implications of clinical relevance. Such honest assessment of preclinical models is necessary for a functional translational pipeline, and worthy of a brief discussion in this editorial. As the majority of the papers in this collection came from TBI research, we will examine preclinical TBI modeling as an example.

Recognizing the Strengths and Limitations of Preclinical TBI Models

Over the past several decades, the neurotrauma field has wrestled with a lamentable translational failure rate. In particular, although hundreds of treatments have shown efficacy in rodent models of TBI, none have shown effectiveness in humans, despite over 30 clinical trials strongly supported by preclinical rodent data [14–17]. The failure of these trials has led to several National Institute of Neurological Disorders and Stroke (NINDS) and Department of Defense (DoD) conferences which have concluded that validated biomarkers of underlying pathological mechanisms are critical for (1) the proper classification of this heterogeneous disease and (2) the development of effective therapies that target specific injury mechanisms [18,19]. Although challenges in clinical trial design and the heterogeneity of human TBI contribute to the lack of positive findings in clinical studies, this dismal translational record also highlights the limitations of rodent models that simply cannot replicate many of the core mechanisms and manifestations of human TBI.

The vast majority of clinical TBIs are closed-head diffuse brain injuries caused by a blow or rapid jolt to the head, causing abrupt motion of the brain within the skull. In moderate-to-severe cases, the acute post-injury phase is marked by rapid cell death and axonal disruption, generally occurring as a direct consequence of biomechanical loading that surpasses cellular/axonal thresholds. In closed-head TBI in humans, the severity of the biomechanical loading is a product of brain mass, neuroanatomy, and head rotational acceleration [20–22]. Therefore, animal models that feature a species with a large brain mass, a gyrencephalic brain architecture, and head rotational acceleration as the inertial loading mechanism are uniquely able to apply biomechanical parameters scaled from those known to occur—and be injurious—in human TBI [20–23]. Indeed, pigs possess a large brain mass and gyrencephalic architecture with a 60:40 white/gray matter ratio as found in the human brain; in contrast, rats and mice present a paucity of white matter, resulting in 14:86 and 10:90 white/gray matter ratios, respectively [24–26]. This is particularly important because diffuse axonal injury (DAI)—predominantly a white matter M been described as the "hallmark" pathology of closed-head diffuse TBI. White matter in the brain consists of long viscoelastic axon tracts that can stretch to accommodate slow shearing forces, but intra-axonal components may snap when exposed to the shear deformation forces generated by rapid acceleration [27,28]. Moreover, the most commonly employed rodent models of TBI rely on either a focal impact to the brain surface (controlled cortical impact—CCI) or a sharp increase in intracranial pressure surrounding the brain (fluid percussion injury—FPI);

however, these loading mechanisms do not reproduce the mechanical forces of inertial TBI in humans, nor do the injuries produce the same clinical manifestations observed after human TBI. For example, when compared to the clinical TBI severity criteria from which they derive their classification terminology, rodent models of "mild" TBI typically result in cortical lesions that do not align with the criteria of a closed skull and an absence of lesions in imaging necessary for clinical classification as mild TBI, and even the most "severe" TBI injuries in rodents do not produce loss of consciousness sufficient even for clinical classification as moderate TBI, let alone severe. The misapplication of these terms to rodent models in the preclinical TBI literature is a major source of confusion, and confusion negatively impacts progress in any field of research. While moderate or severe human TBI may involve pathology from a combination of impact-loaded focal injury and rotational acceleration-loaded diffuse injury, the most obvious manifestation of a moderate-to-severe TBI—protracted loss-of-consciousness >60 min—requires damage from rotational acceleration in closed-head TBI, and cannot be produced via linear acceleration or impact alone [29–32].

Injury via head acceleration has been attempted in rats, but while bleeding was observed, the conclusions were revealed to be flawed in an expert commentary noting that no consideration was given to scaling up acceleration to account for the much smaller brain mass of rats [33,34]. These biomechanical scaling parameters have been extensively studied, and to generate the injurious forces of human TBI in a rat, head acceleration would need to be increased by an unfeasible 8000% relative to human head acceleration [22,34,35]. Despite the commentary associated with the original article that clearly established the fatal flaws of the study and the physical impossibility of administering injury via acceleration in the rodent brain, the original paper was not retracted, and as a result, head acceleration was subsequently revisited in rodent models 14 years later via the deceptively-named CHIMERA (Closed-Head Impact Model of Engineered Rotational Acceleration) system that allows for head movement after impact. Despite the continued spread of misinformation via the CHIMERA model, the fact remains that rodent brain mass is far too small to reach scaled thresholds for generating injurious forces from rotational acceleration, and the pathology—while presenting with a diffuse gradient—emanates from the impact site, as seen in other impact loading models [36,37]. The injurious forces generated by acceleration are highly dependent on brain mass, and so to replicate forces present in a larger human brain during TBI, the acceleration of smaller brains must be scaled up accordingly. Impact loading does not scale with brain mass, so administering an impact powerful enough to scale up the resultant head acceleration would lead to a large discrepancy in the proportion of impact-to-acceleration loading when compared to human TBI. Therefore, even in large gyrencephalic animals like swine with brains closer in size to humans, an unscaled impact and scaled acceleration must be administered separately to replicate the mechanisms of human TBI. Overall, the fundamentally flawed CHIMERA model has created a great deal of confusion in our field, and the scientific record is in dire need of correction. Fortunately, after removing all inaccurate references to and claims of acceleration-loaded injury in small animals, researchers may be left with a valid—if somewhat overcomplicated—CCI model, and therefore many authors duped into utilizing CHIMERA may be able to salvage their publications via errata rather than retraction.

While small animal models cannot reproduce the same biomechanical perturbations as human TBI, they clearly produce a mechanical injury to the rodent brain, and while the clinical manifestations of TBI do not line up with the manifestations of the small animal model injuries, many of the individual endophenotypes of human TBI can be reproduced in some fashion. Therefore, small animal models remain an essential tool for studying the secondary injury mechanisms of TBI, identifying potential therapeutic targets and candidates, and studying the mechanism of action of novel therapeutics. These essential basic and early-phase translational studies require the accessibility and affordability of small animal models—enabling high-throughput testing—while the recognition of limitations and measured interpretation are necessary to improve applicability to clinical TBI. However, a more effective translational pipeline for developing TBI therapeutics should feature pre-IND/IDE studies in large gyrencephalic animal models that incorporate inertial loading with angular rotational acceleration—to better recapitulate the biomechanical causation and pathophysiology of the majority of human TBIs—prior to translation into clinical trials. This will better inform clinical trials (e.g., inclusion criteria, key outcomes, etc.) and allow novel therapeutic candidates to fail early when tested in the context of the mechanisms and manifestations of human injury, thereby avoiding the expensive failures in clinical trials that have stymied industry investment in neurotrauma. The rotational acceleration model is currently only in use at the University of Pennsylvania and Georgia Tech./Emory University, creating a considerable bottleneck for the translation of therapeutics. It will take a concerted effort from funding agencies and research universities to establish additional sites that can utilize this model. Establishing additional sites of rotational acceleration TBI research (which requires extensive training with large gyrencephalic animals and a powerful pneumatic actuator) is a daunting but necessary task if we want a viable translational pipeline for TBI. While TBI is a special case, many of these lessons are applicable to the preclinical study of stroke, spinal cord injury, and other facets of neurotrauma and neurological disorders. Only by recognizing the strengths and limitations of our models can we hope to effectively develop, validate, and translate novel neuroprotective and regenerative therapeutics into clinical practice to maximize functional recovery.

Author Contributions: Writing—original draft preparation, J.C.O.; writing—review and editing, J.C.O. and D.P.; funding acquisition, J.C.O. and D.P. All authors have read and agreed to the published version of the manuscript.

Funding: Financial support was provided by the Department of Veterans Affairs (RR&D IK2-RX003376) (O'Donnell) and the Department of Neurosurgery, Perelman School of Medicine, University of Pennsylvania (Petrov). The opinions, interpretations, conclusions, and recommendations are those of the authors and are not necessarily endorsed by the Department of Veterans Affairs or the University of Pennsylvania.

Conflicts of Interest: The authors declare no conflict of interest.

References

1. Grovola, M.R.; Jinich, A.; Paleologos, N.; Arroyo, E.J.; Browne, K.D.; Swanson, R.L.; Duda, J.E.; Cullen, D.K. Persistence of Hyper-Ramified Microglia in Porcine Cortical Gray Matter after Mild Traumatic Brain Injury. *Biomedicines* **2023**, *11*, 1960. [CrossRef] [PubMed]
2. Oeur, A.; Mull, M.; Riccobono, G.; Arbogast, K.B.; Ciuffreda, K.J.; Joshi, N.; Fedonni, D.; Master, C.L.; Margulies, S.S. Pupillary Light Response Deficits in 4-Week-Old Piglets and Adolescent Children after Low-Velocity Head Rotations and Sports-Related Concussions. *Biomedicines* **2023**, *11*, 587. [CrossRef] [PubMed]
3. Mull, M.; Aderibigbe, O.; Hajiaghamemar, M.; Oeur, R.A.; Margulies, S.S. Multiple Head Rotations Result in Persistent Gait Alterations in Piglets. *Biomedicines* **2022**, *10*, 2976. [CrossRef] [PubMed]
4. Oeur, A.; Torp, W.H.; Arbogast, K.B.; Master, C.L.; Margulies, S.S. Altered Auditory and Visual Evoked Potentials following Single and Repeated Low-Velocity Head Rotations in 4-Week-Old Swine. *Biomedicines* **2023**, *11*, 1816. [CrossRef] [PubMed]
5. O'Donnell, J.C.; Browne, K.D.; Kvint, S.; Makaron, L.; Grovola, M.R.; Karandikar, S.; Kilbaugh, T.J.; Cullen, D.K.; Petrov, D. Multimodal Neuromonitoring and Neurocritical Care in Swine to Enhance Translational Relevance in Brain Trauma Research. *Biomedicines* **2023**, *11*, 1336. [CrossRef] [PubMed]
6. Shin, S.S.; Mazandi, V.M.; Schneider, A.L.C.; Morton, S.; Starr, J.P.; Weeks, M.K.; Widmann, N.J.; Jang, D.H.; Kao, S.-H.; Ahlijanian, M.K.; et al. Exploring the Therapeutic Potential of Phosphorylated *Cis*-Tau Antibody in a Pig Model of Traumatic Brain Injury. *Biomedicines* **2023**, *11*, 1807. [CrossRef] [PubMed]
7. Pavlichenko, M.; Lafrenaye, A.D. The Central Fluid Percussion Brain Injury in a Gyrencephalic Pig Brain: Scalable Diffuse Injury and Tissue Viability for Glial Cell Immunolabeling following Long-Term Refrigerated Storage. *Biomedicines* **2023**, *11*, 1682. [CrossRef] [PubMed]
8. Knibbe, C.A.; Ahmed, R.U.; Wilkins, F.; Sharma, M.; Ethridge, J.; Morgan, M.; Gibson, D.; Cooper, K.B.; Howland, D.R.; Vadhanam, M.V.; et al. SmartPill™ Administration to Assess Gastrointestinal Function after Spinal Cord Injury in a Porcine Model-A Preliminary Study. *Biomedicines* **2023**, *11*, 1660. [CrossRef]

9. Ao, W.; Grace, M.; Floyd, C.L.; Haar, C.V. A Touchscreen Device for Behavioral Testing in Pigs. *Biomedicines* **2022**, *10*, 2612. [CrossRef]
10. Netzley, A.H.; Hunt, R.D.; Franco-Arellano, J.; Arnold, N.; Vazquez, A.I.; Munoz, K.A.; Colbath, A.C.; Bush, T.R.; Pelled, G. Multimodal characterization of Yucatan minipig behavior and physiology through maturation. *Sci. Rep.* **2021**, *11*, 22688. [CrossRef]
11. Netzley, A.H.; Pelled, G. The Pig as a Translational Animal Model for Biobehavioral and Neurotrauma Research. *Biomedicines* **2023**, *11*, 2165. [CrossRef] [PubMed]
12. Purvis, E.M.; Fedorczak, N.; Prah, A.; Han, D.; O'Donnell, J.C. Porcine Astrocytes and Their Relevance for Translational Neurotrauma Research. *Biomedicines* **2023**, *11*, 2388. [CrossRef] [PubMed]
13. Wathen, C.A.; Ghenbot, Y.G.; Ozturk, A.K.; Cullen, D.K.; O'Donnell, J.C.; Petrov, D. Porcine Models of Spinal Cord Injury. *Biomedicines* **2023**, *11*, 2202. [CrossRef] [PubMed]
14. Xiong, Y.; Mahmood, A.; Chopp, M. Animal models of traumatic brain injury. *Nat. Rev. Neurosci.* **2013**, *14*, 128–142. [CrossRef] [PubMed]
15. Vink, R. Large animal models of traumatic brain injury. *J. Neurosci. Res.* **2018**, *96*, 527–535. [CrossRef] [PubMed]
16. Kabadi, S.V.; Faden, A.I. Neuroprotective Strategies for Traumatic Brain Injury: Improving Clinical Translation. *Int. J. Mol. Sci.* **2014**, *15*, 1216–1236. [CrossRef]
17. Loane, D.J.; Faden, A.I. Neuroprotection for traumatic brain injury: Translational challenges and emerging therapeutic strategies. *Trends Pharmacol. Sci.* **2010**, *31*, 596–604. [CrossRef]
18. Levin, H.S.; Diaz-Arrastia, R.R. Diagnosis, prognosis, and clinical management of mild traumatic brain injury. *Lancet Neurol.* **2015**, *14*, 506–517. [CrossRef]
19. Bogoslovsky, T.; Gill, J.; Jeromin, A.; Davis, C.; Diaz-Arrastia, R. Fluid Biomarkers of Traumatic Brain Injury and Intended Context of Use. *Diagnostics* **2016**, *6*, 37. [CrossRef]
20. Holbourn, A.H.S. Mechanics of Head Injuries. *Lancet* **1943**, *242*, 438–441. [CrossRef]
21. Ommaya, A.K.; Yarnell, P.; Hirsch, A.E.; Harris, E.H. *Scaling of Experimental Data on Cerebral Concussion in Sub-Human Primates to Concussion Threshold for Man*; SAE International: Warrendale, PA, USA, 1967. [CrossRef]
22. Margulies, S.S.; Thibault, L.E.; Gennarelli, T.A. Physical model simulations of brain injury in the primate. *J. Biomech.* **1990**, *23*, 823–836. [CrossRef] [PubMed]
23. Wofford, K.L.; Grovola, M.R.; Adewole, O.D.; Browne, K.D.; Putt, E.M.; O'Donnell, J.C.; Cullen, D.K. Relationships between injury kinematics, neurological recovery, and pathology following concussion. *Brain Commun.* **2021**, *3*, fcab268. [CrossRef] [PubMed]
24. Zhang, K.; Sejnowski, T.J. A universal scaling law between gray matter and white matter of cerebral cortex. *Proc. Natl. Acad. Sci. USA* **2000**, *97*, 5621–5626. [CrossRef] [PubMed]
25. Bailey, E.L.; McCulloch, J.; Sudlow, C.; Wardlaw, J.M. Potential animal models of lacunar stroke: A systematic review. *Stroke* **2009**, *40*, e451–e458. [CrossRef] [PubMed]
26. Howells, D.W.; Porritt, M.J.; Rewell, S.S.; O'Collins, V.; Sena, E.S.; van der Worp, H.B.; Traystman, R.J.; Macleod, M.R. Different Strokes for Different Folks: The Rich Diversity of Animal Models of Focal Cerebral Ischemia. *J. Cereb. Blood Flow. Metab.* **2010**, *30*, 1412–1431. [CrossRef] [PubMed]
27. Galbraith, J.A.; Thibault, L.E.; Matteson, D.R. Mechanical and electrical responses of the squid giant axon to simple elongation. *J. Biomech. Eng.* **1993**, *115*, 13–22. [CrossRef] [PubMed]
28. Cullen, D.K.; Vernekar, V.N.; LaPlaca, M.C.; Burda, J.E.; Bernstein, A.M.; Sofroniew, M.V.; Haslach, H.W.; Leahy, L.N.; Hsieh, A.H.; Maneshi, M.M.; et al. Trauma-Induced Plasmalemma Disruptions in Three-Dimensional Neural Cultures Are Dependent on Strain Modality and Rate. *J. Neurotrauma* **2011**, *28*, 2219–2233. [CrossRef] [PubMed]
29. Denny-Brown, D.E.; Russell, W.R. Experimental Concussion: (Section of Neurology). *Proc. R. Soc. Med.* **1941**, *34*, 691–692.
30. Ommaya, A.K.; Gennarelli, T.A. Cerebral concussion and traumatic unconsciousness. Correlation of experimental and clinical observations of blunt head injuries. *Brain* **1974**, *97*, 633–654. [CrossRef]
31. Langlois, J.A.; Rutland-Brown, W.; Wald, M.M. The epidemiology and impact of traumatic brain injury: A brief overview. *J. Head. Trauma Rehabil.* **2006**, *21*, 375–378. [CrossRef]
32. O'Donnell, J.C.; Browne, K.D.; Kilbaugh, T.J.; Chen, H.I.; Whyte, J.; Cullen, D.K. Challenges and demand for modeling disorders of consciousness following traumatic brain injury. *Neurosci. Biobehav. Rev.* **2019**, *98*, 336–346. [CrossRef] [PubMed]
33. Xiao-Sheng, H.; Sheng-Yu, Y.; Xiang, Z.; Zhou, F.; Jian-Ning, Z.; Li-Sun, Y. Diffuse axonal injury due to lateral head rotation in a rat model. *J. Neurosurg.* **2000**, *93*, 626–633. [CrossRef] [PubMed]
34. Meaney, D.F.; Margulies, S.S.; Smith, D.H. Diffuse axonal injury. *J. Neurosurg.* **2001**, *95*, 1108–1110. [PubMed]
35. Meaney, D.F.; Smith, D.H.; Shreiber, D.I.; Bain, A.C.; Miller, R.T.; Ross, D.T.; Gennarelli, T.A.; Johnson, V.E.; Stewart, W.; Weber, M.T.; et al. Biomechanical Analysis of Experimental Diffuse Axonal Injury. *J. Neurotrauma* **1995**, *12*, 689–694. [CrossRef]

36. Namjoshi, D.R.; Cheng, W.H.; McInnes, K.A.; Martens, K.M.; Carr, M.; Wilkinson, A.; Fan, J.; Robert, J.; Hayat, A.; Cripton, P.A.; et al. Merging pathology with biomechanics using CHIMERA (Closed-Head Impact Model of Engineered Rotational Acceleration): A novel, surgery-free model of traumatic brain injury. *Mol. Neurodegener.* **2014**, *9*, 55. [CrossRef]
37. Sauerbeck, A.D.; Fanizzi, C.; Kim, J.H.; Gangolli, M.; Bayly, P.V.; Wellington, C.L.; Brody, D.L.; Kummer, T.T. modCHIMERA: A novel murine closed-head model of moderate traumatic brain injury. *Sci. Rep.* **2018**, *8*, 7677. [CrossRef]

Disclaimer/Publisher's Note: The statements, opinions and data contained in all publications are solely those of the individual author(s) and contributor(s) and not of MDPI and/or the editor(s). MDPI and/or the editor(s) disclaim responsibility for any injury to people or property resulting from any ideas, methods, instructions or products referred to in the content.

Article

Persistence of Hyper-Ramified Microglia in Porcine Cortical Gray Matter after Mild Traumatic Brain Injury

Michael R. Grovola [1,2], Alan Jinich [1,2], Nicholas Paleologos [1,2], Edgardo J. Arroyo [1,2,3], Kevin D. Browne [1,2], Randel L. Swanson [1,2,3], John E. Duda [1,4,5,*] and D. Kacy Cullen [1,2,6,*]

1 Center for Neurotrauma, Neurodegeneration & Restoration, Corporal Michael J. Crescenz VA Medical Center, Philadelphia, PA 19104, USA; mgrovola@pennmedicine.upenn.edu (M.R.G.); ajinich@sas.upenn.edu (A.J.); npale@sas.upenn.edu (N.P.); arroyoe@pennmedicine.upenn.edu (E.J.A.); kbrowne@pennmedicine.upenn.edu (K.D.B.); randel.swanson@pennmedicine.upenn.edu (R.L.S.)
2 Center for Brain Injury & Repair, University of Pennsylvania, Philadelphia, PA 19104, USA
3 Department of Physical Medicine and Rehabilitation, Perelman School of Medicine, University of Pennsylvania, Philadelphia, PA 19104, USA
4 Department of Neurology, Perelman School of Medicine, University of Pennsylvania, Philadelphia, PA 19104, USA
5 Parkinson's Disease Research, Education and Clinical Center, Corporal Michael J. Crescenz VA Medical Center, Philadelphia, PA 19104, USA
6 Department of Bioengineering, School of Engineering and Applied Science, University of Pennsylvania, Philadelphia, PA 19104, USA
* Correspondence: john.duda@va.gov (J.E.D.); dkacy@pennmedicine.upenn.edu (D.K.C.)

Citation: Grovola, M.R.; Jinich, A.; Paleologos, N.; Arroyo, E.J.; Browne, K.D.; Swanson, R.L.; Duda, J.E.; Cullen, D.K. Persistence of Hyper-Ramified Microglia in Porcine Cortical Gray Matter after Mild Traumatic Brain Injury. *Biomedicines* **2023**, *11*, 1960. https://doi.org/10.3390/biomedicines11071960

Academic Editor: Bruno Meloni

Received: 26 May 2023
Revised: 20 June 2023
Accepted: 1 July 2023
Published: 12 July 2023

Copyright: © 2023 by the authors. Licensee MDPI, Basel, Switzerland. This article is an open access article distributed under the terms and conditions of the Creative Commons Attribution (CC BY) license (https://creativecommons.org/licenses/by/4.0/).

Abstract: Traumatic brain injury (TBI) is a major contributor to morbidity and mortality in the United States as several million people visit the emergency department every year due to TBI exposures. Unfortunately, there is still no consensus on the pathology underlying mild TBI, the most common severity sub-type of TBI. Previous preclinical and post-mortem human studies have detailed the presence of diffuse axonal injury following TBI, suggesting that white matter pathology is the predominant pathology of diffuse brain injury. However, the inertial loading produced by TBI results in strain fields in both gray and white matter. In order to further characterize gray matter pathology in mild TBI, our lab used a pig model (n = 25) of closed-head rotational acceleration-induced TBI to evaluate blood-brain barrier disruptions, neurodegeneration, astrogliosis, and microglial reactivity in the cerebral cortex out to 1 year post-injury. Immunohistochemical staining revealed the presence of a hyper-ramified microglial phenotype—more branches, junctions, endpoints, and longer summed process length—at 30 days post injury (DPI) out to 1 year post injury in the cingulate gyrus ($p < 0.05$), and at acute and subacute timepoints in the inferior temporal gyrus ($p < 0.05$). Interestingly, we did not find neuronal loss or astroglial reactivity paired with these chronic microglia changes. However, we observed an increase in fibrinogen reactivity—a measure of blood-brain barrier disruption—predominantly in the gray matter at 3 DPI ($p = 0.0003$) which resolved to sham levels by 7 DPI out to chronic timepoints. Future studies should employ gene expression assays, neuroimaging, and behavioral assays to elucidate the effects of these hyper-ramified microglia, particularly related to neuroplasticity and responses to potential subsequent insults. Further understanding of the brain's inflammatory activity after mild TBI will hopefully provide understanding of pathophysiology that translates to clinical treatment for TBI.

Keywords: mild TBI; neuroinflammation; microglia; large animal models; fibrinogen

1. Introduction

Traumatic brain injury (TBI) is a major contributor to morbidity and mortality in the United States. According to national TBI incidence reports, several million people visit the emergency department each year with unintentional falls, being unintentionally stuck by or against an object, and motor vehicle crashes as the most common mechanisms of injury [1]. Mild TBI, also known as concussion, is much more common than moderate or severe TBI, with the incidence of TBI highest in males among the adult population [2]. Unfortunately, there is still no consensus on the pathology underlying mild TBI as mild TBI is partly defined by an absence of structural changes on clinical neuroimaging, such as hematoma, contusion, and brain swelling [3].

In order to further characterize mild TBI pathology, our lab uses a pig model of closed-head rotational acceleration-induced TBI. This large animal model emulates human TBI by providing more representative biomechanical loading conditions, brain anatomy, and neurophysiology compared to rodent models (see [4] for a review). Previous studies in this porcine model of TBI, as well as post-mortem human TBI studies, have detailed the presence of diffuse axonal injury, suggesting that white matter pathology is the predominant pathology of diffuse brain injury [5–11]. Yet recent neuroimaging studies suggest that gray matter atrophy occurs at acute and chronic timepoints after mild TBI in humans [12–14].

Indeed, the inertial loading produced by TBI results in strain fields in both gray and white matter. One of the consequences of this loading is transient disruption of the neuronal plasma membrane [15,16]. Permeabilized membranes can trigger disruption of normal cell function in a positive feedback manner based on loss of membrane charge, interruption of electrokinetic transport, and osmotic imbalance [17]. Studies suggest that many initially permeabilized cells survive the insult; however, there may be prolonged alteration in physiology or later cell death [18–20]. Previous in vitro and rodent studies suggest that neuronal membrane disruption may occur in separate waves following TBI [21,22]. Moreover, our research team has discovered membrane disruptions in cortical neuron somata paired with significant microglial alterations after TBI [23]. This gray matter pathology may play a pivotal role in post-TBI dysfunction, yet cortical pathology has not been thoroughly characterized in closed-head diffuse TBI.

Here, we sought to evaluate trauma-induced changes to microglia, astrocytes, neurons, and the blood-brain barrier (BBB) within the cerebral cortex using an established large-animal model of closed-head diffuse TBI. We assessed neuroinflammatory and neurodegenerative changes in the cerebral cortex up to 1 year post-injury, with a particular focus on the cingulate gyrus and inferior temporal gyrus. We examined disparate areas of the cortex—one medial gyrus and one lateral gyrus—that may be subjected to variable inertial loading and therefore different pathological patterns and distributions. Specifically, we performed detailed quantification of subtle yet discrete microglial morphological features through skeletal analysis in order to characterize the neuroimmune response to TBI in the cortex. To provide context for these detailed microglial analyses, we also assessed astrocyte reactivity by morphological and cell density measures, cortical neuron dystrophy or loss via cell density measures, and BBB disruption through the presence of blood proteins in the brain parenchyma. This work contributes to a growing body of literature suggesting that TBI-induced gray matter pathology may drive additional neuroinflammatory and neurodegenerative sequela. We hypothesized that microglial morphological changes would coincide with astroglial reactivity, neuronal dystrophy/loss, and BBB disruption in the cerebral cortex after TBI. Understanding these often-overlooked gray matter biophysical responses is crucial to the development of injury prevention strategies and therapeutic interventions following TBI.

2. Materials and Methods

2.1. Animal Subjects

All procedures were approved by the Institutional Animal Care and Use Committees at the University of Pennsylvania and the Michael J. Crescenz Veterans Affairs Medical Center and adhered to the guidelines set forth in the NIH Public Health Service Policy on Humane Care and Use of Laboratory Animals (2015).

For the current study, specimens were obtained from a tissue archive of castrated male pigs subjected to a single mild TBI. This tissue archive was also used in Grovola et al. [9,24]. All pigs were 5–6 months old, sexually mature (considered to be young adult), Yucatan miniature pigs at a mean weight of 34 ± 4 kg (total n = 25, mean ± SD). Pigs were fasted for 16 h, then anesthesia was induced with 20 mg/kg of ketamine and 0.5 mg/kg of midazolam. Following induction, 0.1 mg/kg of glycopyrrolate was subcutaneously administered and 50 mg/kg of acetaminophen was administered per rectum. All animals were intubated with an endotracheal tube and anesthesia was maintained with 2% isoflurane per 2 L O_2. Heart rate, respiratory rate, arterial oxygen saturation, and temperature were continuously monitored throughout the experiment. A forced-air temperature management system was used to maintain normothermia throughout the procedure.

2.2. Head Rotational TBI

In order to attain closed-head diffuse mild TBI in animals, we used a previously described model of head-rotational acceleration in pigs [4,6]. Similar methods were described in Grovola et al. [24]. Briefly, each animal's head was secured to a bite plate, which itself was attached to a pneumatic actuator and a custom assembly that converts linear motion into angular momentum. The pneumatic actuator rotated each animal's head in the coronal plane, reaching an angular velocity between 230–270 rad/sec (n = 15). Any presence of apnea was recorded (maximum apnea time = 45 s), and animals were hemodynamically stabilized if necessary. No animals were excluded from the study due to apnea or hemodynamic instability. Sham animals (n = 10) underwent identical protocols, including being secured to the bite plate, however the pneumatic actuator was not initiated. All animals were transported back to their housing facility, monitored acutely for 3 h, and given access to food and water. Afterwards, the animals were monitored daily for 3 days by veterinary staff.

2.3. Specimen Preparation

At 3 days post-injury (DPI) (n = 4), 7 DPI (n = 5), 30 DPI (n = 3), or 1 year post-injury (YPI) (n = 3), animals were induced and intubated as described above. Sham animals survived for 7 days (n = 4), 30 days (n = 1), or 1 year (n = 5). While under anesthesia, animals were transcardially perfused with 0.9% heparinized saline followed by 10% neutral buffered formalin (NBF). Animals were then decapitated, and tissue stored overnight in 10% NBF at 4 °C. The following day, the brain was extracted, weighed, and post-fixed in 10% NBF at 4 °C for one week. To block the tissue, an initial coronal slice was made immediately rostral to the optic chiasm. The brain was then blocked into 5 mm thick coronal sections from that point by the same investigator. This allowed for consistent blocking and section coordinates across animals. All blocks of tissue were paraffin embedded and 8 μm thick sections were obtained using a rotary microtome. Two sections from each pig—one containing anterior aspects of hippocampal tissue (approximately 10 mm posterior to the optic chiasm) and one containing posterior aspects of hippocampal tissue (approximately 15 mm posterior to the optic chiasm)—were used for histological analysis. Of note, one of the enrolled specimens in the 3 DPI group was excluded from the current study due to an unresolvable error in tissue processing resulting in low-quality and inconsistent tissue staining.

2.4. Immunohistochemical Staining and Microscopy

For 3,3'-Diaminobenzidine (DAB) immunohistochemical labeling, we used a protocol outlined in Johnson et al. [25]. Briefly, slides were dewaxed in xylene, rehydrated in ethanol and de-ionized water. Antigen retrieval was completed in Tris EDTA buffer pH 8.0 using a microwave pressure cooker then blocked with normal horse serum. Slides were incubated overnight at 4 °C using either an anti-rabbit fibrinogen (abcam, Waltham, MA, USA, 183109, 1:5000), an anti-mouse GFAP (SMI-22) (MilliporeSigma, Burlington, MA, USA, NE1015, 1:12,000), or an anti-rabbit Iba-1 (Wako Chemicals, Richmond, VA, USA, 019-19741, 1:4000) primary antibody. The following day, slides were rinsed in PBS and incubated in a horse anti-mouse/rabbit biotinylated IgG secondary antibody (VECTASTAIN Elite ABC Kit, Vector Laboratories, Newark, CA, USA, PK-6200). Sections were rinsed again, then incubated with an avidin/biotinylated enzyme complex (VECTASTAIN Elite ABC Kit), rinsed again, and incubated with the DAB enzyme substrate (Vector Laboratories, SK-4100) for 7 min. Sections were counterstained with hematoxylin, dehydrated in ethanol, cleared in xylene, and finally coverslipped using cytoseal. All sections were stained in the same histological sample run.

For Cresyl Violet (CV) staining, slides were dewaxed in xylene, and rehydrated in ethanol and deionized water. Slides were placed in a solution of 0.1% cresyl violet acetate (MilliporeSigma, C5042) and deionized water for 5 min, rinsed in deionized water, and then differentiated in an acetic acid and 95% ethanol solution. Slides were dehydrated in ethanol, cleared in xylene, and finally coverslipped using cytoseal. All slides were stained in the same histological sample run.

For fluorescence immunohistochemical staining, slides were dewaxed in xylene, and rehydrated in ethanol and deionized water. Antigen retrieval was completed in Tris EDTA buffer pH 8.0 using a microwave pressure cooker then blocked with normal horse serum. Slides were incubated overnight at 4 °C using anti-mouse Neuronal Nuclei (NeuN) (MilliporeSigma, MAB377, 1:500) primary antibody. The following day, sections were rinsed in PBS and incubated in donkey anti-mouse 555 (Thermo Fisher Scientific, Waltham, MA, A31570, 1:500) secondary antibody for 60 min. Sections were rinsed again, then incubated with Hoechst (Thermo Fisher Scientific, H3570, 1:10,000) to visualize DNA, and finally cover slipped using Fluoromount-G (Southern Biotech, Birmingham, AL, USA, 0100-01). All sections were stained on the same histological sample run.

All GFAP, Iba-1, and CV sections were imaged and analyzed at $20\times$ optical zoom using an Aperio CS2 digital slide scanner (Leica Biosystems Inc., Buffalo Grove, IL, USA). All Fibrinogen sections were imaged at $100\times$ using a Keyence VHX-6000 digital microscope (Itasca, IL, USA). All NeuN sections were imaged at $10\times$ using a Keyence BZ-X7000 digital microscope (Itasca, IL, USA).

2.5. Neuropathological Analysis

For Iba-1 skeletal analysis, we employed methods from Grovola et al. [9]. Briefly, we imaged five $40\times$ images per anatomical region for analysis. To conduct skeletal analysis, all Iba-1 positive cells in each $40\times$ field were manually selected, and the image was deconvoluted using Fiji software (version 2.9.0, National Institute of Health, Bethesda, MD, USA). Bandpass filters, unsharp mask, and close plugins were applied before converting the image to binary, skeletonizing, and removing skeletons not overlaid with the manually selected cells. The Analyze Skeleton plugin was then applied to quantify skeletal features such as number of process branches, junctions, process endpoints, and slab voxels in order to measure changes in microglia ramification [26]. For each image, each feature was summed then divided by the total number of cells, thus providing a single field average normalized per cell. Therefore, we examined 5 values from 5 images in the same histological slide for each animal in each anatomical region, which serves as a repeated measure, regional analysis. Slab voxels were then multiplied by the volume of the voxel to calculate the summed process length per cell.

For CV cell counts, a single $20\times$ photomicrograph was taken in layer 2/3 at the gyri apex of the cingulate gyrus for each CV-stained slide. Total cell density was assessed using Fiji software (National Institute of Health); images were converted to grayscale and the Analyze Particles plugin was used to count cells in an automated fashion using an objective set of exclusion parameters [27]. Total neuron density was determined by manual counting; neurons were distinguished from glia by morphology, staining pattern, and sometimes size using the guidelines set by Garcia-Cabezas et al. [28]. All counted neurons had a visible nucleus, and a small rim of visible cytoplasm around the nucleus to distinguish small neurons from astrocytes. All cell densities were normalized according to tissue area. The cell densities were averaged for each specimen in lieu of traditional stereology.

For NeuN neuron counts, the entire left hemisphere cingulate gyrus was stitched at $10\times$ resolution for each specimen. Each image was then cropped to include either cortical layers II/III or all cortical gray matter layers. We assessed total cell density using Fiji software (National Institute of Health); images were converted to grayscale and the Analyze Particles plugin was used to count cells in an automated fashion using an objective set of exclusion parameters [27]. All cell densities were normalized according to tissue area. Neuron density was averaged for each specimen in lieu of traditional stereology.

For astrocyte semi-quantitative analysis, inferior temporal gyrus and cingulate gyrus were assessed in 2 sections per specimen. We adapted a semi-quantitative scale from Sofroniew et al. to histologically classify the progressive severity of reactive astrocytes [29]. Each region was given a 0–3 glial fibrillary acidic protein (GFAP) reactivity score based on cell body size, upregulation of GFAP, and density of GFAP+ cells in the region.

For fibrinogen analysis, we determined percentage area of fibrinogen coverage via the Cavalieri estimator probe within Stereo Investigator software (version 11.06.01, MBF Bioscience, Williston, VT, USA). This probe created a point grid over the tissue and the sum of the number of points that hit fibrinogen-stained regions of interest were used to estimate the area. This fibrinogen area was then divided by the total brain parenchyma area to provide a percentage area of fibrinogen reactivity. Percentage area was averaged between the 2 sections per specimen to provide a more global assessment of fibrinogen reactivity. Different markers were used to label gray versus white matter to allow for comparative analysis in different types of neuronal tissue. Therefore, we assessed total brain, gray matter, and white matter for percentage area fibrinogen reactivity.

2.6. Statistical Analysis

Statistical analysis was performed using GraphPad Prism statistical software (version 9.5.1, GraphPad Software Inc. La Jolla, CA, USA). Any outliers were removed using the ROUT method (Q = 1%). Due to low sample size, GFAP reactivity, CV cell counts, NeuN cell counts, and total fibrinogen reactivity were analyzed with a Kruskal–Wallis test and Dunn's multiple comparisons. Kruskal–Wallis test results are reported as (H (degrees of freedom) = H test statistic, p-value). A T-test was used to analyze gray versus white matter fibrinogen reactivity and reported as (t (degrees of freedom) = t statistic, p-value). The skeletal analysis was statistically assessed via One-way analysis of variance (ANOVA) and Tukey's multiple comparisons test. One-way ANOVA results are reported as (F (degrees of freedom numerator, degrees of freedom denominator) = F value, p-value). Mean, median, standard deviation, and 95% confidence intervals were reported. Differences were considered significant if $p < 0.05$. As this was an archival study, power calculations were not used to determine the number of specimens for each experimental group. The number of images chosen for skeletal analysis mirrors our work from a previous study [9].

3. Results

3.1. Microglia Skeletal Analysis Revealed Chronic Changes in the Cingulate Gyrus

Our previous research has shown that microglia alter their morphology in the white matter in response to a single mild TBI [9]. Therefore, we sought to quantify microglia morphological changes in gray matter structures via automated skeletal analysis. We exclusively analyzed layers II/III of the cerebral cortex in the cingulate gyrus as neurons are preferentially damaged in these cortical layers after mild TBI in our model of injury [30].

There was a significant increase in the number of branches, junctions, endpoints, and summed process length per microglia at 30 DPI and 1 YPI in the anterior cingulate gyrus compared to age-matched sham (Figure 1) (Table 1). There were no statistically significant differences between any of the sham timepoints out to 1 YPI. There was an increase in the number of branches per microglia (F (4114) = 7.199, $p < 0.0001$) at 30 DPI ($p = 0.0006$) and 1 YPI ($p = 0.0023$) compared to sham (Figure 1g). There was an increase in the number of junctions per microglia (F (4114) = 7.235, $p < 0.0001$) at 30 DPI ($p = 0.0006$) and 1 YPI ($p = 0.0022$) compared to sham (Figure 1h). There was an increase in the number of endpoints per microglia (F (4114) = 5.358, $p = 0.0005$) at 30 DPI ($p = 0.0043$) and 1 YPI ($p = 0.0256$) compared to sham (Figure 1i). Finally, there was an increase in the summed process length per microglia ($F(4113) = 8.825$, $p < 0.0001$) at 30 DPI ($p < 0.0001$) and 1 YPI ($p = 0.0062$) compared to sham (Figure 1j).

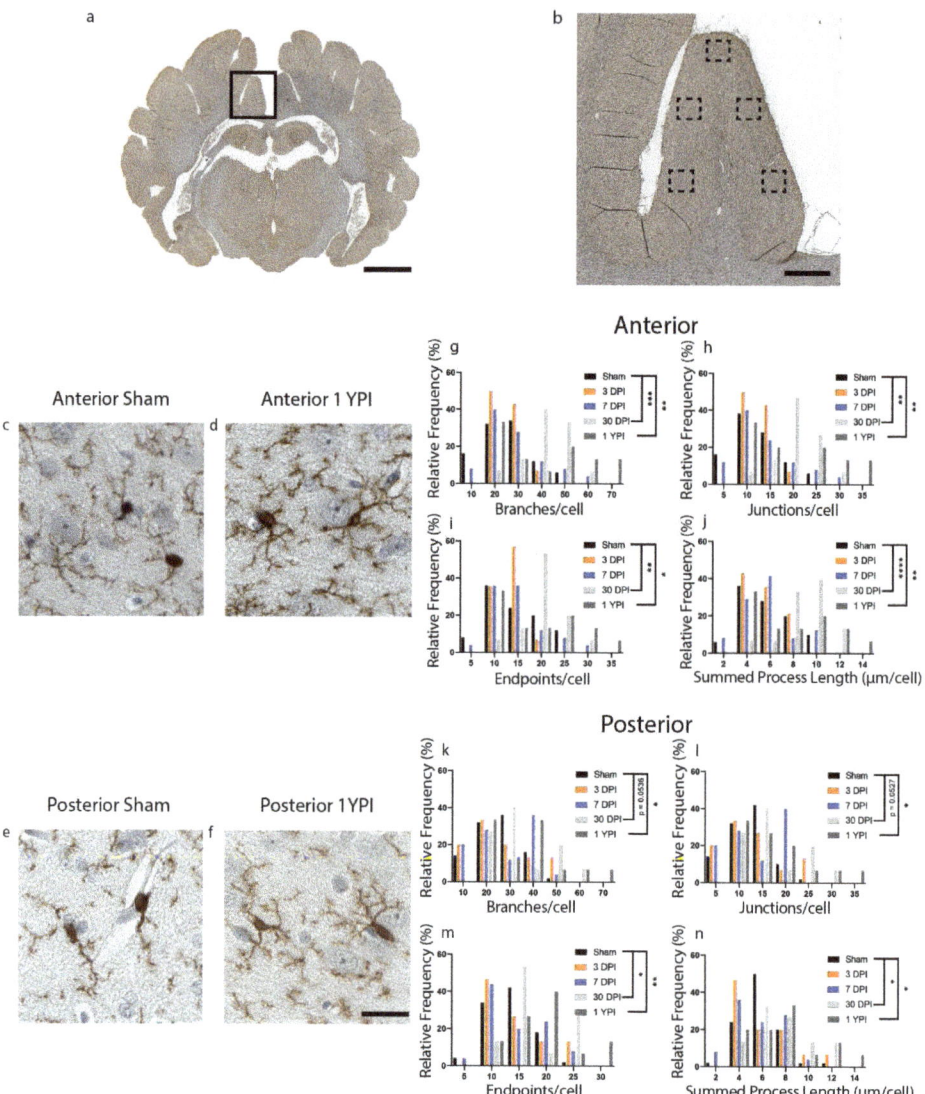

Figure 1. Microglia Become Hyper-ramified in the Cingulate Gyrus at One Month and Persist out to Chronic Time Points. A whole coronal section of Iba-1-stained tissue is shown with a call-out box centered around the cingulate gyrus (**a**, scale = 7 mm). Five call-out boxes in the cingulate gyrus depict the location of the 40× images (image n = 5 per animal) used for skeletal analysis (**b**, scale = 1 mm). Cropped 40× images for sham and 1 YPI are displayed showing the hyper-ramified morphology at 1 YPI (**c–f**, scale = 30 μm). Histograms display the range of data for each experimental group. There is a significant increase in the number of branches, junctions, process endpoints, and summed process length at 30 DPI and 1 YPI compared to sham in anterior sections of cingulate gyrus (**g–j**) and posterior sections of cingulate gyrus (**k–n**) (* $p < 0.05$, ** $p \leq 0.01$, *** $p \leq 0.001$, **** $p \leq 0.0001$).

Table 1. Skeletal analysis measurements in the anterior cingulate gyrus. All values are reported as mean ± standard deviation, 95% confidence interval [lower 95% CI, upper 95% CI].

	Branches	Junctions	Endpoints	Summed Process Length
Sham	26.66 ± 10.03, 95% CI [23.81, 29.51]	12.87 ± 5.09, 95% CI [11.42, 14.31]	14.56 ± 5.36, 95% CI [13.03, 16.08]	5.82 ± 2.07, 95% CI [5.23, 6.41]
3 DPI	26.51 ± 5.70, 95% CI [23.23, 29.80]	12.73 ± 2.89, 95% CI [11.06, 14.40]	14.11 ± 2.57, 95% CI [12.63, 15.60]	5.75 ± 1.31, 95% CI [5.00, 6.51]
7 DPI	28.69 ± 12.15, 95% CI [23.68, 33.70]	13.83 ± 6.19, 95% CI [11.27, 16.38]	14.91 ± 5.37, 95% CI [12.69, 17.12]	5.66 ± 2.00, 95% CI [4.81, 6.50]
30 DPI	40.77 ± 9.92, 95% CI [35.28, 46.27]	20.00 ± 5.01, 95% CI [17.22, 22.78]	20.33 ± 4.45, 95% CI [17.86, 22.79]	8.91 ± 1.95, 95% CI [7.83, 9.99]
1 YPI	39.49 ± 18.73, 95% CI [29.12, 49.87]	19.41 ± 9.53, 95% CI [14.13, 24.69]	19.40 ± 8.15, 95% CI [14.89, 23.91]	8.09 ± 3.56, 95% CI [6.12, 10.06]

In the posterior cingulate gyrus, there was an increase in the number of branches, junctions, endpoints, and summed process length per microglia at 30 DPI and 1 YPI compared to age-matched sham (Figure 1) (Table 2). There were no statistically significant differences between any of the sham timepoints out to 1 YPI. The number of branches per microglia increased ($F(4115) = 3.932$, $p = 0.0050$) at 30 DPI ($p = 0.0536$) and 1 YPI ($p = 0.0147$) compared to sham (Figure 1k). The number of junctions increased ($F(4115) = 3.830$, $p = 0.0058$) at 30 DPI ($p = 0.0527$) and 1 YPI ($p = 0.0177$) compared to sham (Figure 1l). The number of endpoints increased ($F(4115) = 4.261$, $p = 0.0030$) at 30 DPI ($p = 0.0489$) and 1 YPI ($p = 0.0098$) compared to sham (Figure 1m). Finally, the summed process length per microglia increased ($F(4115) = 4.359$, $p = 0.0026$) at 30 DPI ($p = 0.0369$) and 1 YPI ($p = 0.0282$) compared to sham (Figure 1n). These data detail a hyper-ramified microglial morphology at 30 DPI and 1 YPI in both anterior and posterior sections of the cingulate gyrus.

Table 2. Skeletal analysis measurements in the posterior cingulate gyrus. All Values are reported as mean ± standard deviation, 95% confidence interval [lower 95% CI, upper 95% CI].

	Branches	Junctions	Endpoints	Summed Process Length
Sham	26.24 ± 8.18, 95% CI [23.92, 28.57]	12.66 ± 4.10, 95% CI [11.49, 13.82]	13.80 ± 3.79, 95% CI [12.72, 14.88]	5.91 ± 1.75, 95% CI [5.41, 6.41]
3 DPI	26.81 ± 12.01, 95% CI [20.16, 33.46]	12.99 ± 6.14, 95% CI [9.59, 16.39]	13.95 ± 5.31, 95% CI [11.01, 16.88]	5.96 ± 2.63, 95% CI [4.50, 7.42]
7 DPI	28.12 ± 11.28, 95% CI [23.46, 32.78]	13.62 ± 5.78, 95% CI [11.23, 16.00]	14.48 ± 4.98, 95% CI [12.42, 16.53]	5.77 ± 2.12, 95% CI [4.89, 6.64]
30 DPI	34.88 ± 11.58, 95% CI [28.47, 41.30]	17.05 ± 5.91, 95% CI [13.78, 20.33]	17.73 ± 5.20, 95% CI [14.85, 20.61]	7.75 ± 2.29, 95% CI [6.48, 9.02]
1 YPI	36.34 ± 14.42, 95% CI [28.35, 44.32]	17.68 ± 7.31, 95% CI [13.63, 21.73]	18.52 ± 6.44, 95% CI [14.95, 22.08]	7.81 ± 2.84, 95% CI [6.24, 9.38]

3.2. Microglia Skeletal Analysis Revealed Acute and Subacute Changes in the Inferior Temporal Gyrus

Next, we analyzed microglial morphological changes in layers II/III of the inferior temporal gyrus, a lateral cortical structure. In the inferior temporal gyrus, there were significant changes in microglia morphology detected in anterior sections of tissue compared to sham (Figure 2) (Table 3). Specifically, we found an increase in the number of branches, junctions, and endpoints at 7 DPI and 30 DPI, as well as an increase in summed process length per microglia at 3 DPI, 7 DPI, and 30 DPI. The number of branches per microglia increased ($F(4112) = 5.697$, $p = 0.0003$) at 7 DPI ($p = 0.0015$) and 30 DPI ($p = 0.0040$) compared to sham (Figure 2g). The number of junctions per microglia increased ($F(4111) = 6.468$, $p = 0.0001$) at 7 DPI ($p = 0.0006$) and 30 DPI ($p = 0.0018$) compared to sham (Figure 2h). The number of endpoints per microglia increased ($F(4112) = 6.053$, $p = 0.0002$) at 7 DPI ($p = 0.0011$) and 30 DPI ($p = 0.0023$) compared to sham (Figure 2i). Finally, the summed process length per microglia increased ($F(4111) = 5.788$, $p = 0.0003$) at 3 DPI ($p = 0.0272$), 7 DPI ($p = 0.0140$) and 30 ($p = 0.0016$) compared to sham (Figure 2j).

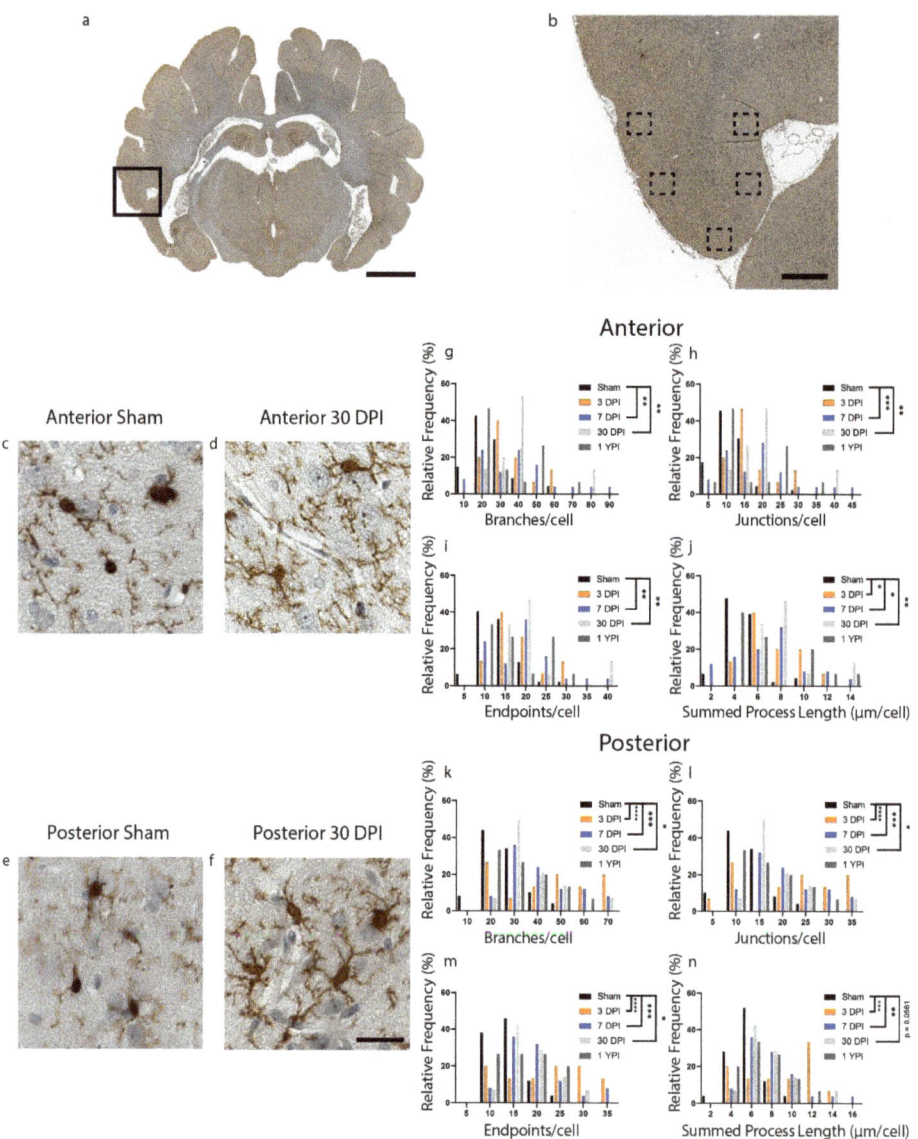

Figure 2. Microglia Become Hyper-ramified in the Inferior Temporal Gyrus at Acute and Subacute Time Points. A whole coronal section of Iba-1-stained tissue is shown with a call-out box centered around the inferior temporal gyrus (**a**, scale = 7 mm). Five call-out boxes in the inferior temporal gyrus depict the location of the 40× images (image n = 5 per animal) used for skeletal analysis (**b**, scale = 1 mm). Cropped 40× images for sham and 30 DPI are displayed showing the comparatively hyper-ramified morphology at 30 DPI in posterior tissue (**c–f**, scale = 30 μm). Histograms display the range of data for each experimental group There is a significant increase in the number of branches (**g**), junctions (**h**), and process endpoints (**i**) at 7 DPI and 30 DPI, as well as a significant increase in summed process length (**j**) at 3 DPI, 7 DPI and 30 DPI in anterior sections of inferior temporal gyrus. There is a significant increase in the number of branches (**k**), junctions (**l**), process endpoints (**m**), and summed process length (**n**) at 3 DPI, 7 DPI, and 30 DPI in posterior sections of inferior temporal gyrus (* $p < 0.05$, ** $p \leq 0.01$, *** $p \leq 0.001$, **** $p \leq 0.0001$).

Table 3. Skeletal analysis measurements in the anterior inferior temporal gyrus. All values are reported as mean ± standard deviation, 95% confidence interval [lower 95% CI, upper 95% CI].

	Branches	Junctions	Endpoints	Summed Process Length
Sham	24.94 ± 10.59, 95% CI [21.84, 28.05]	11.54 ± 4.60, 95% CI [10.18, 12.91]	13.19 ± 4.66, 95% CI [11.82, 14.55]	5.07 ± 1.68, 95% CI [4.58, 5.57]
3 DPI	35.52 ± 13.31, 95% CI [28.15, 42.89]	17.30 ± 6.84, 95% CI [13.51, 21.09]	17.98 ± 5.88, 95% CI [14.72, 21.23]	7.34 ± 2.47, 95% CI [5.98, 8.71]
7 DPI	39.39 ± 19.18, 95% CI [31.48, 47.31]	19.16 ± 9.70, 95% CI [15.16, 23.17]	19.68 ± 8.44, 95% CI [16.19, 23.16]	7.11 ± 3.17, 95% CI [5.81, 8.42]
30 DPI	40.99 ± 17.92, 95% CI [31.06, 50.91]	20.04 ± 9.07, 95% CI [15.02, 25.06]	20.55 ± 7.95, 95% CI [16.15, 24.95]	8.01 ± 2.84, 95% CI [6.44, 9.59]
1 YPI	32.80 ± 17.00, 95% CI [23.38, 42.21]	16.06 ± 8.70, 95% CI [11.24, 20.88]	16.47 ± 7.40, 95% CI [12.38, 20.56]	6.91 ± 3.31, 95% CI [5.08, 8.74]

Importantly, we did find a significant difference between sham specimens which survived for 30 days or less and sham specimens which survived out to 1 year. Specifically, we found a decrease in the number of branches (t (45) = 5.641, $p < 0.0001$), junctions (t (44) = 5.775, $p < 0.0001$), endpoints (t (45) = 5.965, $p < 0.0001$), and summed process length (t (44) = 4.462, $p < 0.0001$) in sham which survived out to one year compared to sham which survived for 30 days or less.

There were also significant differences in branches, junctions, endpoints, and summed process length in posterior sections of tissue at acute and subacute timepoints (Figure 2) (Table 4). The number of branches per microglia increased ($F_{(4,114)} = 9.397$, $p < 0.0001$) at 3 DPI ($p < 0.0001$), at 7 DPI ($p = 0.0001$), and 30 DPI ($p = 0.0293$) compared to sham (Figure 2k). The number of junctions increased ($F_{(4,114)} = 9.323$, $p < 0.0001$) at 3 DPI ($p < 0.0001$), at 7 DPI ($p = 0.0001$), and 30 DPI (p = 0.0290) compared to sham (Figure 2l). The number of endpoints increased ($F_{(4,114)} = 9.458$, $p < 0.0001$) at 3 DPI ($p < 0.0001$), at 7 DPI ($p = 0.0002$), and 30 DPI ($p = 0.0346$) compared to sham (Figure 2m). Finally, the summed process length per microglia increased ($F_{(4,114)} = 7.580$, $p < 0.0001$) at 3 DPI ($p = 0.0001$), at 7 DPI ($p = 0.0013$) and 30 DPI ($p = 0.0561$) compared to sham (Figure 2n). Overall, we found hyper-ramified microglial morphology at 3, 7 and 30 DPI in both anterior and posterior sections of the inferior temporal gyrus.

Table 4. Skeletal analysis measurements in the posterior inferior temporal gyrus. All values are reported as mean ± standard deviation, 95% confidence interval [lower 95% CI, upper 95% CI].

	Branches	Junctions	Endpoints	Summed Process Length
Sham	26.27 ± 8.09, 95% CI [23.97, 28.57]	12.70 ± 4.11, 95% CI [11.53, 13.87]	13.66 ± 3.63, 95% CI [12.63, 14.69]	5.64 ± 1.64, 95% CI [5.18, 6.11]
3 DPI	44.78 ± 19.86, 95% CI [33.78, 55.78]	21.93 ± 9.94, 95% CI [16.43, 27.44]	22.21 ± 9.11, 95% CI [17.16, 27.25]	8.91 ± 3.62, 95% CI [6.91, 10.91]
7 DPI	40.20 ± 14.75, 95% CI [34.11, 46.28]	19.73 ± 7.35, 95% CI [16.69, 22.76]	19.68 ± 6.57, 95% CI [16.97, 22.39]	7.97 ± 2.92, 95% CI [6.76, 9.17]
30 DPI	37.48 ± 11.68, 95% CI [30.73, 44.22]	18.35 ± 5.95, 95% CI [14.92, 21.79]	18.57 ± 5.02, 95% CI [15.68, 21.47]	7.63 ± 2.31, 95% CI [6.30, 8.96]
1 YPI	34.23 ± 12.09, 95% CI [27.53, 40.93]	16.69 ± 6.18, 95% CI [13.27, 20.11]	17.31 ± 5.24, 95% CI [14.40, 20.21]	7.10 ± 2.26, 95% CI [5.85, 8.35]

Finally, we did find a significant difference between sham specimens which survived for 30 days or less and sham specimens which survived out to 1 year. Specifically, we found a decrease in the number of branches (t (48) = 2.724, $p = 0.0090$), junctions (t (48) = 2.661, $p = 0.0106$), endpoints (t (48) = 2.842, $p = 0.0066$), and summed process length (t (48) = 2.129, $p = 0384$) in sham which survived out to one year compared to sham which survived for 30 days or less.

3.3. Total Cell Density and Neuron Density Did Not Change in the Cingulate Gyrus after Mild TBI

As the cingulate gyrus had chronic microglial morphological changes, consistent changes between anterior and posterior aspects of tissue, and no significant differences between aged matched shams, additional pathological analyses focused on this region. Total cell density was next calculated to evaluate any overt glial cell proliferation or loss. However, there were no significant changes to CV total cell density (H (4) = 5.246, p = 0.2630) (Figure 3c). Cohen's d effect sizes are reported below to demonstrate the magnitude of difference between sham and experimental groups and allow for *a priori* power analysis (Table 5).

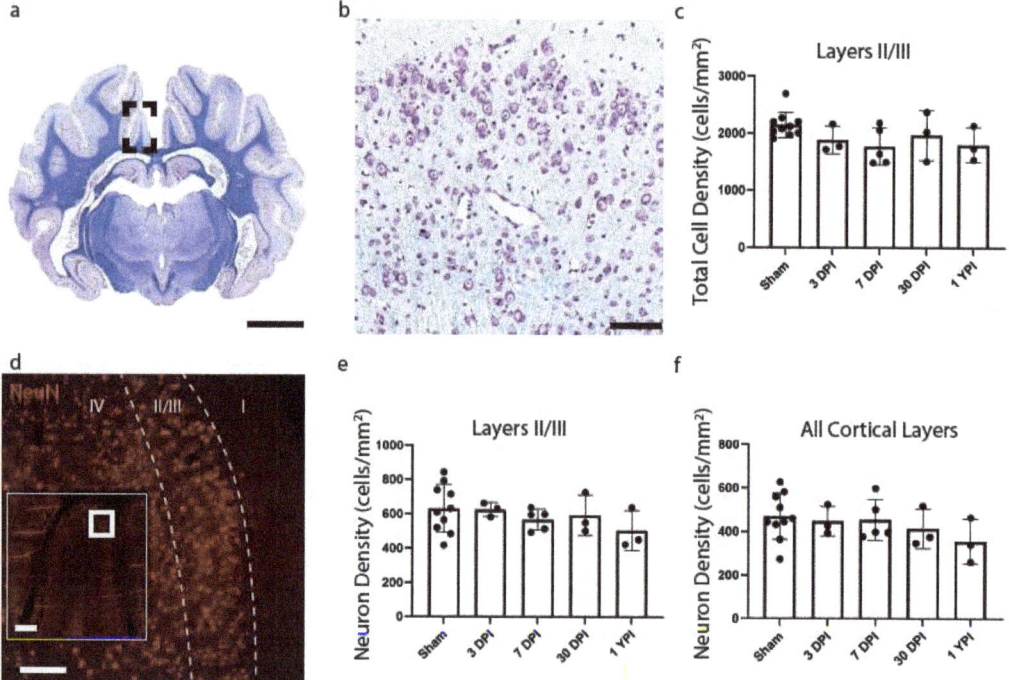

Figure 3. Total Cell Density and Neuronal Density Does Not Change in the Cingulate Gyrus. A whole coronal section of Luxol Fast Blue and Cresyl Violet stained tissue is shown with a call out box centered around the apex of the cingulate gyrus (**a**, scale = 7 mm). A 20× image of cortical cells is shown (**b**, scale = 100 μm). There were no significant changes to total cell density (**c**) as a result of TBI. Whole coronal sections of tissue were also stained with NeuN, a neuronal marker, to count neurons in layers II/III and all cortical layers across the cingulate gyrus. A 10× image of NeuN is shown denoting cortical layers II/III (**d**, scale = 200 μm) with an inset image demonstrating its location along the cingulate gyrus (scale = 500 μm). There were no significant changes in neuron density in layers II/III (**e**) or among all cortical layer (**f**). Each point represents data from an individual animal within each group.

Table 5. Cohen's d effect sizes of cingulate gyrus cell counts compared to sham.

	3 DPI	7 DPI	30 DPI	1 YPI
Total Cell Density	d = 1.13	d = 1.35	d = 0.51	d = 1.28

To assess potential neuronal loss in layers II/III compared to all cortical layers of the cingulate gyrus, we conducted NeuN cell counts in tissue sections adjacent to CV-stained sections. However, there were no significant changes to neuron density in layers II/III (H (4) = 2.699, p = 0.6093) (Figure 3e) or all cortical layers (H (4) = 3.469, p = 0.4826) (Figure 3f) of the cingulate gyrus.

3.4. GFAP Reactivity Did Not Change after Single Mild TBI

To further explore the extent of gray matter pathology after mild TBI and to complement our previous pathological analysis in the white matter, we chose two gray matter structures for closer assessment: the cingulate gyrus, a midline structure, and the inferior temporal gyrus, a lateral structure. We chose these structures based on previous studies in our lab that described permeabilized gray matter neurons after closed-head rotational mild TBI; patches of permeabilized cells were seen in some gyri but absent in others [30]. This pathological distribution suggests that post-TBI gray matter pathology is multi-focal yet may skip some anatomical levels of the cortex. Examination of the cingulate and inferior temporal gyri allows us to thoroughly examine two sample cortices for pathological sequela to TBI.

We began with astrocytic changes as measured by GFAP reactivity as astrocytic endfeet are a component of the BBB, which we have noted was disrupted in our fibrinogen reactivity analysis below. Reactive astrogliosis is often used as a marker for damaged CNS tissue and has been observed colocalizing with fibrinogen blood proteins [29,31]. Yet there was no significant change in GFAP reactivity in the cingulate gyrus (H (4) = 7.00, p = 0.1359) (Figure 4e), nor was there a change in GFAP reactivity in the inferior temporal gyrus (H (4) = 4.60, p = 0.3309) (Figure 4f).

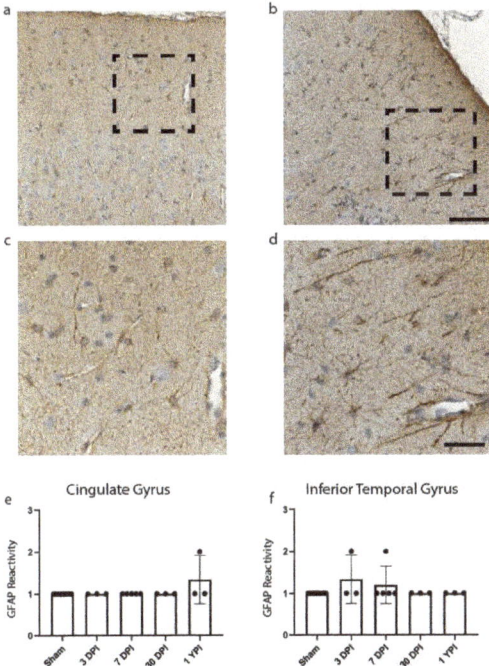

Figure 4. GFAP Reactivity Does Not Change in the Cingulate or Inferior Temporal Gyrus. Examples of GFAP reactivity scored as a 1 (**a**) or 2 (**b**) are shown (scale = 100 μm) with corresponding call-out boxes for scores of 1 (**c**) or 2 (**d**; scale = 50 μm). No images were scored as a 3. There was no significant change in GFAP reactivity in cingulate gyrus (**e**) or inferior temporal gyrus (**f**). Each point represents data from an individual animal within each group.

3.5. Fibrinogen Reactivity Increased at 3 DPI after Mild TBI

Lastly, we investigated BBB disruption, a recently appreciated pathological feature of concussion. Extravasation of the serum protein, fibrinogen, has been used to evaluate BBB disruption at acute timepoints following mild TBI in our model of injury and may be an important pathological feature of concussion [31]. Here, we first assessed fibrinogen reactivity out to chronic timepoints in our cohort. There was an overall increase in percentage area of fibrinogen reactivity in the total brain parenchyma (H (4) = 10.27, p = 0.0360) at 3 DPI (Mean = 4.21, Median = 4.29, SD ± 0.60, 95% CI [2.72, 5.71]) compared to sham (Mean = 0.25, Median = 0.13, SD ± 0.41, 95% CI [−0.04, 0.54]) (p = 0.0198). There were no significant changes to percentage area of fibrinogen reactivity at 7 DPI, 30 DPI, and 1 YPI compared to sham (Figure 5c).

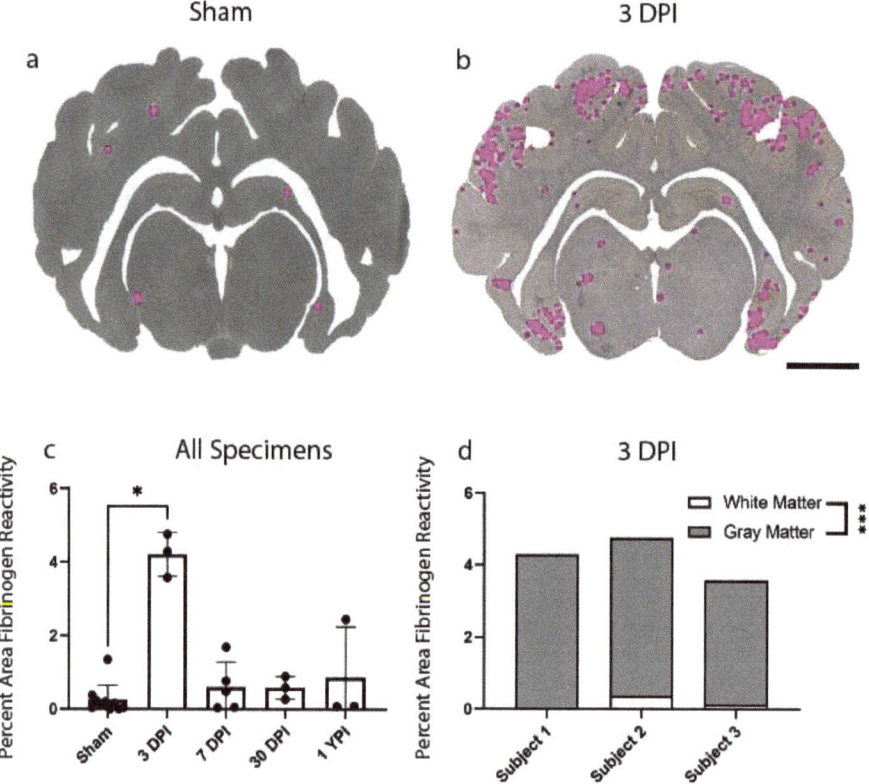

Figure 5. Fibrinogen Reactivity Increases at 3 DPI. Regions of fibrinogen extravasation are shown in sham (**a**) and 3 DPI (**b**) via Stereo Investigator software on whole coronal sections of tissue (scale = 1 cm). The percentage area of fibrinogen reactivity increased at 3 DPI (p = 0.0198) compared to sham before resolving by 7 DPI out to 1 YPI (**c**) (* p < 0.05). Each point represents data from an individual animal within each group. Fibrinogen annotation in gray versus white matter at 3 DPI demonstrated a statistically significant increase in gray matter (p = 0.0003) (**d**) (*** p ≤ 0.001).

Extravasated serum proteins were frequently observed in the gray matter compared to white matter structures at 3 DPI, so we also quantified this gray to white matter reactivity ratio at this timepoint alone (n = 3 per group). There was a statistically significant increase in percentage area of fibrinogen reactivity in gray matter (Mean = 4.04, Median = 4.26, SD ± 0.52, 95% CI [2.75, 5.32]) compared to white matter (Mean = 0.18, Median = 0.13, SD ± 0.18, 95% CI [−0.24, 0.60]) at 3 DPI (t (4) = 12.26, p = 0.0003) (Figure 5d).

4. Discussion

After a single mild TBI in our pig model of closed-head diffuse brain injury, we found chronic changes in microglial morphology in the cingulate gyrus, as well as acute and subacute changes in microglial morphology in the inferior temporal gyrus. Additionally, we found a significant increase in fibrinogen blood proteins extravasated into the brain at 3 DPI, which resolved to sham levels by 7 DPI. Yet we did not find neuronal loss or astrocyte reactivity paired with these pathological changes. We hypothesized that microglial morphological changes would coincide with astroglial reactivity, neuronal dystrophy/loss, and BBB disruption in the cerebral cortex after TBI. Therefore, we are compelled to reject part of our original hypothesis. These microglia morphological changes supplement and support our previous studies in the white matter and hippocampal subregions, which demonstrated microglia morphological changes, microglia density changes, and axonal pathology after a single mild TBI [9,24].

In the current study, we detected chronically hyper-ramified microglia in cingulate gyrus; however, hyper-ramified microglia were only detected at acute and subacute timepoints in inferior temporal gyrus. This discrepancy in pathology may highlight the injury thresholds needed to produce microglial changes after mild TBI in our model. The medial cingulate may be more vulnerable to damage compared to the lateral inferior temporal gyrus at this level of injury, leading to chronic pathology that is consistent along the rostral-caudal axis. Vascular damage or dysfunctional neurons projecting into susceptible white matter may account for this pathology. Additionally, inertial sheer forces in this model of injury may be greater in midline structures versus lateral structures. Follow-up studies are warranted to elucidate the mechanism of region-specific microglial morphological changes.

Absent additional transcriptomic studies, our interpretation of these phenotypic changes on microglia behavior is limited. Cell sorting and RNAseq techniques have highlighted microglia complexity in recent years. Hammond et al. examined individual mouse microglia during development, old age, and after brain injury and detected at least nine transcriptionally distinct microglia subpopulations [32]. Saba et al. characterized several microglia phenotypes following brain injury via multicolor flow cytometry and correlated these subpopulations with long-term cognitive deficits [33]. Furthermore, a 2022 White Paper written by leading microglia researchers urged the field to reject the classic pro- versus anti-inflammatory microglial classifications and provided a new series of recommendations to understand microglial states [34]. Future microglial studies will need to utilize transcriptomic, proteomic, and genetic manipulation techniques to fully characterize microglia subpopulations and determine their functions [35].

Morphological changes in microglia only indicate a change in neuroimmune homeostasis that we are witnessing out to chronic timepoints [9,36]. Importantly, a select few studies have described a hyper-ramified microglia phenotype after injury. After rat midline fluid percussion injury, Morrison et al. conducted microglia skeletal analysis and found a decrease in microglia ramification around the impact site but not in a remote region. Additionally, microglia complexity—measured by fractal analysis—increased in the remote region [37]. Morrison et al. carefully distinguished between ramified microglia and "ramified/hyper-complex" microglia, the former of which is detected most often in sham, while the latter is detected in remote regions at subacute timepoints. Our current findings seem to more closely align with this "hyper-complex" phenotype; our observed microglia have more branches, junctions, endpoints, and summed process lengths that extend far away from the soma. This hyper-ramified phenotype should not be confused with a classic bushy microglia morphology where microglia retract their processes from surrounding tissue and have many short, thickened processes. Finally, studies outside of the TBI literature have reported a hyper-ramified phenotype. One recent study using a mouse model of PTSD reported an increase in hyper-ramified microglia and loss of dendritic spines in specific neuroanatomical regions [38]. Hyper-ramified microglia were also detected in rat prefrontal cortex in a chronic stress model [39]. Overall, microglia becoming de-ramified post-injury is commonly reported while microglia hyper-ramification is observed far less frequently and can indicate stress on the neuroimmune system [37,40].

This hyper-ramified phenotype can be triggered by numerous extracellular signals. For instance, fibrinogen leakage through the BBB is a key change in the extracellular environment that can initiate neuroinflammation (for a review see [41]). Microglia may also be responding to injured neurons and may also be implicated in synaptic modifications [24,40]. Previous research in this model found an acute inflammatory response around permeabilized cortical neurons; microglia density increased and morphology became less ramified 15 min after both mild and severe TBI [23].

Unfortunately, administration of the permeability marker was not performed in this cohort of animals. Yet studies suggest that permeabilized cells initially survive the injury to die at a later timepoint. In particular, Whalen et al. subjected mice to controlled cortical impact and found that many plasma-membrane-damaged neurons recovered their integrity within 24 h, yet disappeared from the brain by 7 DPI [20]. Therefore, we conducted a limited cell density analysis in the cingulate gyrus to determine if neuronal loss coincides with our chronic microglial changes. Neither our total cell density nor our neuronal cell density significantly changed at timepoints post-injury. We initially hypothesized neuronal loss after mild TBI as permeabilized neurons have been prominent in medial gyri and isolated to layers II/III of the cortex [30]. Since neuronal loss was not seen in our cohort, ongoing neuronal dysfunction may be present and potentially driving microglial homeostatic changes and possible microglia-mediated protection. Using a model of cerebral ischemia, Szalay et al. found that microglia reduced excitotoxic damage, as drug-induced knockout of microglia led to dysregulated calcium responses, calcium overload, increased neuronal death, and incidence of spreading depolarization— a process that involves neuronal swelling, injury to dendritic spines, and subsequent dampening of brain electrical activity [42]. Future studies should employ electrophysiological techniques in addition to RNA assays to assess loss of function in cortical neurons in relation to neuroimmune activity.

To further explore the extent of gray matter neuropathology, we utilized other traditional histological markers after TBI. We assessed GFAP reactivity as astrocyte end feet are a critical component of maintaining the BBB, and when disturbed, could promote pathology [43]. We examined both the cingulate gyrus and the inferior temporal gyrus yet did not find any indication of increased astrocyte reactivity after mild TBI. This is consistent with our previous examination of astrocyte reactivity in the white matter [9]. Astrocytes have been documented colocalizing with fibrinogen after mild TBI and reactive astrocytes are considered a hallmark pathology of injured CNS tissue, yet overt astrocyte reactivity does not seem to occur at these injury levels at a level detectable with our methodology [29,31]. Astrocytes may undergo genomic changes in response to injury, so future studies may want to employ RNA-based assays to detect any subtle changes in astrocyte activity post-TBI [44]. Moreover, astrocytes undergo proliferative changes that can vary depending on injury severity so astrocytes should still be examined at higher injury levels or after repetitive mild injuries in this model [44].

We also assessed fibrinogen reactivity as a global measure of blood-brain barrier breakdown and neuropathology post TBI. Previous studies in our injury model have found an increase in fibrinogen after a single mild TBI. Utilizing a different strain of pigs, Johnson et al. (2018) conducted postmortem histology at 6 h and 3 DPI and found an increase in fibrinogen extravasation at both timepoints. They noted fibrinogen was specifically found in gray-white matter boundaries, periventricular, and subpial regions [31]. Our analysis found fibrinogen at gray-white matter boundaries as well as in middle layers of the cortex. The different pig strains and potentially different vasculature may account for the disparities in fibrinogen's pattern and distribution in the brain parenchyma. Our analysis also found an increase in fibrinogen reactivity at 3 DPI and that this reactivity resolved by 7 DPI out to chronic timepoints. In contrast, Hay et al. (2015) found that 47% of humans that survived a moderate or severe TBI for a year or more showed multifocal parenchymal fibrinogen immunoreactivity [45]. This may suggest that fibrinogen-measured BBB disruption is a common pathological finding after moderate or severe TBI but does not persist beyond the acute phase of mild TBI. Additionally, we examined the extent of fibrinogen reactivity in gray versus white matter at 3 DPI and found a significant presence in gray matter compared to white matter, underscoring the importance of gray matter pathology after mild TBI.

Finally, we found a statistically significant difference in age-matched sham microglia morphology at one year in the inferior temporal gyrus compared to 30-day or less age-matched sham. These differences are specific to the inferior temporal gyrus, which has shown heterogenous reactivity to TBI in our model. This is interesting for future studies in the intersection of aging and TBI, however that is not the focus of this study. Other regions had no detectable differences in sham over time.

There are several limitations in this study. First, the sample size of injured specimens was small, which reduces the power of the study. Large animal subjects are costly and require specialized housing and trained veterinary staff. Additionally, this was an archival study of formalin-fixed paraffin-embedded tissue, which constrained our options for experimental methodology. Also, many commercially available antibodies do not cross-react with pig tissue, thus also restricting our immunohistochemical staining and pathological analysis.

In conclusion, we have detailed a temporal sequence of neuropathology in the cerebral cortex, an often-overlooked area in diffuse brain injury. Fibrinogen-measured BBB disruption increases at 3 DPI but resolves by 7 DPI. Meanwhile, microglia adopt a hyper-ramified morphology from 3 DPI to 30 DPI in the inferior temporal gyrus and are delayed until 30 DPI but persist out to 1 YPI in the cingulate gyrus. These delayed consequences may be an effect of distal diffuse axonal injury and warrant follow-up studies. We have demonstrated that a single, closed-head, mild TBI is associated with chronic alterations to microglia homeostasis. Microglia activity has become increasingly implicated in neurodegenerative disease pathogenesis though precisely how they contribute to disease progression remains elusive [46,47]. Further experimentation is needed to supplement our histological analysis. Microglia modulate synaptic plasticity, therefore presynaptic and postsynaptic markers as well as electrophysiological recordings could examine the impact of hyper-ramified microglia after TBI. Compounds used to transiently knock out microglia should be used in our model to assess potential neuron loss and cell physiological dysfunction that may be microglia-mediated in gray matter. Additionally, neuroimaging would allow us to track microglia activity in a single specimen over time, providing details in between the terminal timepoints in the current study. Finally, no changes in neuronal density were observed but subtle structural and/or functional changes in neurons, axons, and/or dendrites are possible, similar to our recent report of alterations in electrophysiological function absent signs of neuronal or axonal pathology in the hippocampal formation post-TBI in pigs [48]. Golgi staining or dyes used for neuronal tracing can be performed to detect neurostructural changes related to critical functional parameters such as dendritic number, length, and spine density. It is hoped that further understanding of the brain's inflammatory activity after mild TBI will provide us with knowledge of neuronal health and cognitive integrity, and that this will translate to treatment of TBI in people.

Author Contributions: M.R.G., R.L.S., J.E.D. and D.K.C. designed the experiments; M.R.G., N.P., E.J.A. and K.D.B. performed research; M.R.G., A.J. and N.P. analyzed data; M.R.G. and D.K.C. wrote the paper. All authors have read and agreed to the published version of the manuscript.

Funding: This work was made possible through financial support provided by the Department of Veterans Affairs [RR&D Merit Review I01-RX001097 (Duda), BLR&D Merit Review I01-BX005017 (Cullen), and RR&D Career Development Award IK2-RX003651 (Swanson)], the National Institutes of Health [NINDS R03-NS116301 (Cullen), NINDS R01-NS117757 (Cullen)], and the Pennsylvania Department of Health [Cure Award 4100077083 (Swanson)]. None of the funding sources aided in the collection, analysis, and interpretation of data, in the writing of the report, or in the decision to submit the paper for publication. Disclaimer: The contents of this work do not represent the views of the Department of the Veterans Affairs or the United States Government.

Institutional Review Board Statement: All procedures were approved by the Institutional Animal Care and Use Committees at the University of Pennsylvania and the Michael J. Crescenz Veterans Affairs Medical Center (UPenn IACUC Protocol # 803165 last approved on 10 November 2022) and adhered to the guidelines set forth in the NIH Public Health Service Policy on Humane Care and Use of Laboratory Animals (2015).

Informed Consent Statement: Not applicable.

Data Availability Statement: The datasets used during the current study are available from the corresponding author on reasonable request.

Acknowledgments: The authors thank John O'Donnell for his guidance.

Conflicts of Interest: No competing financial interests exist. The authors have no conflict of interest related to this work to disclose.

References

1. Centers for Disease Control and Prevention. *Surveillance Report of Traumatic Brain Injury-Related Emergency Department Visits, Hospitalizations, and Deaths-United States, 2014*; Centers for Disease Control and Prevention (United States): Atlanta, GA, USA, 2019; Volume 24.
2. Nguyen, R.; Fiest, K.M.; McChesney, J.; Kwon, C.-S.; Jette, N.; Frolkis, A.D.; Atta, C.; Mah, S.; Dhaliwal, H.; Reid, A.; et al. The international incidence of traumatic brain injury: A systematic review and meta-analysis. *Can. J. Neurol. Sci.* **2016**, *43*, 774–785. [CrossRef]
3. Levin, H.S.; Diaz-Arrastia, R.R. Diagnosis, prognosis, and clinical management of mild traumatic brain injury. *Lancet Neurol.* **2015**, *14*, 506–517. [CrossRef]
4. Cullen, D.K.; Harris, J.P.; Browne, K.D.; Wolf, J.A.; Duda, J.E.; Meaney, D.F.; Margulies, S.S.; Smith, D.H. A porcine model of traumatic brain injury via head rotational acceleration. *Methods Mol. Biol.* **2016**, *1462*, 289–324. [CrossRef]
5. Chen, X.-H.; Meaney, D.; Xu, B.-N.; Nonaka, M.; McIntosh, T.K.; Wolf, J.A.; Saatman, K.E.; Smith, D.H. Evolution of Neurofilament Subtype Accumulation in Axons Following Diffuse Brain Injury in the Pig. *J. Neuropathol. Exp. Neurol.* **1999**, *58*, 588–596. [CrossRef] [PubMed]
6. Smith, D.H.; Chen, X.-H.; Xu, B.-N.; McIntosh, T.K.; Gennarelli, T.A.; Meaney, D. Characterization of Diffuse Axonal Pathology and Selective Hippocampal Damage following Inertial Brain Trauma in the Pig. *J. Neuropathol. Exp. Neurol.* **1997**, *56*, 822–834. [CrossRef]
7. Chen, X.-H.; Siman, R.; Iwata, A.; Meaney, D.F.; Trojanowski, J.Q.; Smith, D.H. Long-Term Accumulation of Amyloid-β, β-Secretase, Presenilin-1, and Caspase-3 in Damaged Axons Following Brain Trauma. *Am. J. Pathol.* **2004**, *165*, 357–371. [CrossRef] [PubMed]
8. Cullen, D.K.; Harris, J.P.; Browne, K.D.; Wolf, J.A.; Duda, J.E.; Meaney, D.F.; Margulies, S.S.; Smith, D.H. Injury Models of the Central Nervous System. *Methods Mol. Biol.* **2016**, *1462*, 289–324. [CrossRef] [PubMed]
9. Grovola, M.R.; Paleologos, N.; Brown, D.P.; Tran, N.; Wofford, K.L.; Harris, J.P.; Browne, K.D.; Shewokis, P.A.; Wolf, J.A.; Cullen, D.K.; et al. Diverse changes in microglia morphology and axonal pathology during the course of 1 year after mild traumatic brain injury in pigs. *Brain Pathol.* **2021**, *31*, e12953. [CrossRef] [PubMed]
10. Johnson, V.E.; Stewart, W.; Smith, D.H. Axonal pathology in traumatic brain injury. *Exp. Neurol.* **2012**, *246*, 35–43. [CrossRef]
11. Browne, K.D.; Chen, X.-H.; Meaney, D.F.; Smith, D.H. Mild Traumatic Brain Injury and Diffuse Axonal Injury in Swine. *J. Neurotrauma* **2011**, *28*, 1747–1755. [CrossRef]
12. Song, J.; Li, J.; Chen, L.; Lu, X.; Zheng, S.; Yang, Y.; Cao, B.; Weng, Y.; Chen, Q.; Ding, J.; et al. Altered gray matter structural covariance networks at both acute and chronic stages of mild traumatic brain injury. *Brain Imaging Behav.* **2021**, *15*, 1840–1854. [CrossRef] [PubMed]
13. Mayer, A.R.; Meier, T.B.; Dodd, A.B.; Stephenson, D.D.; Robertson-Benta, C.R.; Ling, J.M.; Reddy, S.P.; Zotev, V.; Vakamudi, K.; Campbell, R.A.; et al. Prospective Study of Gray Matter Atrophy Following Pediatric Mild Traumatic Brain Injury. *Neurology* **2023**, *100*, e516–e527. [CrossRef] [PubMed]
14. Muller, A.M.; Panenka, W.J.; Lange, R.T.; Iverson, G.L.; Brubacher, J.R.; Virji-Babul, N. Longitudinal changes in brain parenchyma due to mild traumatic brain injury during the first year after injury. *Brain Behav.* **2021**, *11*, e2410. [CrossRef]

15. LaPlaca, M.C.; Prado, G.R.; Cullen, D.K.; Simon, C.M. Plasma membrane damage as a marker of neuronal injury. *Annu. Int. Conf. IEEE Eng. Med. Biol. Soc.* **2009**, *2009*, 1113–1116. [CrossRef]
16. Keating, C.E.; Cullen, D.K. Mechanosensation in traumatic brain injury. *Neurobiol. Dis.* **2020**, *148*, 105210. [CrossRef] [PubMed]
17. Bartoletti, D.C.; Harrison, G.I.; Weaver, J.C. The number of molecules taken up by electroporated cells: Quantitative determination. *FEBS Lett.* **1989**, *256*, 4–10. [CrossRef]
18. Farkas, O.; Lifshitz, J.; Povlishock, J.T. Mechanoporation induced by diffuse traumatic brain injury: An irreversible or reversible response to injury? *J. Neurosci.* **2006**, *26*, 3130–3140. [CrossRef]
19. Singleton, R.H.; Povlishock, J.T. Identification and Characterization of Heterogeneous Neuronal Injury and Death in Regions of Diffuse Brain Injury: Evidence for Multiple Independent Injury Phenotypes. *J. Neurosci.* **2004**, *24*, 3543–3553. [CrossRef]
20. Whalen, M.J.; Dalkara, T.; You, Z.; Qiu, J.; Bermpohl, D.; Mehta, N.; Suter, B.; Bhide, P.G.; Lo, E.H.; Ericsson, M.; et al. Acute Plasmalemma Permeability and Protracted Clearance of Injured Cells after Controlled Cortical Impact in Mice. *J. Cereb. Blood Flow Metab.* **2007**, *28*, 490–505. [CrossRef]
21. Hernandez, M.L.; Chatlos, T.; Gorse, K.M.; Lafrenaye, A.D. Neuronal Membrane Disruption Occurs Late Following Diffuse Brain Trauma in Rats and Involves a Subpopulation of NeuN Negative Cortical Neurons. *Front. Neurol.* **2019**, *10*, 1238. [CrossRef]
22. Cullen, D.K.; Vernekar, V.N.; LaPlaca, M.C.; Burda, J.E.; Bernstein, A.M.; Sofroniew, M.V.; Haslach, H.W.; Leahy, L.N.; Hsieh, A.H.; Maneshi, M.M.; et al. Trauma-Induced Plasmalemma Disruptions in Three-Dimensional Neural Cultures Are Dependent on Strain Modality and Rate. *J. Neurotrauma* **2011**, *28*, 2219–2233. [CrossRef] [PubMed]
23. Wofford, K.L.; Harris, J.P.; Browne, K.D.; Brown, D.P.; Grovola, M.R.; Mietus, C.J.; Wolf, J.A.; Duda, J.E.; Putt, M.E.; Spiller, K.L.; et al. Rapid neuroinflammatory response localized to injured neurons after diffuse traumatic brain injury in swine. *Exp. Neurol.* **2017**, *290*, 85–94. [CrossRef]
24. Grovola, M.R.; Paleologos, N.; Wofford, K.L.; Harris, J.P.; Browne, K.D.; Johnson, V.; Duda, J.E.; Wolf, J.A.; Cullen, D.K. Mossy cell hypertrophy and synaptic changes in the hilus following mild diffuse traumatic brain injury in pigs. *J. Neuroinflamm.* **2020**, *17*, 44. [CrossRef]
25. Johnson, V.E.; Stewart, W.; Weber, M.T.; Cullen, D.K.; Siman, R.; Smith, D.H. SNTF immunostaining reveals previously undetected axonal pathology in traumatic brain injury. *Acta Neuropathol.* **2016**, *131*, 115–135. [CrossRef] [PubMed]
26. Arganda-Carreras, I.; Fernández-González, R.; Muñoz-Barrutia, A.; Ortiz-De-Solorzano, C. 3D reconstruction of histological sections: Application to mammary gland tissue. *Microsc. Res. Tech.* **2010**, *73*, 1019–1029. [CrossRef] [PubMed]
27. Schindelin, J.; Arganda-Carreras, I.; Frise, E.; Kaynig, V.; Longair, M.; Pietzsch, T.; Preibisch, S.; Rueden, C.; Saalfeld, S.; Schmid, B.; et al. Fiji: An open-source platform for biological-image analysis. *Nat. Methods* **2012**, *9*, 676–682. [CrossRef]
28. García-Cabezas, M.; John, Y.J.; Barbas, H.; Zikopoulos, B. Distinction of Neurons, Glia and Endothelial Cells in the Cerebral Cortex: An Algorithm Based on Cytological Features. *Front. Neuroanat.* **2016**, *10*, 107. [CrossRef]
29. Sofroniew, M.V.; Vinters, H.V. Astrocytes: Biology and pathology. *Acta Neuropathol.* **2010**, *119*, 7–35. [CrossRef]
30. Harris, J.P.; Mietus, C.J.; Browne, K.D.; Wofford, K.L.; Keating, C.E.; Brown, D.P.; Johnson, B.N.; Wolf, J.A.; Smith, D.H.; Cohen, A.S.; et al. Neuronal Somatic Plasmalemmal Permeability and Dendritic Beading Caused By Head Rotational Traumatic Brain Injury in Pigs—An Exploratory Study. *Front. Cell. Neurosci.* **2023**, *17*, 1055455. [CrossRef]
31. Johnson, V.E.; Weber, M.T.; Xiao, R.; Cullen, D.K.; Meaney, D.F.; Stewart, W.; Smith, D.H. Mechanical disruption of the blood–brain barrier following experimental concussion. *Acta Neuropathol.* **2018**, *135*, 711–726. [CrossRef]
32. Hammond, T.R.; Dufort, C.; Dissing-Olesen, L.; Giera, S.; Young, A.; Wysoker, A.; Walker, A.J.; Gergits, F.; Segel, M.; Nemesh, J.; et al. Single-Cell RNA Sequencing of Microglia throughout the Mouse Lifespan and in the Injured Brain Reveals Complex Cell-State Changes. *Immunity* **2019**, *50*, 253–271.e6. [CrossRef] [PubMed]
33. Saba, E.S.; Karout, M.; Nasrallah, L.; Kobeissy, F.; Darwish, H.; Khoury, S.J. Long-term cognitive deficits after traumatic brain injury associated with microglia activation. *Clin. Immunol.* **2021**, *230*, 108815. [CrossRef] [PubMed]
34. Paolicelli, R.; Sierra, A.; Stevens, B.; Tremblay, M.E.; Aguzzi, A.; Ajami, B.; Amit, I.; Audinat, E.; Bechmann, I.; Bennett, M.; et al. Microglia states and nomenclature: A field at its crossroads. *Neuron* **2022**, *110*, 3458–3483. [CrossRef] [PubMed]
35. Grovola, M.R.; von Reyn, C.; Loane, D.J.; Cullen, D.K. Understanding microglial responses in large animal models of traumatic brain injury: An underutilized resource for preclinical and translational research. *J. Neuroinflamm.* **2023**, *20*, 67. [CrossRef] [PubMed]
36. Salter, M.W.; Stevens, B. Microglia emerge as central players in brain disease. *Nat. Med.* **2017**, *23*, 1018–1027. [CrossRef]
37. Morrison, H.; Young, K.; Qureshi, M.; Rowe, R.K.; Lifshitz, J. Quantitative microglia analyses reveal diverse morphologic responses in the rat cortex after diffuse brain injury. *Sci. Rep.* **2017**, *7*, 13211. [CrossRef] [PubMed]
38. Smith, K.L.; Kassem, M.S.; Clarke, D.J.; Kuligowski, M.P.; Bedoya-Pérez, M.A.; Todd, S.M.; Lagopoulos, J.; Bennett, M.R.; Arnold, J.C. Microglial cell hyper-ramification and neuronal dendritic spine loss in the hippocampus and medial prefrontal cortex in a mouse model of PTSD. *Brain Behav. Immun.* **2019**, *80*, 889–899. [CrossRef]
39. Hinwood, M.; Tynan, R.J.; Charnley, J.L.; Beynon, S.B.; Day, T.; Walker, F.R. Chronic Stress Induced Remodeling of the Prefrontal Cortex: Structural Re-Organization of Microglia and the Inhibitory Effect of Minocycline. *Cereb. Cortex* **2013**, *23*, 1784–1797. [CrossRef]
40. Vidal-Itriago, A.; Radford, R.A.W.; Aramideh, J.A.; Maurel, C.; Scherer, N.M.; Don, E.K.; Lee, A.; Chung, R.S.; Graeber, M.B.; Morsch, M. Microglia morphophysiological diversity and its implications for the CNS. *Front. Immunol.* **2022**, *13*, 997786. [CrossRef]

41. Petersen, M.A.; Ryu, J.K.; Akassoglou, K. Fibrinogen in neurological diseases: Mechanisms, imaging and therapeutics. *Nat. Rev. Neurosci.* **2018**, *19*, 283–301. [CrossRef]
42. Szalay, G.; Martinecz, B.; Lénárt, N.; Környei, Z.; Orsolits, B.; Judák, L.; Császár, E.; Fekete, R.; West, B.L.; Katona, G.; et al. Microglia protect against brain injury and their selective elimination dysregulates neuronal network activity after stroke. *Nat. Commun.* **2016**, *7*, 11499. [CrossRef] [PubMed]
43. Zhao, Z.; Nelson, A.R.; Betsholtz, C.; Zlokovic, B.V. Establishment and Dysfunction of the Blood-Brain Barrier. *Cell* **2015**, *163*, 1064–1078. [CrossRef] [PubMed]
44. Burda, J.E.; Bernstein, A.M.; Sofroniew, M.V.; Angeles, L. Astrocyte roles in TBI. *Exp. Neurol.* **2016**, *275*, 305–315. [CrossRef]
45. Hay, J.R.; Johnson, V.E.; Young, A.M.; Smith, D.H.; Stewart, W. Blood-Brain Barrier Disruption Is an Early Event That May Persist for Many Years After Traumatic Brain Injury in Humans. *J. Neuropathol. Exp. Neurol.* **2015**, *74*, 1147–1157. [CrossRef] [PubMed]
46. Zhang, B.; Gaiteri, C.; Bodea, L.-G.; Wang, Z.; McElwee, J.; Podtelezhnikov, A.A.; Zhang, C.; Xie, T.; Tran, L.; Dobrin, R.; et al. Integrated Systems Approach Identifies Genetic Nodes and Networks in Late-Onset Alzheimer's Disease. *Cell* **2013**, *153*, 707–720. [CrossRef] [PubMed]
47. Liddelow, S.A.; Marsh, S.E.; Stevens, B. Microglia and Astrocytes in Disease: Dynamic Duo or Partners in Crime? *Trends Immunol.* **2020**, *41*, 820–835. [CrossRef] [PubMed]
48. Wolf, J.A.; Johnson, B.N.; Johnson, V.E.; Putt, M.E.; Browne, K.D.; Mietus, C.J.; Brown, D.P.; Wofford, K.L.; Smith, U.H.; Grady, M.S.; et al. Concussion Induces Hippocampal Circuitry Disruption in Swine. *J. Neurotrauma* **2017**, *34*, 2303–2314. [CrossRef]

Disclaimer/Publisher's Note: The statements, opinions and data contained in all publications are solely those of the individual author(s) and contributor(s) and not of MDPI and/or the editor(s). MDPI and/or the editor(s) disclaim responsibility for any injury to people or property resulting from any ideas, methods, instructions or products referred to in the content.

Article

Pupillary Light Response Deficits in 4-Week-Old Piglets and Adolescent Children after Low-Velocity Head Rotations and Sports-Related Concussions

Anna Oeur [1], Mackenzie Mull [1], Giancarlo Riccobono [1], Kristy B. Arbogast [2,3], Kenneth J. Ciuffreda [4], Nabin Joshi [5], Daniele Fedonni [2], Christina L. Master [2,3,6] and Susan S. Margulies [1,*]

1. Wallace H. Coulter Department of Biomedical Engineering, Emory University and Georgia Institute of Technology, Atlanta, GA 30332, USA
2. Center for Injury Research and Prevention, Children's Hospital of Philadelphia, Philadelphia, PA 19146, USA
3. Perelman School of Medicine, University of Pennsylvania, Philadelphia, PA 19104, USA
4. College of Optometry, State University of New York, New York, NY 10036, USA
5. Tesseract Health Inc., 530 Old Whitfield St., Guilford, CT 06437, USA
6. Sports Medicine and Performance Center, Children's Hospital of Philadelphia, Philadelphia, PA 19104, USA
* Correspondence: susan.margulies@emory.edu; Tel.: +1-(404)-727-9827

Abstract: Neurological disorders and traumatic brain injury (TBI) are among the leading causes of death and disability. The pupillary light reflex (PLR) is an emerging diagnostic tool for concussion in humans. We compared PLR obtained with a commercially available pupillometer in the 4 week old piglet model of the adolescent brain subject to rapid nonimpact head rotation (RNR), and in human adolescents with and without sports-related concussion (SRC). The 95% PLR reference ranges (RR, for maximum and minimum pupil diameter, latency, and average and peak constriction velocities) were established in healthy piglets (N = 13), and response reliability was validated in nine additional healthy piglets. PLR assessments were obtained in female piglets allocated to anesthetized sham (N = 10), single (sRNR, N = 13), and repeated (rRNR, N = 14) sagittal low-velocity RNR at pre-injury, as well as days 1, 4, and 7 post injury, and evaluated against RRs. In parallel, we established human PLR RRs in healthy adolescents (both sexes, N = 167) and compared healthy PLR to values obtained <28 days from a SRC (N = 177). In piglets, maximum and minimum diameter deficits were greater in rRNR than sRNR. Alterations peaked on day 1 post sRNR and rRNR, and remained altered at day 4 and 7. In SRC adolescents, the proportion of adolescents within the RR was significantly lower for maximum pupil diameter only (85.8%). We show that PLR deficits may persist in humans and piglets after low-velocity head rotations. Differences in timing of assessment after injury, developmental response to injury, and the number and magnitude of impacts may contribute to the differences observed between species. We conclude that PLR is a feasible, quantifiable involuntary physiological metric of neurological dysfunction in pigs, as well as humans. Healthy PLR porcine and human reference ranges established can be used for neurofunctional assessments after TBI or hypoxic exposures (e.g., stroke, apnea, or cardiac arrest).

Keywords: oculomotor; animal models; mild traumatic brain injury (mTBI); diagnostic methods; pupillary light response; visual processing; porcine

1. Introduction

Problems with vision are among commonly reported symptoms after a concussion [1], with 70% of children ages 11 to 17 years old reporting at least one problem, typically vergence disorders (the inability of both eyes to track an object up close (converge) or far away (diverge)) and accommodative disorders (the inability to maintain focus on an object as it changes in distance) [2]. Visual deficits affect children's ability to participate in school-, sports-, and work-related activities and may have long-term impacts on development, learning, and behavior [3].

Clinical pupillary light reflex (PLR) assessments have been incorporated as a diagnostic tool in neurodegenerative diseases such as traumatic brain injury (TBI) [1,4], Alzheimer's disease [5], Parkinson's disease [6], and autism [7]. The PLR controls pupil constriction and dilation in response to changes in light intensity to moderate the amount of light that reaches the retina [8]. As an assessment of the integrated response of retinal photoreceptor circuit that is innervated by (via the oculomotor nerve and hypothalamus/brainstem) the parasympathetic and sympathetic pathways to effect constriction of radial and/or circular muscles of the iris [1,9,10], pupillary response evaluations are powerful, noninvasive, and cost-effective assessments of autonomic nervous system function and can be used as quantitative biomarkers for diagnosis of a wide variety of neurological disorders [1,4,6,7]. Typical clinical pupillary dynamics metrics include pupil diameter, pupillary latency (the onset of pupil movement in reaction to the onset of light stimulation), change in pupil diameter, peak pupillary constriction velocity, and peak pupillary dilation velocity [1]. More recent clinical use of the PLR has employed a neurological pupil index (NPi) score derived from a patented algorithm using pupil size, constriction, and dilation parameters [11]. NPi values range from 0 to 5, where scores >3 are healthy and scores <3 are abnormal; this index has demonstrated some clinical utility in predicting TBI patient outcomes for those admitted to hospital [12,13] and identifying patients requiring surgical interventions at triage [14]. The potential to use these assessments in both clinical human settings and preclinical animal models presents a powerful tool for understanding the mechanistic basis for neurological dysfunction.

Thiagarajan and Ciuffreda [15] summarized the human literature and reported that a cohort of military adults with diagnosed concussions had significantly slower constriction and dilation velocity compared with controls when assessed several months after injury. In contrast, high-school-aged children with sports-related concussions (SRC) had opposite findings, where the injured cohorts had greater pupil diameters (maximum and minimum) and constriction velocities (average and peak) than the healthy controls [16,17] when evaluated on average 2 weeks post injury. In addition to the differences in age, the timing of the assessments (chronic vs. acute/subacute) may account for some of these differences.

Animal models provide invaluable opportunities to address the heterogeneity of causation of neurological deficits in humans that may in part lead to the incongruent PLR findings between children and adults at different timepoints after mild TBI (mTBI). Due to the shape of iris muscles, the pupil of mammals such as primates, piglets, and rodents remains circular, similar to humans, at different light intensities, in contrast to domestic cats and seals that have pupils that contract to create a vertical slit or sheep and reindeer that have horizontal pupils [18]. In animals, PLR is typically assessed in the laboratory using devices that are customized for small animals such as mice [19], rats [20], guinea pigs [21], large animals such as dogs [22], and rhesus monkeys [23]. To the authors' knowledge, there has only been one study examining pupil responses in animals after severe TBI, in which these authors studied head impacts delivered to monkeys and used the presence or absence of the PLR as a confirmation of sustaining a neurological insult [24].

The piglet model is increasingly used as a preclinical model to study a wide range of neurological diseases [25], such as ischemic stroke [26] and Alzheimer's disease [27], due to the similarities of its morphology and gray- and white-matter distributions that parallel human features [28,29]. The 4 week old piglet is a model of the adolescent brain and has been used to study severe TBI, including diffuse axonal injuries and brain hemorrhages [30–32]. In addition, piglet models have been used for mTBI studies for brain biomechanics [33,34], as well as diagnosis [35] and treatment [36]. With particular relevance to this study, an additional advantage of the large animal piglet model is that the conventional clinical equipment used in the clinical setting for humans can be used to study pigs [30], thereby facilitating the translation between species.

The aim of this study was to establish quantifiable PLR biomarkers of mTBI in 4 week old piglets using a clinically available handheld pupillometer designed for humans. First, we evaluated the feasibility of PLR measurements and utility of a pupillometer in 4 week old piglets by establishing a methodology for assessment of pupillary response in a cohort of healthy animals. After confirming feasibility, we established 95% reference ranges (RR) from these healthy animals and validated the RR for each PLR metric with a separate set of healthy animals. Secondly, we used the RR to evaluate a cohort of experimental animals in the acute phase (up to 7 days post injury) that experienced a single or repeated mild sagittal head rotations at load levels scaled from measured human soccer headers in the sagittal direction [37]. We hypothesized that porcine PLR metrics would be affected by load condition (single or repeated) and post-injury timepoint. In parallel, we examined human PLR metrics, using the same assessment tool, in both healthy adolescents and those within 28 days of injury from an SRC. We established similar RR for PLR metrics in the healthy human cohort and examined these metrics in adolescent humans with SRC. Comparison across species allows a more robust understanding of the neurological response.

2. Materials and Methods

2.1. Piglet Model

Fifty-nine 4 week old Yorkshire piglets (Sus scrofa) were obtained (Palmetto Research Swine, Reevesville, SC and Oak Hill Genetics, Ewing, IL, USA). Animals were housed together on a 7 a.m.–7 p.m. light–dark cycle and permitted to freely eat and drink as desired in the housing pen (LabDiet 5080, St. Louis, MO, USA). Prior to the start of each study cohort, the eyelashes of the animals were trimmed to aid in visualization of the pupil, and animals were acclimated to the behavior study room and commercially available piglet sling restraint with a metal frame for at least 2 days (Lomir Biomedical Inc., Notre-Dame-de-l'Île-Perrot, QC, Canada).

Pupillary dynamics in humans are measured by several technologies such as infrared videography [21], high-speed cameras [38], and smartphones [21]. A number of handheld automated devices, including PLR-200™ and PLRTM-3000 (NeurOptics Inc., Laguna Hills, CA, USA), RAPDx® (Konan Medical Inc., Irvine, CA, USA), and NeuroLight (IDMED, Marseille, France), are commercially available to measure pupillary response [39] and provide quick, quantitative, and repeatable measurements of pupillary dynamics for diagnostic and prognostic purposes. In the current study, one of the handheld infrared pupilometers for humans (PLRTM-3000, NeurOptics, Laguna Hills, CA, USA) was used to capture the pupillary responses to light in piglets [40,41]. The pupillometer (Figure 1) emits a light flash with fixed intensity to stimulate the pupil, captures the pupil via infrared camera (32 frames/s), and determines the pupil diameter with a ±0.03 mm precision. While the top and bottom eyelids were held gently by fingers, the pupillometer eye cup was placed in front of each eye, and the pupilometer was held steady. The eye cup gently encloses the animal's eye as the pupillometer captures an image of the eye and pupil in response to a light stimulus emitted from the device. A standard positive white-light pulse stimulus, which consists of a bright pulse over a dimmer background that triggers pupil constriction, was used in this study. This white-light stimulus (0.8 s duration, 180 µW) was preceded and followed by a dark (0 µW) background signal. All protocols were approved by Emory University Institutional Animal Care and Use Committee.

On the day of study, animals were fed before 9 a.m., and PLR measurements were performed between 9 a.m. and 1 p.m. Feeding animals before the experiments increased compliance with the assessment. During acclimation and data collection, food rewards, such as dried apple, yogurt chips, and fresh banana, were provided to reward head position and attention to the testing [42,43]. Anesthesia has been known to depress PLR responses such as pupil size, latency, and constriction velocity compared with the pre-anesthesia [22]; therefore, the PLR response of piglets was measured while animals were fully awake and stationary (Figure 2).

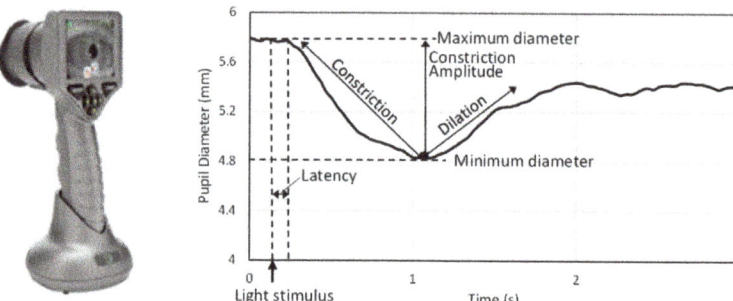

Figure 1. Handheld PLR-3000 device (**left**) and a typical pupil diameter time history and pupillary response parameters (**right**).

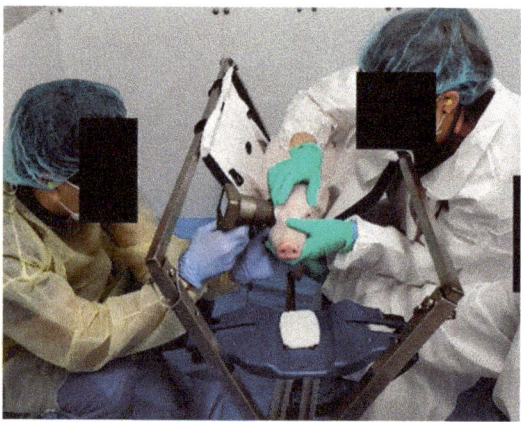

Figure 2. The experimental setup used in this study showing the position of the piglets in the sling and placement of the pupillometer on the animal's eye.

The pupillary response was recorded for 5 s to measure the constriction and the subsequent re-dilation of the pupil after the light flash. Repeated measurements of the same eye were performed at least 15 s apart. Figure 1 shows a typical time history of pupillary response to light and illustrates parameters associated with pupillary response. Analysis focused on metrics related to PLR dynamics: latency after light stimulus to constriction initiation (ms), average velocity of pupil diameter constriction (mm/s), and peak velocity of pupil diameter constriction (mm/s). Dilation velocity was not captured consistently and, therefore, was excluded. Abnormal readings were flagged by the device and were not included in data analysis. Additionally, trials with non-negative constriction velocity values, typically associated with spurious pupil diameter readings, were also excluded from the analysis.

2.2. Establishing the Healthy Reference Range

From the 59 subjects, we designated a healthy reference cohort (N = 13 females), in which the left and right eyes were tested three times over a 1 week period with ≥ 3 trials each day. The healthy reference cohort was studied over multiple days to establish the middle 95% healthy reference range (RR) for five metrics: maximum pupil diameter, minimum pupil diameter, pupil latency, average constriction velocity, and peak constriction velocity. An additional cohort of healthy piglets (N = 9, two males and seven females) were allocated to a validation data set. The two male piglets and three female piglets were tested ≥ 3 times a day for 3 days, and the four female piglets were tested ≥ 3 times in 1 day, but testing of this set was terminated prematurely due to unplanned facility closures. Therefore, the 19 testing days of this validation cohort was used to evaluate consistency and reliability of the healthy RR generated from the first set.

Because some subjects had usable data from only one eye, we sought to determine if healthy animals' responses differed between right and left eyes to evaluate the hypothesis that data from one eye could be representative of responses from both eyes to create the 95% RR. Pooling data from healthy reference cohort and healthy validation groups, the results of all trials for each animal were averaged within each day, handling the left and right eye separately. Using paired t-tests (JMP®, Version 15. SAS Institute Inc., Cary, NC, USA), we found no significant difference between measurements of the left and right eye ($p > 0.05$).

Next, we evaluated the effect of test day in the healthy animals (one-way repeated-measure ANOVA) and determined no significant effect of test day. Therefore, to determine the healthy RR, each healthy animal's data were first averaged across trials (within a day), then across the eyes (when only one eye passed quality control, that value was used), and then averaged again across test days, resulting in a single value for each animal. The subsequent values from all 13 animals were again averaged, and the 2.5th and 97.5th percentiles were calculated to generate the 95% RR for healthy animals. Data from each of the validation animals were then compared to the healthy RR to determine the percentage of validation animals that fell within the healthy RR for each PLR metric. For the validation cohort, each animal's data were first averaged across trials, and then across eyes (when only 1 eye passed quality control, that value was used); next, each day was treated as a separate validation data point resulting in 19 validation values.

2.3. Experimental RNR Animals

The remaining (N = 37) animals were allocated to experimental groups: anesthetized shams (N = 10 females); single (sRNR) and repeated, rapid, nonimpact head rotation (rRNR). The sRNR (N = 13 females) consisted of a single sagittal rotation, and the rRNR group (N = 14 females) experienced five sagittal head rotations all within 1 h. The sRNR group experienced a single "high" rotation and the rRNR group experienced one "high" load followed by four "medium" loads spaced 8.4 ± 1.1 min apart (Table 1). The "high" and "medium" levels for piglet TBI loading conditions were scaled from soccer headers in high-school players instrumented with wearable head impact sensors. A total of 267 frontal headers resulting in primarily sagittal head loading were verified via video footage [37]. The kinematic data from the head impact sensors were input into a finite element model of the human brain, and the maximum axonal strain (MAS) was estimated for each head impact [44,45]. From the human data, the 50th and 90th percentile MAS values were identified. These human MAS values were scaled to determine the peak angular velocity and angular acceleration for piglets that would produce a strain-based deformation of the same magnitude [44,45]. Scaled loading conditions for pigs were such that the target load for the single (sRNR) cohort was one scaled 90th percentile "header" (100 rad/s and 36 krad/s^2) and the repeated (rRNR) cohort target loads were one 90th percentile scaled "header", followed by four 50th percentile "headers" (60 rad/s and 13 krad/s^2). From the same human soccer heading data, we found that boys typically experienced six impacts per hour and girls four impacts per hour, both spaced 8 min apart [37].

Table 1. Mean ± standard error of angular velocity and angular accelerations for single and multiple RNR loading groups.

Injury Group	Load Level	Angular Velocity (rad/s)	Angular Acceleration (rad/s^2)
Single	High	104.3 (±0.49)	38,444 (±1272)
Multiple	Medium	61.18 (±0.20)	15,033 (±169.0)
	High	104.5 (±0.40)	38,394 (±509.3)

Head rotations were performed in animals that were sedated with ketamine/xylazine/midazolam (4/2/0.2 mg/kg IM), anesthetized with isoflurane (1–5%) via mask, and administered the analgesic buprenorphine (0.1 mg/kg IM). Rapid nonimpact head rotations (RNR) were delivered using a pneumatic HYGE device (HYGE, Inc., Kittanning, PA, USA) described elsewhere [30]. Angular transducers were affixed to a side-arm linkage to capture angular kinematics of the head rotational events (Applied Technology Associates (ATA), Inc., Arlington, VA, USA and Diversified Technical Systems, Inc., Seal Beach, CA, USA). Measured velocity and calculated accelerations are provided in Table 1. For sham, sRNR and rRNR cohorts, at least three left and right eye PLR measurements were taken on a pre-injury day, as well as on days 1, 4, and 7 post RNR injury.

2.4. Human Adolescent Participants

A secondary analysis was performed on PLR data collected from a prospective cohort of adolescents between ages 12 and 19 years with pupillary light reflex (PLR) assessment conducted between 1 August 2017 and 11 May 2021, recruited from a specialty concussion program and private suburban high school, where some results from this cohort have been previously published [16]. The prospective observational cohort study was approved by the Children's Hospital of Philadelphia institutional review board. Adolescents and/or their parents/legal guardians provided written assent/written informed consent. Pupillary light reflex metrics were measured via the same Neuroptics PLR-3000 handheld, infrared, automated, monocular pupillometer model used for the piglet portion of this study [46]. The pupillometer is approved by the US Food and Drug Administration and has been used in similar studies of mTBI in human adults and adolescents. The diagnosis of SRC was made by trained sports medicine pediatricians on the basis of the most recent Consensus Statement on Concussion in Sports [47], and all adolescents with concussion were assessed with pupillometry within 28 days of injury. Overall, PLR data from 167 healthy controls were used to establish a healthy RR for humans. PLR metrics for 177 concussed cases were obtained and compared to the healthy RR.

2.5. Statistics

For each PLR metric, comparisons between the experimental animals and the healthy RR were completed. Data for each injury group (sham, sRNR, and rRNR) and study day (preinjury, as well as days 1, 4, and 7 post injury) were evaluated to determine the proportion of animals that fell within the healthy RR. To test if the proportion of animals that fell within the healthy RR on each day was significantly different by loading group (group effect at each day), a Fisher's exact test was conducted. To test if the proportion of injured animals that fell within the healthy RR for each loading group differed by day (day effect per loading group), a Cochran's Q test was performed followed by a McNemar's post hoc test within each injury group.

We did not assume that the effects of sagittal loading on the visual pathway were axisymmetric. Therefore, a three-way ANOVA with repeated measures for day was run to test the effect of eye (left or right), day, and loading group (sham, sRNR, and rRNR). For metrics where eye was not a significant factor, a two-way ANOVA with repeated measures was performed to determine the effect of study day and loading group (sham, sRNR, and rRNR) with Bonferroni post hoc tests for each PLR measure. All statistical analyses were conducted using SPSS Statistical software (V 28, IBM), and significance was accepted at $p < 0.05$.

For human PLR metrics, the proportion of concussed cases with PLR metrics within the healthy RR was compared to the proportion of healthy controls within the healthy reference range . The proportion of PLR metrics from cases at initial visit that were within the healthy reference range were compared to healthy controls using χ^2 with a significance level of 0.05. Additionally, the proportion of concussed cases with PLR metrics within health RR was compared by sex and previous history of concussion using χ^2 with a significance level of 0.05.

3. Results

3.1. Animal Results

For each of the five PLR metrics, 95% of the validation animals fell within the healthy RR for maximum diameter and latency, 84% of validation animals fell within the RR for minimum diameter, and 68% of validation animals fell within the RR for average constriction velocity. However, 0% of validation animals fell within the RR for peak constriction velocity; thus, peak constriction velocity was excluded from further analysis, resulting in statistical comparisons completed for the remaining four metrics (maximum and minimum diameter, latency, and average constriction velocity). The values for mean, standard deviation, 2.5th, 50th, and 97.5th percentiles of pupillary response metrics from healthy animals (N = 13) are provided in the Appendix A Table A1. Maximum and minimum diameter were the only metrics that had significant findings for the reference range statistical analysis, where day 1 post RNR responses were significantly below the healthy RR for rRNR compared with shams and their own pre-injury baselines (Figure 3A,C).

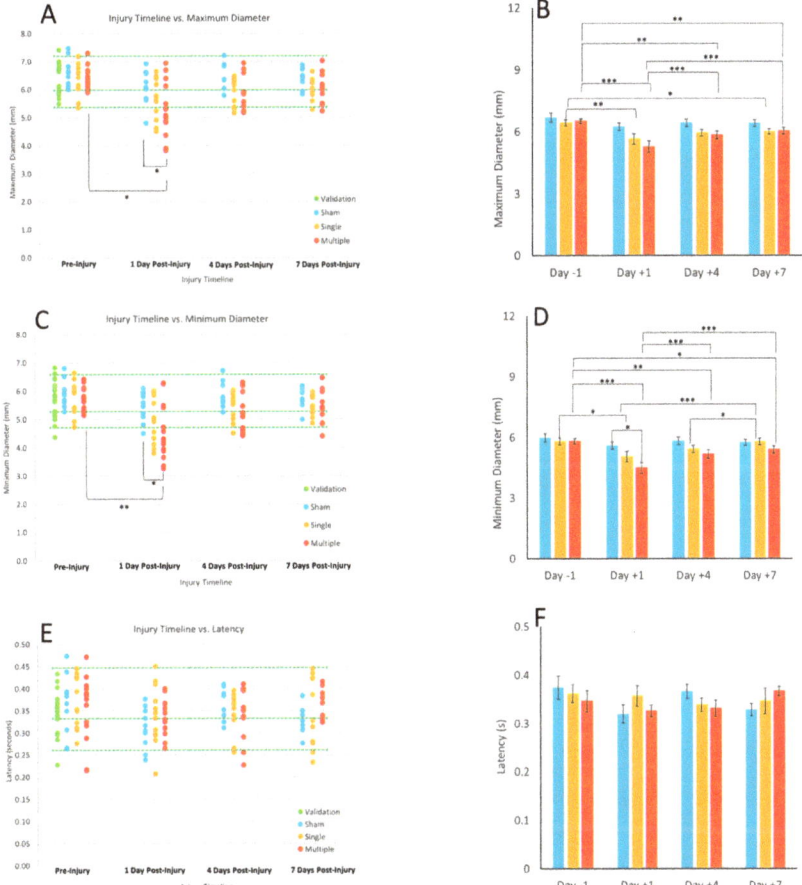

Figure 3. Proportion of validation and experimental animals in the healthy reference range (**A,C,E**) and two-way repeated-measure ANOVA results for the effect of injury group and study day (**B,D,F**) for maximum diameter (**top row**), minimum diameter (**middle row**), and latency (**bottom row**). Blue bars represents sham, yellow bars represents sRNR, and red bars represents rRNR groups. The healthy RR is demarcated with green dotted lines in (**A,C,E**). Overlaying cross bars show significant comparisons with * $p < 0.05$, ** $p < 0.01$, and *** $p < 0.001$.

The three-way ANOVA did not yield significant results for maximum diameter, minimum diameter, and latency; therefore a two-way ANOVA was conducted for these metrics. Two-way repeated-measure ANOVA results demonstrate that sRNR had reduced maximum diameters at days 1 and 7 post TBI in comparison to pre-injury levels (Figure 3B). The rRNR group had reduced maximum diameters relative to pre-injury on all days studied, where values on day 1 post TBI were significantly smaller than those on days 4 and 7 (Figure 3B). There was a main effect of load group on minimum diameter on day 1 post TBI, where the rRNR had reduced minimum diameters compared to sham ($p = 0.025$). Furthermore, sRNR animals had significantly reduced minimum diameters at day 1 relative to pre-injury levels, where day 1 remained significantly reduced compared to day 4 and 7 (Figure 3D). The rRNR animals had significantly reduced minimum diameters on all post-injury days relative to pre-injury, where day 1 values remained significantly lower than days 4 and 7 (Figure 3D). There were no significant findings for the reference range analysis or two-way ANOVA findings for latency (Figure 3E,F).

For average constriction velocity, the three-way ANOVA showed a significant interaction effect between side and loading group. Figure 4A illustrates the average constriction velocities for both the left eye (solid dot) and right eye (patterned dot) of each animal per loading group (sham, sRNR, and rRNR). Fisher's exact test for the effect of group at each day was significant for the left eye on all days; however, results for the right eye did not reach significance. Because the Cochran's Q test and McNemar's test were not significant for either eye, we found no overall or day-by-day relationship between groups for the proportion of animals within the average constriction velocity healthy RR. Figure 4B shows the post hoc tests from the ANOVA that employed paired t-tests for eye per loading group, and it was found that the left eye had faster constriction velocities than the right eye in the sham group at pre-injury and in the sham and sRNR groups at day 1 ($p < 0.05$). Furthermore, significant differences were found for right eyes in the single group between pre-injury and day 1, for right eyes between single and repeated RNR groups on day 1, and for left eyes between sham and sRNR on day 4 (Figure 4B).

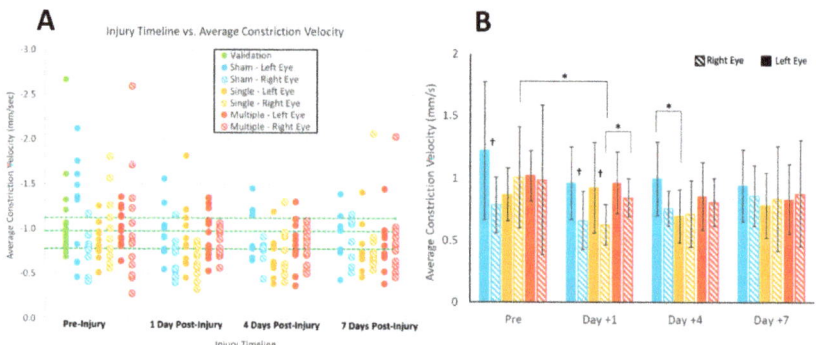

Figure 4. Proportion of validation and experimental animals in the healthy reference range (**A**) and three-way repeated-measure ANOVA for side, injury group, and study day on average constriction velocity (**B**). The effect of side was found to be significant for average constriction velocity; therefore, left and right eye data are plotted (**B**). Blue bars represents sham, yellow bars represents sRNR, and red bars represents rRNR groups. The healthy RR is demarcated with green dotted lines in (**A**). Overlaying cross bars show significant comparisons with * $p < 0.05$. † in (B) illustrates significant differences between left and right eyes ($p < 0.05$).

3.2. Human Results

The patient data and patient characteristics for the human adolescent cohort of healthy and concussed individuals is described in Tables 2 and 3.

Table 2. Percentage within 2.5th and 97.5th percentile reference range comparison for human study cohort.

Metric	Cases at Initial Visit (N = 177)
Latency (ms)	94.9
Average constriction velocity (mm/s)	92.6
Maximum constriction velocity (mm/s)	93.2
Maximum pupil diameter (mm)	85.8 *
Minimum pupil diameter (mm)	99.4

*Adolescents with concussion had a significantly lower proportion within the healthy reference range for this parameter compared to healthy adolescents.

Table 3. Demographic and clinical characteristics of the study cohort.

Demographics	Healthy Controls (N = 167)	Concussed Cases (N = 177)
Sex		
Female	91 (54.5%)	97 (54.8%)
Male	76 (45.5%)	80 (45.2%)
Race/ethnicity		
Non-Hispanic White	125 (77.6%)	144 (81.8%)
Non-Hispanic Black	19 (11.8%)	14 (7.9%)
Hispanic	6 (3.7%)	6 (3.4%)
Non-Hispanic Asian	5 (3.1%)	3 (1.7%)
Non-Hispanic mixed race	5 (3.1%)	1 (0.6%)
Non-Hispanic other	1 (0.6%)	8 (4.6%)
Age (Median, IQR)	15.6 (14.3–17.2)	15.4 (14.2–16.6)
Previous concussion	40 (23.95)	81 (45.76)

Overall, PLR data from 167 healthy controls were used to establish a healthy RR for humans (Table A2 in Appendix A). The 95% reference ranges for latency (180.0–243.3 ms), average constriction velocity (1.26–3.91 mm/s), peak constriction velocity (2.00–6.21 mm/s), maximum pupil diameter (2.67–5.57 mm), and minimum pupil diameter (2.02–3.4 mm) were established from the PLR metrics from the sub-cohort of healthy adolescents.

PLR metrics for 177 concussed cases were obtained and compared to the healthy RR (Table 2). Adolescents with concussion had a significantly lower proportion (85.8%) within the healthy reference range for maximum pupil diameter compared to healthy adolescents (95.8%). The proportions of all other PLR metrics within the healthy reference ranges for concussed did not have significant differences from the proportion of healthy adolescents within the reference range compared to the corresponding PLR metrics.

In contrast to the piglet data, there was no difference in PLR metrics between adolescents with multiple concussions compared to those with only one concussion. In addition, there was no significant difference in PLR metrics between female and male adolescents (see Tables A3 and A4 in Appendix A).

4. Discussion

This is a first report of the pupillary light reflex in a porcine preclinical model of TBI. Importantly, a commercially available pupillometer, commonly used in clinical assessments, could obtain reliable and consistent PLR measurements in 4 week old piglets to obtain maximum and minimum pupil diameter, highlighting interesting nuances with average constriction velocity. We established healthy and validated reference ranges which can serve as baseline comparisons for use in the preclinical and animal sciences literature. We show that head rotations, characteristic of typical heading kinematics in soccer, yielded significantly reduced maximum and minimum diameter on day 1 post sRNR in comparison to pre-injury. In addition, the rRNR group had faster average constriction velocities than the sRNR group on day 1 post injury. Interestingly, pupil diameters (maximum and minimum) tended to approach recovery on days 4 and 7; however, these metrics remained significantly lower than at pre-injury, which suggests that PLR deficits may persist beyond the acute (7 day) period for mTBI in piglets. The injury paradigm employed in the animal model illustrated the effect of severity when comparing a single head rotation to repeated head rotations. There were a greater number of significant pupil deficits in the rRNR group than the sRNR group, as well as a greater magnitude of those differences on each study day (e.g., maximum and minimum diameter, Figure 3A,B and Figure 3C,D).

In addition, to compare across species, we examined human adolescent PLR metrics, from both healthy and concussed individuals (Figure 5). We established an RR from healthy adolescents for each PLR metric and compared the proportion of concussed and healthy adolescents falling within the RR for each group. In our investigation of the human adolescent PLR data, we found that maximum pupil diameter was the only PLR metric to have a significantly fewer concussed adolescents falling within the RR, whereas the other metrics yielded similar proportions of healthy and concussed adolescents within the RR. This may be, in part, due to the mild nature of SRC along the spectrum of TBI, likely representing the mildest form of mTBI and, as such, resulting in subtle physiological perturbations. Quantitative PLR metrics remain a promising target outcome measure for identifying and monitoring SRC; our findings in this investigation indicate that a narrower RR with more stringent criteria for "healthy" may be useful, and future studies should investigate optimizing sensitivity of these measures to an mTBI such as SRC, while maximizing specificity.

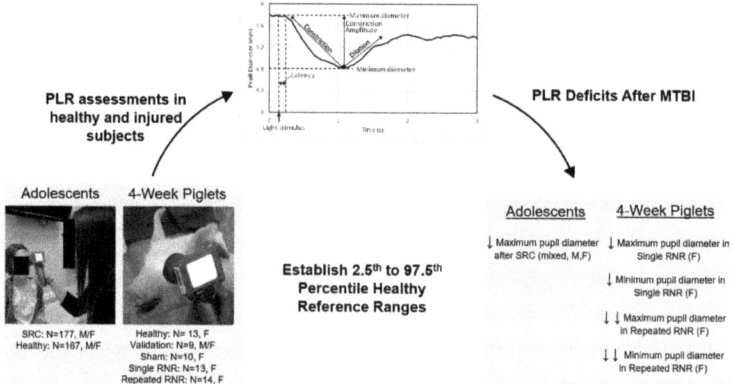

Figure 5. Illustration of PLR cohorts and comparison of main findings for adolescent children with and without an SRC and healthy and experimental RNR 4 week old piglets. M = male, F = female. Mixed refers to mixed head injury mechanisms causing the SRC. A greater number of PLR deficits were observed in controlled piglet RNR studies, with repeated RNR having more severe deficits.

Interestingly, our comparative study of piglet and human PLR metrics revealed differences in the direction of post-mTBI alterations. Specifically, while the piglets PLR metrics were diminished acutely (during the first week) after mTBI, the PLR metrics in the concussed adolescents were enhanced sub-acutely (<28 days) after mTBI. The athletic adolescent population may be a special cohort in comparison with their nonathletic counterparts, potentially exhibiting unique pupillary responses as a result of sports and skill demands. With the assumption that the developmental stages of the piglet and human adolescents are similar, the main remaining explanation for the different findings could be the timing of assessment and similarity of the injury severity and mechanism (total number, direction of head loading, and magnitude).

In a study examining 200 adult military personnel in the acute phase of concussion (<72 h) after non-blast-related mTBI (i.e., from aerial jumps, motor vehicle accidents, falls, sports, and recreation), the authors found that those with an mTBI had the following decreased PLR metrics: maximum and minimum diameter, percentage constriction, average constriction velocity, and average dilation velocity in comparison to control (N = 100) [48]. A separate study investigated military patients diagnosed with blast-induced mTBI (N = 20) compared to age-matched controls (N = 20) in the subacute period post injury (15–45 days), and the results also reported that maximum and minimum pupil diameters, average constriction velocity, and maximum constriction velocity were significantly lower in the mTBI group [49]. Another report in military subjects with chronic (<1 year) non-blast mTBI (N = 17) compared to healthy controls (N = 15) also found decreased maximum constriction velocity, average constriction velocity, and maximum and minimum pupil diameter among others [46]. These acute and chronic results in adults are similar to our piglet results.

Age has a demonstrable influence on injury responses. In contrast to adults, we (Master and Podolak [16]) examined a younger human population and conducted a study of high-school athletes (12–18 years old) with a sports-related concussion (N = 110) and measured PLR in the subacute period (within 28 days). We found that the injured cohorts had greater pupil diameters (maximum and minimum) and constriction velocities (average and peak) than the healthy controls (N = 143). We hypothesized that concussion may cause a traumatically induced autonomic dysfunction in the sympathetic pathway, the primary driver influencing the dilation PLR response. Greater maximum pupil diameters allow more light to enter the eye during light stimulation, affecting subsequent constriction velocity and dilation dynamics [16]. In a separate study, Hsu and Stec [17] obtained PLR in children seen in a concussion clinic (N = 92) with an average date from injury of almost 2 months and reported similar findings, where average and peak constriction velocities were greater in the concussed pediatric cohort than in controls (N = 192).

Differences in these PLR findings between adults and children have been attributed to age as PLR measures have been known to be sensitive to age and change throughout development [1,21]. Additionally, the time post TBI when the PLR measurements were taken represents another factor that could lead to incongruous findings. Lastly, the mechanism of injury is an additional potential contributing factor for the differences observed. Pupil diameters (minimum and maximum) were not found to be different between groups of children in Hsu and Stec's [17] study; however, the values were larger (6 mm) than those reported in Master and Podolak's [16] study (4.8 mm). A larger maximum pupil diameter at the beginning of the PLR test would result in larger response amplitudes, as well as greater pupil dynamics and constriction velocities (average and peak), which could help to explain the findings observed in these pediatric studies [15]. Future longitudinal studies are needed in order to delineate the trajectory of PLR metrics over the clinical course after injury, to determine the pattern of changes in adults and children, which may better characterize a pattern with diminished PLR metrics both acutely and chronically, perhaps demonstrating a subacute enhancement of PLR measures in the first weeks after injury.

Injury setting may contribute to the differences in PLR responses, in addition to age and timing. Studies involving military personnel had subjects with concussions from blast injury and blunt force head trauma from aerial maneuvers, falls, and combat training. In contrast, the pediatric studies were primarily injuries due to sport and recreation (falls onto various surfaces and player-to-player contact). The head loading conditions, comprising various combinations of surfaces, inbound velocities, impact locations, and directions, may play an important role in how energy is translated to the head and brain resulting in trauma. Variations in loading conditions have been demonstrated to result in distinct patterns of head kinematics [50], stresses and strains from computational modelling [51], neurobehaviors in animals [52], and histopathology [51]. The studies involving concussions in children and adults did not control for biomechanical differences in head loading conditions, which may further contribute to the disparate findings.

Our piglet study had several limitations. First, we studied head loading in the sagittal direction only, which applies the biomechanical loads equally across the two hemispheres and visual pathways. Despite restricting the study to the sagittal direction to limit the influences of asymmetric kinematics on the eyes, we observed differences between the left and right eyes, albeit only for average constriction velocity and in the sham and sRNR groups (Figure 4D). Other directions may reveal more pronounced sidedness. Most studies in the literature treated patient PLR data as an average between the left and right eye [16,17]; however, in a more recent clinical study involving stroke and TBI patients, a difference in the Npi score of the right and left eye was associated with poorer outcomes and greater levels of disability at discharge [53]. Therefore, future studies changing loading direction should continue to conduct analyses specific to the left or right eye. Second, only female piglets were examined in this study for the injury cohort, which may limit extrapolations to male piglets. In a study examining males and females and tracking the time to recovery after a concussion, females were found to take longer to recover, with a longer time for symptoms to abate and to return to scholastic and sports activities than males [54]. Therefore, the trajectory of PLR deficits and recovery may not only be sex-dependent but also affected by age, as well as injury severity and mechanism. Future work should include changing head loading direction, including male animals, and studying animals beyond 7 days to track when PLR deficits return to baseline to better understand injury severity (single and repeated RNR) and predict recovery trajectories on the basis of biomechanical conditions.

5. Conclusions

In this study, we obtained consistent and reliable pupillary light reflex responses in 4 week old piglets using a commercially available pupillometer commonly used in clinical assessments and established healthy reference ranges for a substantial number of animals. Although this study highlighted some species differences in pupillary light metrics, conducting parallel experimental studies using the same device, methodology, and metrics in human and animals, as introduced in this study, can facilitate preclinical–clinical translation of objective, involuntary diagnostics and treatment efficacy metrics. We applied these ranges to study PLR in animals subject to single and repeated mild head rotations characteristic of non-injurious heading in soccer during the acute time period (within 7 days); we found that the most severe deficits occurred on day 1 after injury, and that repeated injuries tended to be more severe than a single mild head rotation as reflected in a greater number of PLR deficits and a larger magnitude of those decreases. Furthermore, pupil diameters (maximum and minimum) remained significantly decreased 7 days after low-velocity head rotations. We conclude that PLR holds promise as a feasible involuntary quantitative physiological metric of neurological dysfunction in piglets.

Author Contributions: Conceptualization, S.S.M. and A.O.; methodology, M.M., A.O., G.R., C.L.M. and K.B.A.; validation, M.M., A.O., G.R. and C.L.M.; formal analysis, M.M., A.O., C.L.M. and D.F.; investigation, S.S.M., A.O., K.B.A. and C.L.M.; resources, S.S.M., K.B.A. and C.L.M.; data curation, M.M., A.O., C.L.M. and D.F.; writing—original draft preparation, A.O., G.R., S.S.M., K.B.A., C.L.M. and D.F.; writing—review and editing, A.O., M.M., G.R., K.B.A., K.J.C., N.J., D.F., C.L.M. and S.S.M.; visualization, A.O., M.M. and C.L.M.; supervision, S.S.M.; project administration, S.S.M.; funding acquisition, S.S.M., K.B.A. and C.L.M. All authors have read and agreed to the published version of the manuscript.

Funding: This research was supported by the National Institutes of Health (R01NS097549) and the Georgia Research Alliance.

Institutional Review Board Statement: All procedures were approved by the Institutional Animal Care and Use Committee (IACUC) at Emory University School of Medicine. Experiments were carried out in an Assessment and Accreditation of Laboratory Animal Care (AAALAC)-accredited facility. All human studies were approved by the Children's Hospital of Philadelphia Institutional Review Board.

Informed Consent Statement: Not applicable.

Data Availability Statement: The datasets used and/or analyzed during the current study are available from the corresponding author on reasonable request.

Acknowledgments: This work was supported by the National Institutes of Health [NIH-R01NS097549] and the Georgia Research Alliance. The authors would like to thank Rebecca Grant, Brittany Davis, Hayden Thomas, Gbemi Aderibigbe, Akshara Thakore, and Morteza Seidi for assistance with the animal studies.

Conflicts of Interest: The authors declare no conflict of interest.

Appendix A

Table A1. Mean, standard deviation, and 2.5th, 50th, and 97.5th percentiles of pupillary response metrics from healthy animals (N = 13).

Metric	Mean	STD	2.5th %	50th %	97.5th %
Latency (s)	0.34	0.087	0.26	0.33	0.45
Average constriction velocity (mm/s)	−1.0	0.21	−1.1	−0.98	−0.78
Peak constriction velocity (mm/s)	−2.2	0.40	−2.6	−2.2	−1.9
Maximum diameter (mm)	6.0	0.58	5.4	5.8	7.2
Minimum diameter (mm)	5.4	0.57	4.7	5.3	6.6

Table A2. Mean, standard deviation, and 2.5th, 50th, and 97.5th percentiles of pupillary response metrics of healthy controls.

Metric	Mean	STD	2.5th %	50th %	97.5th %
Latency (s)	0.21	0.017	0.18	0.20	0.24
Average constriction velocity (mm/s)	2.6	0.71	1.3	2.7	3.9
Maximum constriction velocity (mm/s)	4.1	1.1	2.0	4.1	6.2
Maximum pupil diameter (mm)	4.1	0.78	2.7	4.1	5.6
Minimum pupil diameter (mm)	2.6	0.38	2.0	2.6	3.4

Table A3. Percentage of cases with a history of concussion compared with those without a history (total concussed, N = 177). No significant differences were found between groups for each PLR metric.

Metric	History of Concussion	No History of Concussion
Latency (s)	94.21%	96.41%
Average constriction velocity (mm/s)	91.74%	95.07%
Maximum constriction velocity (mm/s)	91.74%	95.52%
Maximum pupil diameter (mm)	87.60%	92.38%
Minimum pupil diameter (mm)	100.00%	97.76%

Table A4. Percentage of male and female cases (total concussed, N = 177). No significant differences were found between sexes for each PLR metric.

Metric	History of Concussion	No History of Concussion
Latency (s)	95.88%	93.75%
Average constriction velocity (mm/s)	91.75%	93.75%
Maximum constriction velocity (mm/s)	92.78%	93.75%
Maximum pupil diameter (mm)	84.54%	87.50%
Minimum pupil diameter (mm)	98.97%	100.00%

References

1. Ciuffreda, K.J.; Joshi, N.R.; Truong, J.Q. Understanding the effects of mild traumatic brain injury on the pupillary light reflex. *Concussion* **2017**, *2*, CNC36. [CrossRef]
2. Scheiman, M.; Grady, M.F.; Jenewein, E.; Shoge, R.; Podolak, O.E.; Howell, D.H.; Master, C.L. Frequency of oculomotor disorders in adolescents 11 to 17 years of age with concussion, 4 to 12 weeks post injury. *Vis. Res.* **2021**, *183*, 73–80. [CrossRef]
3. Haarbauer-Krupa, J.; Arbogast, K.B.; Metzger, K.B.; Greenspan, A.I.; Kessler, R.; Curry, A.E.; Bell, J.M.; DePadilla, L.; Pfeiffer, M.R.; Zonfrillo, M.R.; et al. Variations in Mechanisms of Injury for Children with Concussion. *J. Pediatr.* **2018**, *197*, 241–248.e1. [CrossRef] [PubMed]
4. Truong, J.Q. Mild Traumatic Brain Injury (mTBI) and Photosensitivity: Objective Pupillometric Findings. 2016. Available online: http://hdl.handle.net/20.500.12648/1145 (accessed on 4 September 2021).
5. Bittner, D.M.; Wieseler, I.; Wilhelm, H.; Riepe, M.W.; Mueller, N.G. Repetitive pupil light reflex: Potential marker in Alzheimer's disease? *J. Alzheimer's Dis.* **2014**, *42*, 1469–1477. [CrossRef] [PubMed]
6. Giza, E.; Fotiou, D.; Bostantjopoulou, S.; Katsarou, Z.; Karlovasitou, A. Pupil Light Reflex in Parkinson's Disease: Evaluation with Pupillometry. *Int. J. Neurosci.* **2011**, *121*, 37–43. [CrossRef] [PubMed]
7. Nyström, P.; Gliga, T.; Jobs, E.N.; Gredebäck, G.; Charman, T.; Johnson, M.H.; Bölte, S.; Falck-Ytter, T. Enhanced pupillary light reflex in infancy is associated with autism diagnosis in toddlerhood. *Nat. Commun.* **2018**, *9*, 1678. [CrossRef] [PubMed]
8. Park, J.C.; Moura, A.L.; Raza, A.S.; Rhee, D.W.; Kardon, R.H.; Hood, D.C. Toward a clinical protocol for assessing rod, cone, and melanopsin contributions to the human pupil response. *Investig. Ophthalmol. Vis. Sci.* **2011**, *52*, 6624–6635. [CrossRef]
9. Truong, J.Q.; Ciuffreda, K.J. Comparison of pupillary dynamics to light in the mild traumatic brain injury (mTBI) and normal populations. *Brain Inj.* **2016**, *30*, 1378–1389. [CrossRef]
10. Sand, A.; Schmidt, T.M.; Kofuji, P. Diverse types of ganglion cell photoreceptors in the mammalian retina. *Prog. Retin. Eye Res.* **2012**, *31*, 287–302. [CrossRef]
11. Chen, J.W.; Vakil-Gilani, K.; William, C.J.; Cecil, S. Infrared pupillometry, the Neurological Pupil index and unilateral pupillary dilation after traumatic brain injury: Implications for treatment paradigms. *Springerplus* **2014**, *3*, 548. [CrossRef]
12. Butt, A.A.; Atem, F.D.; Stutzman, S.E.; Aiyagari, V.; Venkatachalam, A.M.; Olson, D.M.; Yokobori, S. Contribution of pupillary light reflex assessment to Glasgow Coma Scale for prognostication in patients with traumatic brain injury. *J. Neurocrit. Care* **2021**, *14*, 29–35. [CrossRef]
13. Teixeira, T.L.; Peluso, L.; Banco, P.; Njimi, H.; Abi-Khalil, L.; Pillajo, M.C.; Schuind, S.; Creteur, J.; Bouzat, P.; Taccone, F.S. Early Pupillometry Assessment in Traumatic Brain Injury Patients: A Retrospective Study. *Brain Sci.* **2021**, *11*, 1657. [CrossRef] [PubMed]
14. El Ahmadieh, T.Y.; Bedros, N.; Stutzman, S.E.; Nyancho, D.; Venkatachalam, A.M.; MacAllister, M.; Ban, V.S.; Dahdaleh, N.S.; Aiyagari, V.; Figueroa, S.; et al. Automated Pupillometry as a Triage and Assessment Tool in Patients with Traumatic Brain Injury. *World Neurosurg.* **2021**, *145*, e163–e169. [CrossRef] [PubMed]
15. Thiagarajan, P.; Ciuffreda, K.J. Accommodative and pupillary dysfunctions in concussion/mild traumatic brain injury: A Review. *Neurorehabilitation* **2022**, *50*, 261–278. [CrossRef] [PubMed]
16. Master, C.L.; Podolak, O.E.; Ciuffreda, K.J.; Metzger, K.B.; Joshi, N.R.; McDonald, C.C.; Margulies, S.S.; Grady, M.F.; Arbogast, K.B. Utility of Pupillary Light Reflex Metrics as a Physiologic Biomarker for Adolescent Sport-Related Concussion. *JAMA Ophthalmol.* **2020**, *138*, 1135–1141. [CrossRef]
17. Hsu, J.; Stec, M.; Ranaivo, H.R.; Srdanovic, N.; Kurup, S.P. Concussion Alters Dynamic Pupillary Light Responses in Children. *J. Child Neurol.* **2020**, *36*, 195–202. [CrossRef]
18. Douglas, R.H. The pupillary light responses of animals; a review of their distribution, dynamics, mechanisms and functions. *Prog. Retin. Eye Res.* **2018**, *66*, 17–48. [CrossRef]
19. Kircher, N.; Crippa, S.V.; Martin, C.; Kawasaki, A.; Kostic, C. Maturation of the Pupil Light Reflex Occurs Until Adulthood in Mice. *Front. Neurol.* **2019**, *10*, 56. [CrossRef]
20. Schriver, B.J.; Bagdasarov, S.; Wang, Q. Pupil-linked arousal modulates behavior in rats performing a whisker deflection direction discrimination task. *J. Neurophysiol.* **2018**, *120*, 1655–1670. [CrossRef]
21. Chang, L.Y.-L.; Turuwhenua, J.; Qu, T.Y.; Black, J.M.; Acosta, M.L. Infrared Video Pupillography Coupled with Smart Phone LED for Measurement of Pupillary Light Reflex. *Front. Integr. Neurosci.* **2017**, *11*, 6. [CrossRef]
22. Kim, J.; Heo, J.; Ji, D.; Kim, M.-S. Quantitative assessment of pupillary light reflex in normal and anesthetized dogs: A preliminary study. *J. Veter-Med. Sci.* **2015**, *77*, 475–478. [CrossRef] [PubMed]
23. Gamlin, P.; Zhang, H.; Clarke, R. Luminance neurons in the pretectal olivary nucleus mediate the pupillary light reflex in the rhesus monkey. *Exp. Brain Res.* **1995**, *106*, 177–180. [CrossRef] [PubMed]
24. Masuzawa, H.; Nakamura, N.; Hirakawa, K.; Sano, K.; Matsuno, M.; Sekino, H.; Mii, K.; Abe, Y. Experimental Head Injury & Concussion in Monkey Using Pure Linear Acceleration Impact. *Neurol. Med.-Chir.* **1976**, *16 Pt 1*, 77–90. [CrossRef]
25. Hoffe, B.; Holahan, M.R. The Use of Pigs as a Translational Model for Studying Neurodegenerative Diseases. *Front. Physiol.* **2019**, *10*, 838. [CrossRef]
26. Kuluz, J.W.; Prado, R.; He, D.; Zhao, W.; Dietrich, W.D.; Watson, B. New pediatric model of ischemic stroke in infant piglets by photothrombosis: Acute changes in cerebral blood flow, microvasculature, and early histopathology. *Stroke* **2007**, *38*, 1932–1937. [CrossRef]

27. Lee, S.-E.; Hyun, H.; Park, M.-R.; Choi, Y.; Son, Y.-J.; Park, Y.-G.; Jeong, S.-G.; Shin, M.-Y.; Ha, H.-J.; Hong, H.-S.; et al. Production of transgenic pig as an Alzheimer's disease model using a multi-cistronic vector system. *PLoS ONE* **2017**, *12*, e0177933. [CrossRef]
28. Sauleau, P.; Lapouble, E.; Val-Laillet, D.; Malbert, C.-H. The pig model in brain imaging and neurosurgery. *Animal* **2009**, *3*, 1138–1151. [CrossRef]
29. Ryan, M.C.; Sherman, P.; Rowland, L.; Wijtenburg, S.A.; Acheson, A.; Fieremans, E.; Veraart, J.; Novikov, D.; Hong, L.E.; Sladky, J.; et al. Miniature pig model of human adolescent brain white matter development. *J. Neurosci. Methods* **2017**, *296*, 99–108. [CrossRef]
30. Cullen, D.K.; Harris, J.P.; Browne, K.D.; Wolf, J.A.; Duda, J.E.; Meaney, D.F.; Margulies, S.S.; Smith, D.H. A Porcine Model of Traumatic Brain Injury via Head Rotational Acceleration. *Inj. Model. Cent. Nerv. Syst.* **2016**, *1462*, 289–324. [CrossRef]
31. Weeks, D.; Sullivan, S.; Kilbaugh, T.; Smith, C.; Margulies, S.S. Influences of Developmental Age on the Resolution of Diffuse Traumatic Intracranial Hemorrhage and Axonal Injury. *J. Neurotrauma* **2014**, *31*, 206–214. [CrossRef]
32. Ryan, M.C.; Kochunov, P.; Sherman, P.M.; Rowland, L.M.; Wijtenburg, S.A.; Acheson, A.; Hong, L.E.; Sladky, J.; McGuire, S. Miniature pig magnetic resonance spectroscopy model of normal adolescent brain development. *J. Neurosci. Methods* **2018**, *308*, 173–182. [CrossRef] [PubMed]
33. Hajiaghamemar, M.; Seidi, M.; Margulies, S. Head Rotational Kinematics, Tissue Deformations, and Their Relationships to the Acute Traumatic Axonal Injury. *J. Biomech. Eng.* **2020**, *142*, 031006. [CrossRef] [PubMed]
34. Hajiaghamemar, M.; Wu, T.; Panzer, M.B.; Margulies, S.S. Embedded axonal fiber tracts improve finite element model predictions of traumatic brain injury. *Biomech. Model. Mechanobiol.* **2019**, *19*, 1109–1130. [CrossRef] [PubMed]
35. Hajiaghamemar, M.; Kilbaugh, T.; Arbogast, K.B.; Master, C.L.; Margulies, S.S. Using Serum Amino Acids to Predict Traumatic Brain Injury: A Systematic Approach to Utilize Multiple Biomarkers. *Int. J. Mol. Sci.* **2020**, *21*, 1786. [CrossRef] [PubMed]
36. Armstead, W.M.; Riley, J.; Vavilala, M. Dopamine prevents impairment of autoregulation after TBI in the newborn pig through inhibition of upregulation of ET-1 and ERK MAPK. *Pediatr. Crit. Care Med.* **2013**, *14*, e103. [CrossRef] [PubMed]
37. Patton, D.A.; Huber, C.M.; Margulies, S.S.; Master, C.L.; Arbogast, K.B. Comparison of Video-Identified Head Contacts and Sensor-Recorded Events in High School Soccer. *J. Appl. Biomech.* **2021**, *37*, 573–577. [CrossRef]
38. Espinosa, J.; Roig, A.B.; Pérez, J.; Mas, D. A high-resolution binocular video-oculography system: Assessment of pupillary light reflex and detection of an early incomplete blink and an upward eye movement. *Biomed. Eng. Online* **2015**, *14*, 22. [CrossRef]
39. Schallenberg, M.; Bangre, V.; Steuhl, K.-P.; Kremmer, S.; Selbach, J.M. Comparison of the Colvard, Procyon, and Neuroptics Pupillometers for Measuring Pupil Diameter Under Low Ambient Illumination. *J. Refract. Surg.* **2010**, *26*, 134–143. [CrossRef]
40. Vassilieva, A.; Olsen, M.H.; Peinkhofer, C.; Knudsen, G.M.; Kondziella, D. Automated pupillometry to detect command following in neurological patients: A proof-of-concept study. *Peerj* **2019**, *7*, e6929. [CrossRef]
41. Kinney, M.; Johnson, A.D.; Reddix, M.; McCann, M.B. Temporal Effects of 2% Pilocarpine Ophthalmic Solution on Human Pupil Size and Accommodation. *Mil. Med.* **2020**, *185*, 435–442. [CrossRef] [PubMed]
42. Fosse, T.K.; Toutain, P.L.; Spadavecchia, C.; Haga, H.A.; Horsberg, T.E.; Ranheim, B. Ketoprofen in piglets: Enantioselective pharmacokinetics, pharmacodynamics and PK/PD modelling. *J. Veter-Pharmacol. Ther.* **2011**, *34*, 338–349. [CrossRef] [PubMed]
43. Zeltner, A. *Handling, Dosing and Training of the Göttingen Minipig*; Educational Package; Ellegaard Göttingen Minipigs: Dalmose, Denmark, 2013; 44p.
44. Wu, T.; Hajiaghamemar, M.; Giudice, J.S.; Alshareef, A.; Margulies, S.S.; Panzer, M.B. Evaluation of Tissue-Level Brain Injury Metrics Using Species-Specific Simulations. *J. Neurotrauma* **2021**, *38*, 1879–1888. [CrossRef] [PubMed]
45. Hajiaghamemar, M.; Margulies, S. *Traumatic Brain Injury: Translating Head Kinematics Outcomes between Pig and Human*; International Research Council on Biomechanics of Injury (IRCOBI): Munich, Germany, 2020; pp. 605–607.
46. Thiagarajan, P.; Ciuffreda, K.J. Pupillary responses to light in chronic non-blast-induced mTBI. *Brain Inj.* **2015**, *29*, 1420–1425. [CrossRef] [PubMed]
47. McCrory, P.; Meeuwisse, W.; Dvorak, J.; Aubry, M.; Bailes, J.; Broglio, S.; Cantu, R.; Cassidy, D.; Echemendia, R.; Castellani, R.; et al. Consensus statement on concussion in sport—The 5th international conference on concussion in sport held in Berlin, October 2016. *Br. J. Sport. Med.* **2017**, *51*, 838–847.
48. E Capó-Aponte, J.; A Beltran, T.; Walsh, D.V.; Cole, W.R.; Dumayas, J.Y. Validation of Visual Objective Biomarkers for Acute Concussion. *Mil. Med.* **2018**, *183*, 9–17. [CrossRef] [PubMed]
49. Capó-Aponte, J.E.; Urosevich, T.G.; Walsh, D.V.; Temme, L.A.; Tarbett, A.K. Pupillary light reflex as an objective biomarker for early identification of blast-induced mTBI. *J. Spine* **2013**, *4*, 1–5.
50. Oeur, R.A.; Gilchrist, M.D.; Hoshizaki, T.B. Interaction of impact parameters for simulated falls in sport using three dif-ferent sized Hybrid III headforms. *Int. J. Crashworthiness* **2018**, *24*, 326–335. [CrossRef]
51. Sullivan, S.; Eucker, S.A.; Gabrieli, D.; Bradfield, C.; Coats, B.; Maltese, M.R.; Lee, J.; Smith, C.; Margulies, S.S. White matter tract-oriented deformation predicts traumatic axonal brain injury and reveals rotational direction-specific vulnerabilities. *Biomech. Model. Mechanobiol.* **2015**, *14*, 877–896. [CrossRef]
52. Eucker, S.A.; Smith, C.; Ralston, J.; Friess, S.H.; Margulies, S.S. Physiological and histopathological responses following closed rotational head injury depend on di-rec-tion of head motion. *Exp. Neurol.* **2011**, *227*, 79–88. [CrossRef]

53. Privitera, C.M.; Neerukonda, S.V.; Aiyagari, V.; Yokobori, S.; Puccio, A.M.; Schneider, N.J.; Stutzman, S.E.; Olson, D.M.; Hill, M.; DeWitt, J.; et al. A differential of the left eye and right eye neurological pupil index is associated with discharge modified Rankin scores in neurologically injured patients. *BMC Neurol.* **2022**, *22*, 273. [CrossRef]
54. Desai, N.; Wiebe, D.J.; Corwin, D.J.; Lockyer, J.E.; Grady, M.F.; Master, C.L. Factors Affecting Recovery Trajectories in Pediatric Female Concussion. *Clin. J. Sport Med.* **2019**, *29*, 361–367. [CrossRef] [PubMed]

Disclaimer/Publisher's Note: The statements, opinions and data contained in all publications are solely those of the individual author(s) and contributor(s) and not of MDPI and/or the editor(s). MDPI and/or the editor(s) disclaim responsibility for any injury to people or property resulting from any ideas, methods, instructions or products referred to in the content.

Article

Multiple Head Rotations Result in Persistent Gait Alterations in Piglets

Mackenzie Mull [1], Oluwagbemisola Aderibigbe [1], Marzieh Hajiaghamemar [2], R. Anna Oeur [1] and Susan S Margulies [1,*]

[1] Wallace H. Coulter Department of Biomedical Engineering, Georgia Institute of Technology and Emory University, Atlanta, GA 30322, USA
[2] Department of Biomedical Engineering and Chemical Engineering, University of Texas at San Antonio, San Antonio, TX 78249, USA
* Correspondence: susan.margulies@emory.edu; Tel.: +1-(404)-727-9827

Abstract: Multiple/repeated mild traumatic brain injury (mTBI) in young children can cause long-term gait impairments and affect the developmental course of motor control. Using our swine model for mTBI in young children, our aim was to (i) establish a reference range (RR) for each parameter to validate injury and track recovery, and (ii) evaluate changes in gait patterns following a single and multiple (5×) sagittal rapid non-impact head rotation (RNR). Gait patterns were studied in four groups of 4-week-old Yorkshire swine: healthy ($n = 18$), anesthesia-only sham ($n = 8$), single RNR injury ($n = 12$) and multiple RNR injury ($n = 11$). Results were evaluated pre-injury and at 1, 4, and 7 days post-injury. RR reliability was validated using additional healthy animals ($n = 6$). Repeated mTBI produced significant increases in gait time, cycle time, and stance time, as well as decreases in gait velocity and cadence, on Day One post-injury compared to pre-injury, and these remained significantly altered at Day Four and Day Seven post-injury. The gait metrics of the repeated TBI group also significantly fell outside the healthy RR on Day One, with some recovery by Day Four, while many remained altered at Day Seven. Only a bilateral decrease in hind stride length was observed at Day Four in our single RNR group compared to pre-injury. In sum, repeated and single sagittal TBI can significantly impair motor performance, and gait metrics can serve as reliable, objective, quantitative functional assessments in a juvenile porcine RNR TBI model.

Keywords: concussion; gait; pediatric; swine; rapid non-impact head rotation

1. Introduction

Traumatic brain injury (TBI) is a leading cause of morbidity and mortality among children in the United States [1]. Globally, the estimated annual incidence of pediatric TBI ranges between 47 and 280 per 100,000 children, with the United States estimating about 70 per 100,000 children [2,3]. Between 1997 and 2017, there were over 95,000 TBI-induced deaths in children, known to be largely caused by motor vehicle accidents, child abuse, and falls [1,4]. In the developing brain, the extent, location, and mechanism of injury can cause poor neurological outcomes and functional disabilities that impact the somatic, cognitive, and emotional aspects of a child's life [5,6]. Many studies on children have focused on the cognitive or behavioral changes caused by TBI, resulting in limited focus on the functional outcomes of gait velocity and balance post-injury.

Although persistent motor impairments in children have been reported after moderate to severe TBI [7–10], children with mild TBI also exhibit slower gait velocity or dual-task gait impairments that persist even after concussion symptoms have resolved [11–13], perhaps due to their efforts to avoid falling and to maintain stability [14]. Furthermore, TBI in younger children affects the developmental course of motor control [15].

Younger children are also known to have increased risk of sustaining repeated/multiple mild brain injuries [16–18]. Between 2002 and 2006, 51% of brain injuries reported each year

occurred during the period of cerebral development (ages 0–24 years) and the estimated incidence of repeated injuries for this population ranges from 5.6% to 36% [4,19]. Repeated brain injuries in younger children are usually attributed to child abuse or falls [4,20,21]. Slower recovery in balance deficits, increased difficulties in memory and concentration, as well as increases in learning disabilities have been observed in adult and junior athletes who experienced multiple brain injuries [22–24]. However, the effects of repeated brain injury on gait are understudied in the pediatric population.

Animal models are essential in better understanding and treating motor impairments after TBI. An important consideration when choosing an appropriate animal model is replicating the injury type, neuropathology, and mechanisms in human TBI and applying proper biomechanical loadings to the head that can cause brain tissue deformations in animal models similar to humans. Although studies have utilized small animal models (i.e., rodents) to evaluate the impact of injury severity on histopathological changes and motor impairments, many of these animal studies have been unable to efficiently mimic the biomechanics of TBI observed in children due to differences in skull thickness, brain anatomy, and physiology. Additionally, scaling inertial and impact kinematics from adults to children are inaccurately captured because adult skull and brain properties cannot be linearly extrapolated to represent the infant and child head [25–28]. In addition, when compared to the human brain, the rodent brain is smaller in size, smooth, and possesses a lower white-to-grey matter ratio [29,30]. These structural differences may be responsible for the substantially different responses to trauma between rodents and humans [31,32]. It may also contribute to the failure of clinical trials for neuroprotective drugs that were identified as being effective in rodent TBI models [31]. In contrast, piglets are popular large animal models that are similar in brain anatomy, physiology, and development to children [33–35]. Compared to children's brains, piglet's brains have similar patterns of post-natal neurogenesis, similar time course of myelination, and similar white matter volume [36–40]. These similarities make piglets an appropriate animal model for evaluating and studying TBI in the pediatric population. Many gait studies in piglets have utilized focal injury models like the controlled cortical impact (CCI) model, which mimics focal contusions [38]. However, there are no piglet studies that have focused on the effect of diffuse TBI on gait. In this present study, we used a rapid non-impact head rotation (RNR) model that mimics the inertial diffuse injuries resulting from high translational and rotational accelerations of the head with or without impact [41]. These RNR injuries are usually caused by falls or low speed motor vehicle accidents, which account for most TBIs noted in young children [27,42]. Particularly, we concentrated on the sagittal RNR head movement known to injure the brainstem, which plays a crucial role in balance, posture, and locomotion [43]. In addition, most studies have utilized rodent animals to study the effects of single and repetitive brain injury on gait [25,44–46]. At this time, there are no studies that have evaluated the effects of repeated brain injury on gait in a large animal model.

Therefore, in this study, we used a piglet RNR model of TBI to study potential gait deficits due to single and repeated brain injuries in pediatric populations. We hypothesized that (1) gait time, velocity, cycle time, cadence, number of stances, stance time, and stride length are reproducible motor performance metrics in young pigs; (2) mild levels of rapid head rotations acutely affect gait; and (3) gait deficits are exacerbated with multiple head rotations. We studied four piglet cohorts: (i) healthy, (ii) anesthetized, uninjured sham, (iii) single RNR, and (iv) multiple RNR. Healthy animals were utilized to develop and validate performance reference ranges used as baselines for identifying important gait changes after TBI. These healthy and injured piglet data will provide a platform that can be used in the future to evaluate the influence of therapeutics and interventions on motor function following single and repeated TBI.

2. Materials and Methods

2.1. Animals

Forty-two 18–19-day old female and two 18–19-day old uncastrated male Yorkshire swine were received from Palmetto Research Swine (Reevesville, SC, USA), two 18–19-day old female Yorkshire swine were received from Oak Hill Genetics (Franklin County, IL, USA), and three 18–19-day old female Yorkshire swine were received from Premier BioSource (Ramona, CA, USA). Animals were given physical exams by the Emory University Division of Animal Resources (DAR) upon arrival to ensure no abnormalities were present, such as hoof deformations. Animals were received in cohorts of 2–3 littermates and housed together for the duration of the study. Housing consisted of a 12-h light-dark cycle with ad libitum access to pellets and water.

2.2. Accreditation

The protocol used in this research was approved by the Emory University Institutional Animal Care and Use Committee (IACUC). All lab space and animal records were inspected by the United States Department of Agriculture (USDA), the Association for Assessment and Accreditation of Laboratory Animal Care (AAALAC), and the Emory University IACUC.

2.3. Acclimation

Prior to data collection, animals were acclimated to research staff and equipment for a minimum of three days. A Tekscan Strideway™ pressure system (Tekscan Inc., MA, USA) was used to assess gait. The total area of the mat was 10.7 feet by 3.0 feet; the total area of active sensitivity was 6.4 feet by 2.1 feet. On the first day of acclimation, animals were exposed to the Tekscan Strideway™ mat as a cohort. Animals were placed on one end of the mat and encouraged to ambulate to the opposite end through the presentation of an auditory stimulus (e.g., a clicker). After successfully reaching the end of the gait mat, animals were rewarded with food enrichment (e.g., yogurt on a tongue depressor). This was repeated several times to allow for learning of the behavior as well as modelling the behavior to observing littermates. On the second and third day of acclimation, animals were exposed individually to the Tekscan Strideway™ mat using the same techniques described above.

2.4. Design of Animal Experiments Based on On-Field Head Impact Measurements in Soccer

The piglet TBI experiments in this study were designed based on the head impact kinematics that occur in high school soccer games. Measurements of video-confirmed frontal head-ball impact header kinematics during two seasons of high school competitions (rotated dominantly in sagittal plane) [47] were selected for the purpose of this study. Fortunately, the sagittal plane cerebrum kinematics have an anatomic fidelity between the quadruped and the biped. Also, our previous porcine studies showed that this direction can cause more severe axonal pathology compared to other rotational directions given the same peak head angular velocity [48]. The ball-head impact kinematic data from the high school athlete subjects ($n = 267$) were scaled for the pig TBI experiments in this study, as previously described in detail [49].

To summarize, the ratio of peak angular acceleration (α) to peak angular velocity (ω) (α/ω ratio) relates the frequency/duration of head impact rotational motion and represents the characteristics of head impact kinematics, and this ratio directly influences the intracranial axonal deformations [48]. Therefore, the α/ω ratio was calculated for each head impact ($n = 267$) from this high school data, and the 50th percentile of this ratio was calculated to represent the mean head impact kinematic characteristic in human soccer. The spectrum of diffuse brain injuries (including concussion, diffuse axonal injury, and coma [50]) result from the deformation of white matter tissues in the brain [51]. Therefore, the maximum axonal strain (MAS) value for each head impact ($n = 267$) was estimated using previously published MAS surface contours that relate MAS and head kinematics (peak angular velocity and peak angular acceleration) through brain finite element modeling [52]. Next, the 50th percentile and 90th percentile of MAS values in this human data were calculated and the intersection of these MAS curves with the 50th percentile α/ω were selected to represent the 'median' and the 'high' head kinematic loadings in humans. Then, a tissue deformation-based optimal scaling method [52,53] was used to identify scaled sagittal peak angular acceleration (α) and peak angular velocity (ω) values that produce the same MAS values in pigs as the 'median' and the 'high' head kinematic loadings in human soccer [53]. Following these steps, the peak angular velocity and acceleration that needed to be applied to the pig heads to replicate the 'median' and the 'high' head impacts experienced in human soccer games were found to be $\omega = 60$ rad/s and $\alpha = 20$–30 krad/s^2 for the 'median' or 50th percentile, and $\omega = 100$ rad/s and $\alpha = 50$–60 krad/s^2 for the 'high' or 90th percentile head kinematic loadings.

From the same on-field soccer head impact dataset [54], we evaluated the rate and interval between multiple headers on the field and found that at the 95th percentile, there were 6 and 4 impacts per hour for boys and girls, respectively, with an interval of 8 min. The most typical pattern for repeated impact was a single 'high', and 4 to 5 'median' level loads. Therefore, our repeated head rotation group (multiple) received one 'high' followed by four 'median' level sagittal head rotations, spaced 8 min apart. The single head rotation group received one 'high' level rotation.

2.5. Head Rotation Methodology

A well-established, rapid non-impact rotational (RNR) injury model was used in this study to produce mild TBI in piglets similar to that of a sports-related concussion in adolescent humans.

Subjects were distributed into a naïve group and an experimental group. The naïve group had healthy animals with no anesthesia experience ($n = 16$, female; $n = 2$, male). The experimental group consisted of animals allocated to multiple RNR ($n = 11$, female), single RNR ($n = 9$, female), anesthesia-only shams ($n = 8$, female). The single RNR group experienced one 'high' level rotation followed by 32 min of anesthesia, and the repeated RNR group experienced one 'high' level rotation followed by four 'median' level rotations with 8 min intervals between rotations, totaling to 32 min of anesthesia. Anesthesia-only shams experienced no rotations and received 32 min of anesthesia. Within the naive group, a few animals were used to create a healthy reference range ($n = 12$, female), and the remaining were used to validate the created reference range ($n = 4$, female; $n = 2$, male). Healthy animals without an anesthetic experience were used to control for the effects of anesthesia while also establishing a healthy reference range. Healthy animals were studied for three non-consecutive days, and experimental group animals were studied for one day pre-injury and at 1, 4, and 7 days post-injury.

Prior to injury, animals were sedated with Ketamine (4 mg/kg), Xylazine (2 mg/kg), and Midazolam (0.2 mg/kg) via intramuscular (IM) injection. Animals were subsequently anesthetized with 5% isoflurane and 1.5–2.0 L/min of oxygen via gas mask. Once a surgical plane of anesthesia was achieved, characterized by a lack of toe pinch reflexes, the animal was intubated, placed on the ventilator (10–15 mL/kg), and secured to the pneumatic actuator by a padded snout clamp. Maintenance anesthesia was administered for the duration of the procedure at 3% isoflurane. Prior to the first rotational injury, Buprenorphine (0.1 mg/kg) was administered via IM injection and toe pinch reflex was checked again to confirm the depth of anesthesia. Isoflurane was then withdrawn, and the animal was removed from ventilator for less than 2 min. The head of the animal was rotated 60–70° in the sagittal plane at a target level of 100 rad/s (high level) over 15 milliseconds by inertial loading of the pneumatic device, with the center of rotation occurring in the cervical spine. Immediately post-injury, the animal was placed back on the ventilator with maintenance anesthesia provided. For single injury animals, ventilation and anesthesia were provided for 32 min post-injury. For multiple injury animals, ventilation and anesthesia were provided for 8 min and withdrawn again before the next rotation at a target level of 60 rad/s (low level) over 20 milliseconds. The low-level rotation was repeated four times, all occurring approximately 8 min apart with anesthesia being withdrawn prior to rotation. Actual sagittal angular velocity and acceleration are provided in Table 1.

Table 1. Summary of Angular velocity and accelerations of single and multiple RNR injuries. Values represent the calculated average ± standard error.

	Rotation Type	Angular Velocity (rad/s)	Angular Acceleration (rad/s^2)
Single	High	104.5 ± 0.47	40,052 ± 1559
Multiple	Median	61.3 ± 0.18	15,010 ± 169
	High	104.6 ± 0.41	38,368 ± 499

After rotations and anesthesia were completed, all animals were checked for tongue lacerations, then removed from the ventilator and isoflurane. Once the animal was respiring independently and maintaining oxygen levels > 95%, the animal was extubated and transported to housing for recovery. Animals were considered fully recovered once they were able to eat and drink, able to ambulate to food and water, and maintain stable vitals (oxygen saturation levels, rectal temperature, heart rate, and respirations per minute). For the remainder of the study until euthanasia, wellness checks were performed twice daily by lab members to observe physical and cognitive status.

2.6. Gait Assessment

Injury animals underwent gait assessment once at least 1 day prior to injury, then at 1, 4 and 7 days post-injury, and healthy animals underwent gait assessment on three non-consecutive days. For all animal study days, including both experimental and healthy groups, animal weights were recorded in kilograms and used to calibrate the Tekscan Strideway™ to a similar pressure. After calibration, the animal was placed at one end of the Tekscan Strideway™ mat and encouraged to walk across using techniques described previously. An ELP camera (ELP-USBFHD05MT) with infrared (IR) and light emitting diode (LED) was used to record video of gait assessment trials at an acquisition speed of 250 frames per second (ELP, Shenzhen Ailipu Technology Co., Ltd, Shenzhen, China). A trial was defined as one attempt by the animal to cross the mat. A minimum of three trials was collected per animal per study day. Trials were considered to be unacceptable if any of the following occurred: galloping, pause in ambulation, not ambulating directly towards the opposite end of mat, slipping/sliding, or exceeding 25 s to cross mat.

2.7. Data Processing

There were two components to a gait assessment trial: (1) a pressure recording collected by the Tekscan Strideway™ gait mat, and (2) a video recording from a camera mounted at the end of the mat. Both recordings were collected through the Strideway™ software. A trial was considered to be successful and acceptable if the animal crossed the full length of the mat in less than 25 s and did not exhibit any unacceptable ambulatory behaviors described previously (galloping, pausing, slipping, sliding, not ambulating directly to the opposite end of the mat). Gait trials were selected for processing through the review of pressure recordings, video recordings, and observation notes taken during data collection. Trials were also validated by confirming that the video and pressure recordings were synchronized. For each animal on each study day, the first three acceptable and validated trials were selected and imported into the Strideway™ software for data extraction. Parameters extracted by the software included the number of stances, gait time, gait distance, gait velocity, cycles per minute (CPM), cycle time, stance time, swing time, stride time, and stride velocity (Table 2). Gait distance was not examined due to the requirement of all animals to cross the entirety of the mat. A data table was created by the Strideway software containing the parameter averages for each trial collected as well as the averages of the trials combined. Data tables were then exported to Microsoft Excel (Microsoft Corporation, Redmon, WA, USA), where standard deviation and standard error were calculated for individual animal daily averages. The animal daily averages were then combined per experimental group (single, multiple, and sham) with subsequent standard deviation and standard error calculated.

Table 2. Parameter definitions.

Parameter Type	Parameter	Definition
Single-Value Parameters	Number of Stances	Also known as 'Number of Strikes'; how many total stances an animal makes during trial; stances in quadrupeds can involve 2- or 3-legged support
	Gait Time	Time, in seconds, that it takes for animal to cross the gait mat; begins with contact of the first left or right front stance and ends with the time of contact of the last left or right front stance registered on the sensor
	Gait Velocity	Gait distance divided by gait time; centimeters per second
	Cycle Time	Average time, in seconds, to complete a gait cycle
	Cycles Per Minute	Also known as "cadence"; number of complete gait cycles per minute
Individual Hoof Parameters (left front, right front, left hind, right hind)	Swing Time	Elapsed time between the last contact of a preceding hoof and first contact of the next step of that same hoof, in seconds
	Stride Time	Elapsed time between the first contacts of two consecutive hoof falls, in seconds
	Stance Time	Average time from first contact to last contact of each hoof, in seconds
	Stride Length	Distance measured parallel to the line of progression, between the posterior heel points of two consecutive hoof falls, in centimeters
	Stride Velocity	Stride length divided by stride time for each hoof; centimeters per second

3. Statistics and Analysis

All statistical analyses were performed using the SPSS Statistics (IBM, Armonk, NY, USA) software. Of the forty-six animals received, eleven were not able to have data from all study days collected. These animals were not included in the analysis of variance tests (ANOVA) with repeated measures but were used in the reference range validations (naive group) and non-parametric tests (experimental group: single, repeated, sham). Figures were generated using Microsoft Excel.

3.1. Healthy Reference Range

First, each parameter was categorized as either a single-value parameter or an individual hoof parameter (Table 2). The single-value parameters were gait time, gait velocity, number of stances, cycle time, and cycles per minute and were represented with one value per trial. The individual hoof parameters were stance time, swing time, stride time, stride velocity, and stride length and were collected for each hoof (left front, right front, left hind, right hind) for each trial.

For the single-value parameters, a one-way ANOVA with repeated measures and a post-hoc Bonferroni were performed in the healthy group to determine if there was an effect of day. Parameters were excluded if there was found to be an effect of day in the healthy group due to the potential for significance found post-injury in the experimental groups to be unrelated to anesthesia or injury. No effect of day was found; therefore, gait time, gait velocity, number of stances, cycle time, and cycles per minute were studied in the experimental groups.

For the individual hoof parameters, a two-way ANOVA with repeated measures and a post-hoc Bonferroni were used in the healthy group to understand if there was an effect of day or hoof. Stance time and stride length were found to have no effect of day or hoof in the healthy group and were studied in the experimental groups. If the day was a significant factor, the parameter was excluded.

In the experimental groups, a three-way ANOVA with repeated measures and a post-hoc Bonferroni were performed on stance time and stride length to understand if hoof had an effect post-injury. Stance time was found to have no effect of hoof post-injury; therefore, the hoofs were averaged for each study day per each animal in both the healthy and experimental groups. Stride length was found to have an effect of hoof post-injury, so hoofs were studied and reported separately for all groups.

In total, gait time, gait velocity, number of stances, cycle time, and cycles per minute (single-value parameters), along with stance time and stride length (individual hoof parameters), were selected for further statistical analyses. The 2.5th and 97.5th percentiles were calculated for each parameter to establish reference range (RR) interval values in piglets that followed a method congruent with reference intervals established for healthy patients in the clinical setting [55]. To establish the reproducibility of the calculated reference ranges, a separate validation group of healthy animals that did not complete all three study days were compared against the ranges ($n = 6$). Animals were studied on one or two days, and data from each day was considered to be a single data point ($n = 10$). The percentage of data points that fell in the healthy range was evaluated to determine whether the reference range was an accurate representation of healthy values. If the percentage of validation data points in the healthy range was below 75%, the parameter was deemed not reliable. All seven parameters satisfied the criteria for reliability, with at least 75% of the validation group falling within the healthy range. These reliable parameters were then evaluated in the experimental groups to determine the percentage of animals that fell within the healthy reference range on each day of study.

3.2. Experimental Groups

The influence of experimental group (sham, single, multiple) and study day (pre-injury, Day One, Day Four, Day Seven) were examined for six parameters (number of stances, gait time, gait velocity, cycle time, cycles per minute, and stance time) using a two-way ANOVA with repeated measures, followed by a post-hoc Bonferroni. For stride length, a three-way ANOVA (group, day, hoof) was performed along with a post-hoc Bonferroni. For all evaluations, significance was defined as $p < 0.05$.

The percentage of animals in each experimental group whose trial values fell within the healthy reference range was determined on each study day to determine if there were significant variations between these proportions by experimental group and by day. For all evaluations, significance was defined as $p < 0.05$. To examine if the experimental group had a significant effect, a Fisher Exact Test was performed. If a group was found to be significant, the Fisher Exact Test was then repeated and restricted to comparing two groups against each other for the post-hoc analysis. To determine if study day had a significant effect within an experimental group, a Cochran Q test was performed, along with a post-hoc McNemar's test.

4. Results

4.1. Overview

For the pre-injury study day, there was found to be no significant effect of study group ($p > 0.05$) for any of the parameters (Figures 1–7, Table S1). On Day One post-injury, the multiple injury group was found to have significantly increased gait time, cycle time, and stance time, and decreased gait velocity and cadence, relative to the sham group. For all the above parameters except gait time, differences between the multiple injury and sham groups persisted on to Days Four and Seven post-injury. The only significance found for the single injury group was decreased cadence on Day Seven post-injury, and decreased stride length on both hind limbs (left hind and right hind) on Day Four post-injury, relative to the sham group.

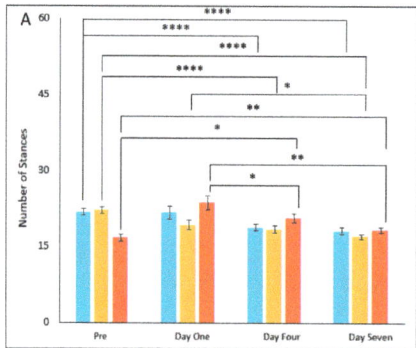 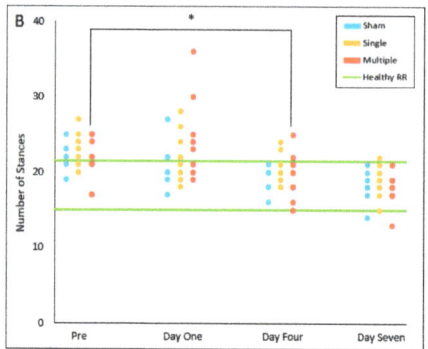

Figure 1. Number of Stances (**A**) Presents daily group average ± standard error and statistical comparisons noted from the two-way ANOVA with repeated measures analysis; (**B**) Presents individual animal daily averages and statistical comparisons are noted from the Fisher Exact Test and McNemar's analysis (* $p < 0.05$, ** $p < 0.01$, **** $p < 0.001$).

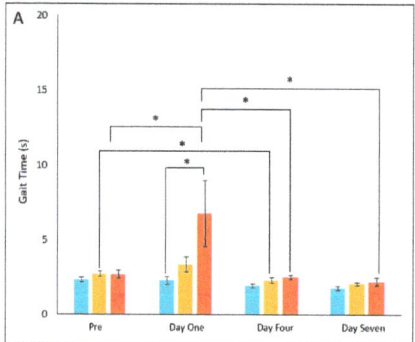 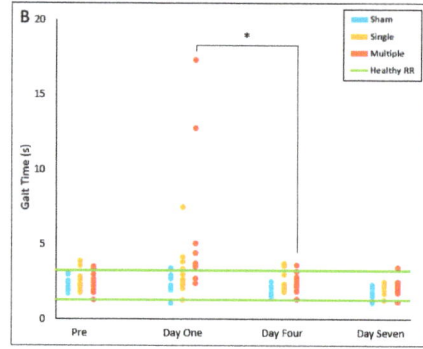

Figure 2. Gait Time (**A**) Presents daily group average ± standard error and statistical comparisons noted from the two-way ANOVA with repeated measures analysis; (**B**) Presents individual animal daily averages with statistical comparisons noted from the Fisher Exact Test and McNemar's analysis (* $p < 0.05$).

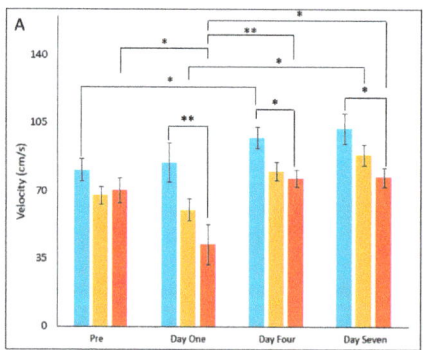 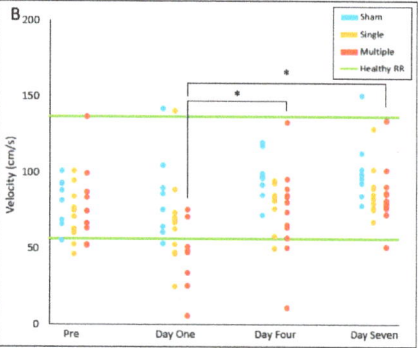

Figure 3. Gait Velocity (**A**) Presents daily group average ± standard error and statistical comparisons noted from the two-way ANOVA with repeated measures analysis; (**B**) Presents individual animal daily averages with statistical comparisons noted from the Fisher Exact Test and McNemar's analysis (* $p < 0.05$, ** $p < 0.01$).

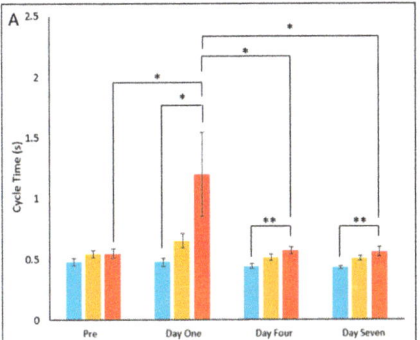 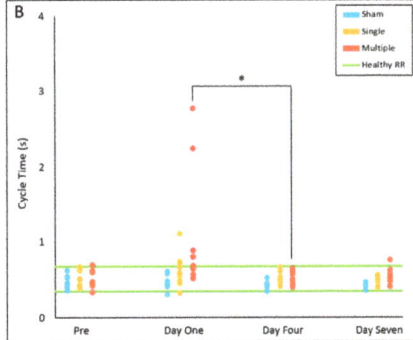

Figure 4. Cycle Time (**A**) Presents daily group average ± standard error and statistical comparisons noted from the two-way ANOVA with repeated measures analysis; (**B**) Presents individual animal daily averages with statistical comparisons noted from the Fisher Exact Test and McNemar's analysis (* $p < 0.05$, ** $p < 0.01$).

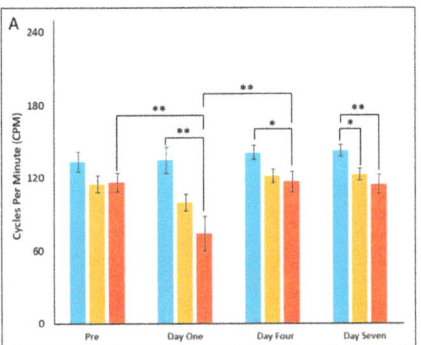 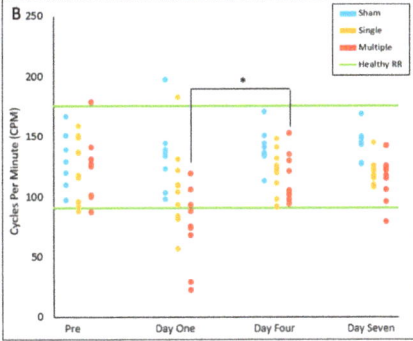

Figure 5. Cycles per Minute (cadence) (**A**) Presents daily group average ± standard error and statistical comparisons noted from the two-way ANOVA with repeated measures analysis; (**B**) Presents individual nimal daily averages with statistical comparisons noted from the Fisher Exact Test and McNemar's analysis (* $p < 0.05$, ** $p < 0.01$).

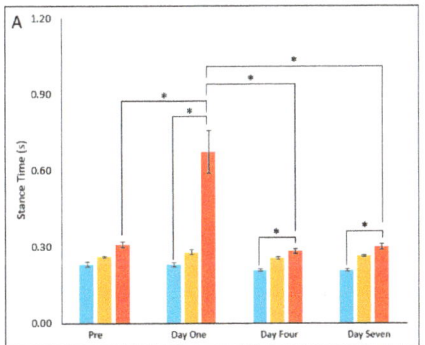 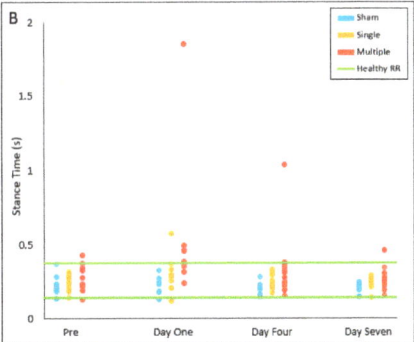

Figure 6. Stance Time (**A**) Presents daily group average ± standard error and statistical comparisons noted from the two-way ANOVA with repeated measures analysis; (**B**) Presents individual animal daily averages with statistical comoparisons noted from the Fisher Exact Test and McNemar's analysis (* $p < 0.05$).

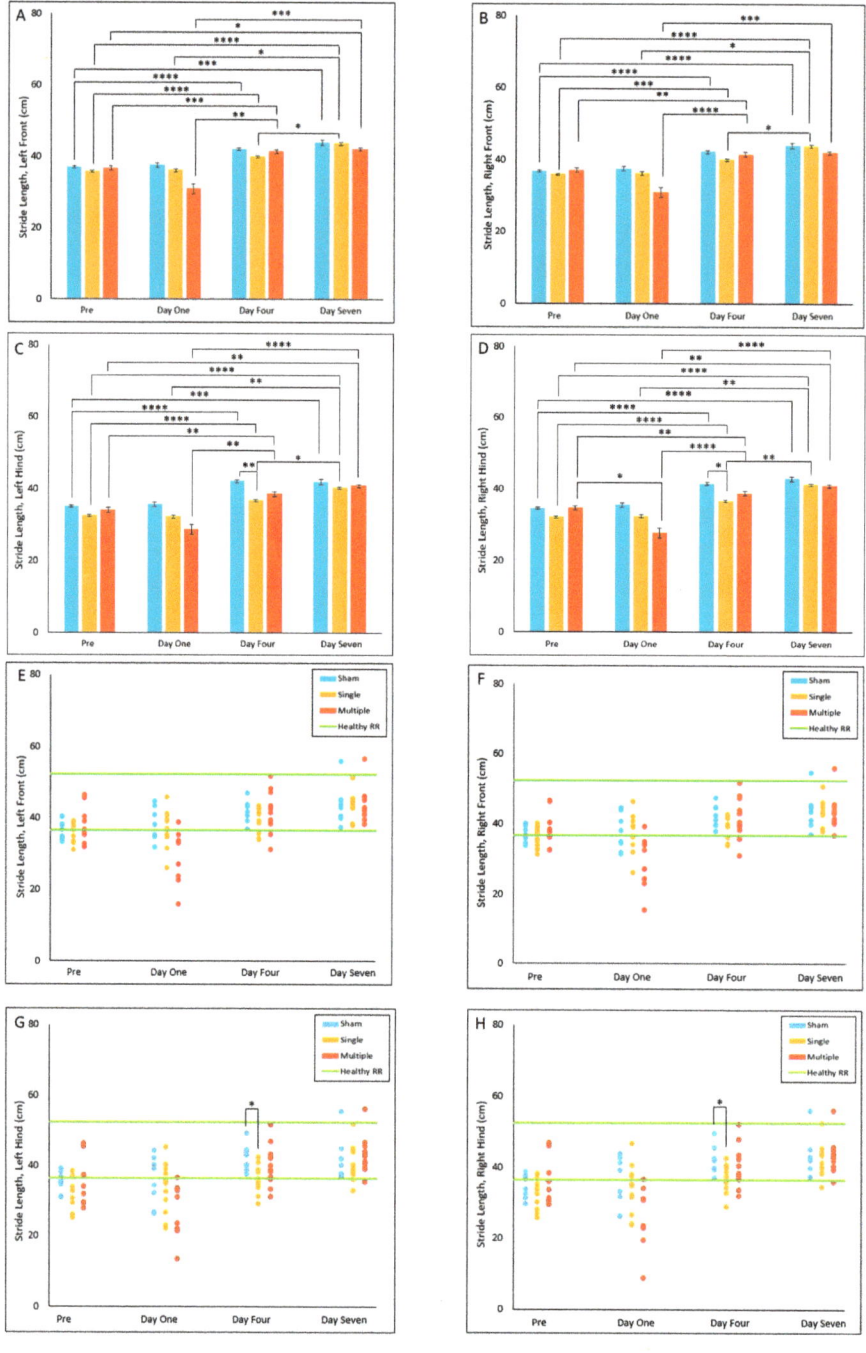

Figure 7. Stride Length by Limb (**A–D**) Presents daily group average ± standard error and statistical comparisons noted from the two-way ANOVA with repeated measures analysis; (**E–H**) Presents individual animal daily averages with statistical comparisons noted from the Fisher Exact Test and McNemar's analysis (* $p < 0.05$, ** $p < 0.01$, *** $p = 0.001$, **** $p < 0.001$).

Within the multiple injury group, gait time, cycle time, and stance time were found to be significantly increased on Day One post-injury relative to all other study days, and gait velocity was found to be decreased on Day One relative to all other study days. Cadence was only found to be decreased on Day One relative both to pre-injury and Day Four. Number of stances was found to be increased on Day One relative to both Day Four and Day Seven.

Within the single injury group, there was a delayed increase in gait velocity from Day One to Day Seven, and for number of stances, there was a decrease in stances from pre-injury to Day Four and Day Seven, as well as from Day One to Day Seven. Within the sham group, there was an increase in gait velocity from pre-injury to Day Four, and a decrease in stances from pre-injury to Day Four and Day Seven.

When applying the reference range to the groups, a significant decrease in the proportion of animals in the healthy reference range was found in the multiple injury group on Day One relative to Day Four for gait time, gait velocity, cycle time, and cadence. Gait velocity also experienced this decrease on Day One relative to Day Seven. For number of stances, the multiple injury group experienced an increase in the proportion of animals in the reference range from pre-injury to Day Four. For stride length, only the hind limbs in the single injury group experienced a significant proportion of animals outside the reference range on Day Four relative to the sham group.

4.2. Gait Parameters

4.2.1. Number of Stances

The number of stances was found to demonstrate no significant differences between groups for all study days. Within the sham group, the number of stances was found to be significantly decreased on Day Four ($p < 0.001$) and Day Seven ($p < 0.001$) compared to pre-injury. Within the single injury group, Day Four ($p < 0.001$) and Day Seven ($p < 0.001$) were found to have decreased stances compared to pre-injury. Day Seven was also found to have decreased stances relative to Day One ($p = 0.032$). Within the multiple injury group, Day Four ($p = 0.014$) and Day Seven ($p = 0.007$) were found to have decreased stances relative to pre-injury values. Day One was found to have increased stances relative to Day Four ($p = 0.011$) and Day Seven ($p = 0.005$). For reference range comparisons, there was no effect of group on any study day; however, it was found that the multiple injury group had an increase in animals in the reference range on Day Four relative to pre-injury ($p = 0.031$) (Figure 1A,B).

4.2.2. Gait Time

Gait time was found to be increased in the multiple injury group compared to the sham group on Day One post-injury ($p = 0.042$). Within the single injury group, Day Four post-injury was found to have decreased gait time relative to pre-injury ($p = 0.035$). Within the multiple injury group, gait time on Day One post-injury was found to be significantly increased compared to pre-injury ($p = 0.02$), Day Four ($p = 0.015$), and Day Seven ($p = 0.012$). For reference range comparisons, the multiple injury group's proportion of animals in the healthy reference range was significantly decreased on Day One relative to Day Four ($p = 0.031$), with all out-of-range animals exhibiting longer gait times than the healthy RR. (Figure 2A,B).

4.2.3. Gait Velocity

The multiple injury group had decreased gait velocity on Day One ($p = 0.009$), Day Four ($p = 0.032$), and Day Seven ($p = 0.041$) post-injury relative to the sham group. Within the sham group, gait velocity on Day Four had significantly increased ($p = 0.014$) from pre-injury values. Within the single injury group, gait velocity on Day Seven was significantly increased from Day One ($p = 0.028$). Within the multiple injury group, gait velocity on Day One was significantly decreased compared to pre-injury ($p = 0.031$), Day Four ($p = 0.008$), and Day Seven ($p = 0.023$). For reference range comparisons, the multiple injury group's proportion of animals in the healthy range was significantly decreased on Day One relative to Day Four ($p = 0.031$) and Day Seven ($p = 0.031$), with all out-of-range gait velocities slower than the healthy RR (Figure 3A,B).

4.2.4. Cycle Time

The multiple injury group was found to have significantly decreased gait cycle time on Day One ($p = 0.030$), Day Four ($p = 0.005$) and Day Seven ($p = 0.005$) post-injury relative to the sham group. Within the multiple injury group, cycle time on Day One was significantly decreased compared to pre-injury ($p = 0.013$), Day Four ($p = 0.017$), and Day Seven ($p = 0.014$). The multiple injury group had a significantly decreased proportion of animals in the healthy reference range on Day One relative to Day Four ($p = 0.031$), with all out-of-range cycle times longer than the healthy RR (Figure 4A,B).

4.2.5. Cycles per Minute (Cadence)

The multiple injury group had a significantly decreased cadence on Day One ($p = 0.003$), Day Four ($p = 0.015$), and Day Seven ($p = 0.006$) post-injury relative to the sham group. The single injury group was found to have decreased cadence on Day Seven ($p = 0.045$) post-injury relative to the sham group on Day Seven. Within the multiple injury group, cadence on Day One was found to be decreased relative to pre-injury ($p = 0.006$) and Day Four ($p = 0.005$) post-injury. The multiple injury group had a significantly decreased proportion of animals in the healthy reference range on Day One relative to Day Four ($p = 0.031$), with all out-of-range cadences slower than the healthy RR. (Figure 5A,B).

4.2.6. Stance Time

The multiple injury group stance time values were found to be significantly increased on Day One ($p = 0.035$), Day Four ($p = 0.046$) and Day Seven ($p = 0.012$) post-injury relative to the sham group. Within the multiple injury group, stance time was significantly increased on Day One relative to pre-injury ($p = 0.035$), Day Four ($p = 0.021$) and Day Seven ($p = 0.022$) post-injury. No significant differences were found for reference range comparisons (Figure 6A,B).

4.2.7. Stride Length

In both hind limbs (left hind; LH and right hind; RH), the single injury group was found to have significantly decreased stride length on Day Four post-injury ($p < 0.017$) relative to the sham group. However, no significant differences were found between groups in either front limb (left front; LF and right front; RF, Figure 7A–D).

In all four limbs, the sham and single injury group experienced an increase in stride length from pre-injury to Day Four ($p \leq 0.001$) and Day Seven ($p \leq 0.001$). Within the single injury group, we also noted a significant increase in stride length of all limbs on Day One ($p \leq 0.016$) and Day Four ($p < 0.04$) relative to Day Seven (Figure 7A–D).

In all four limbs, the multiple injury group experienced an increase in stride length from pre-injury to Day Four ($p \leq 0.007$). Additionally, in the multiple group a significant increase from pre-injury to Day Seven ($p \leq 0.031$) was observed in all limbs except for the right front (RF). Only the right hind (RH) limb experienced a significant decrease in stride length from pre-injury to Day One ($p = 0.029$) in the multiple injury group. All four limbs also displayed an increase in stride length on Day Four ($p \leq 0.003$) and Day Seven ($p \leq 0.002$) relative to Day One (Figure 7A–D).

No significant differences were found for reference range comparisons in either front limb, i.e., left front (LF) and right front (RF). However, in both hind limbs, i.e., left hind (LH) and right hind (RH), the single injury group had a decreased proportion of animals in the reference range relative to the sham group ($p = 0.018$) (Figure 7E–H).

5. Discussion
5.1. General Summary

In this study, we identified persistent changes in gait patterns following a sagittal RNR injury and the exacerbating influence of multiple head rotations at Day One post-injury. Our findings suggest that at Day One post-injury, our multiple RNR group walked significantly slower (higher gait time and lower velocity), had fewer step cycles per minute (lower cadence), and spent longer time with their feet on the ground (higher cycle time and stance time). Additionally, in our multiple RNR group, velocity, cycle time, cadence, and stance time were not only affected at Day One post-TBI, but deficits in these parameters significantly persisted at Day Four and Day Seven post-injury, suggesting that multiple injuries have long-term effects on gait. Similarly, the literature shows that children walk slower, take fewer steps per minute, and have difficulty maintaining balance months or years post-injury [7,9,56–58]. In contrast, multiple RNR injury did not seem to significantly increase the total stances (number of stances) or shorten the distance between steps (stride length). While we did not notice a significant persistent reduction in the distance between steps (stride length), studies in children have identified significant reductions in stride length and shorter step length caused by TBI [7,9,10]. These findings in children highlight the long-term effect that TBI has on gait, which we also found in our piglet multiple RNR group but with different gait parameter alterations following TBI, as discussed above.

5.2. Relationship with Previous Pediatric Studies

Children who sustain moderate to severe TBI show evidence of decreased velocity that persists for years post-injury (Table S2). Two studies carried out 3–12 months after brain injury in young children indicated a 27.7% and 20% decrease in velocity [9,59]. Another study noted a 50% reduction in gait speed 3.5 years after severe TBI in adolescents [10]. Based on our findings, our multiple RNR group also showed a 50% significant reduction in velocity at Day One post-injury and maintained a 24% significant decrease in velocity by Day Seven post-injury compared to sham animals. We noted that at Day Seven post-injury, the percentage decrease in velocity observed in our multiple RNR group was similar to those documented in young children by 1 year post-injury. This significant reduction in the velocity of our multiple RNR group may be due to the piglets attempts in maintaining balance and stability, as noted in pediatric TBI and older adult studies [14,59,60].

Studies in pediatric TBI have also identified decreases in cadence that persist for up to 7.8 months following moderate to severe TBI (Table S2). At 2.8 months post-injury, a TBI study in young children noted a 13% decrease in cadence, and although there was an improvement by 7.8 months post-injury, a 5.6% decrease in cadence persisted [7]. This study also noted a 23% and 7.1% decrease in velocity at 2.8 months and 7.8 months post-injury, respectively. The percentage decrease observed in young children at both time points were quite similar for velocity and cadence [7]. Compared to the sham group, our single RNR group exhibited a 14% significant decrease in cadence at Day Seven post-injury. Our multiple RNR group also displayed a 45% significant decrease in cadence at Day One post-injury, and a 20% significant decrease in cadence by Day Seven post-injury compared to sham animals. Similar to observations in young children, the decrease in cadence was also quite similar to the decrease in velocity in our multiple RNR group. It is important to note that velocity is a product of cadence and stride length, and it can be significantly affected by changes in either one of these parameters [61]. The similarity observed in changes to cadence and velocity enables us to conclude that multiple RNR injury may not have an effect on stride length, and this is reflected by the non-significant changes observed in stride length for this group. However, studies in children are quite different from our findings because they show that velocity, cadence, and stride length seem to decrease and improve together post-injury [7]. These differences may be due to children being bipedal, and piglets being quadrupedal, which is a distinction that should be put into consideration when assessing gait patterns from both species. Cadence has also been shown to alter balance, and the significant decrease in cadence in our multiple RNR group may be responsible for the difficulty that this group experienced in maintaining balance during gait trials [62].

Variability in gait patterns and decrease in stride length are other impairments that have been identified in children and piglets post-TBI [37,63–65]. A pediatric TBI study noted increased step variability in children who suffered severe TBI [10]. Another study showed that despite significant improvements by 7.8 months post-injury, differences in stride length were still present in children who had suffered moderate to severe TBI compared to those who had not [7]. This study noted a 16% decrease in stride length at 2.8 months post-injury, with a slight improvement to about 7.9% decrease by 7.8 months post-injury. Kinder and colleagues also noted a decrease in hind reach in their pediatric controlled cortical impact (CCI) piglet model and described it as the lagging behind of the hind limbs compared to the front limbs [36]. They proposed that this decrease in hind reach may be caused by an overall decrease in stride length and an increase in percentage stance. Our single RNR injury group had significantly shorter stride length in the left and right hind limbs at Day Four post-injury compared to the sham group. This significant difference was not observed in the front limbs, which may signify some level of variability in gait patterns of the front (LF, RF) and hind (LH, RH) limbs of this group. Although noted at Day Four post-injury in the single RNR group, we did not identify gait variability in the front and hind limbs of the single RNR group at Day One and Seven post-injury and in the multiple RNR group at Days One, Four, and Seven post-injury compared to the sham group. Recent studies in children and piglets also mention that a decrease in hind reach and stride length may be responsible for directly influencing an increase in cycle time, decrease in cadence, and decrease in velocity [36,66,67]. Our multiple RNR group displayed a 152% significant increase in cycle time at Day One post-injury, and a 30% significant increase in cycle time by Day Seven post-injury. Although we noted a significant increase in cycle time and significant decreases in cadence and velocity, no decrease in stride length was observed in this group, perhaps due to a concurrent significant group-independent weight gain from pre-injury (7.11 ± 0.6 kg mean ± SE) to Day Seven (9.33 ± 0.6 kg, 2-way ANOVA with repeated measures) (Table 3).

Table 3. Summary of animal weight averages (in kilograms) for pre-injury and Day Seven post-injury and analysis results. Values represent the calculated group average ± standard error and the p-value for Paired-Sample t-test.

	Pre-Injury	Day Seven	p-Value
Sham	6.34 ± 0.601	9.20 ± 0.728	<0.001
Single	6.60 ± 0.221	8.90 ± 0.444	<0.001
Multiple	7.11 ± 0.556	9.33 ± 0.625	<0.001

Additionally, although several studies indicate that children with TBI improve post-injury, many of these studies also show significant differences in velocity, cadence, and stride length months or years post-injury compared to healthy controls [7,9,10,59,68]. Kuhtz-Buschbeck and colleagues noted that velocity and cadence improved in about 80% of pediatric patients, but differences between controls and injured children persisted months post-injury [7]. An improvement of 67% in gait time, 83% in velocity, 54% in cycle time, 54% in cadence, and 55% in stance time were observed in our multiple RNR injury group by Day Seven post-injury compared to Day One post-injury. However, irrespective of these improvements in gait, significant differences in gait time, velocity, cycle time, cadence, and stance time persisted by Day Seven post-injury. Similarly, a focal TBI piglet study also noted transient impairments in their cycle time, cadence, and stride length [36]. Significant differences in cycle time and cadence of their CCI piglets persisted by Day Seven post-injury compared to their baseline measures. These findings suggest that our sagittal RNR piglet models exhibit similar gait abnormalities as seen in both humans and CCI piglet models during the early and later phases post-injury.

In summary, this is the first study to (i) utilize a piglet diffuse TBI model to evaluate changes in gait patterns, (ii) compare gait changes in injured groups to both healthy and sham groups, and (iii) evaluate changes in gait patterns following repeated mild TBI.

5.3. Limitations and Future Work

A distinction of this diffuse TBI study is that the levels of rotational head loads applied to the piglets are associated with active sport participation in heading the ball in soccer and with common recreational behavior in children. In previous studies published by our group as a model of abusive head trauma in infants, younger 3–5-day old piglets (35–40 gm brain, typically) experienced single or repeated cyclical "trains" of sagittal head rotations with velocities of 20–40 rad/s and accelerations of 600–700 rad/s^2 scaled from reconstructions of human infant shaking [69,70]. Using traditional mass-scaling laws [71] to determine equivalent kinematics in a 4-week old piglet used in the current study (60 gm brain, typically), the equivalent rotational loads for vigorous shaking correspond to 15–35 rad/s and 450–530 rad/s^2. Based on these scaled levels, the mTBI loads in the current study for the "median headers" were three times higher rotational velocities and 30 times higher rotational accelerations than in vigorous shaking (Table 1). While we observed prolonged gait deficits following head rotations at levels associated with recreational play in children, future work should expand on the functional deficits associated with much lower level head rotations indicative of shaking without impact. Another limitation of this study was the use of only female piglets in our experimental group. Gait analysis in both sexes is needed to better understand sex variability in gait impairments. In humans, women tend to report longer and worse outcomes than men, yet pre-clinical studies [72] show females subjects are known to have lower comorbidities, implicating female sex hormones may have neuroprotective effects [31]. This limitation was mitigated by using pre-pubertal female piglets. A third limitation is the use of only a sagittal RNR injury model. Additional studies in axial and coronal directions should be explored to provide a more complete understanding of how different injury directions can affect motor function. Lastly, another limitation was the exclusion of cognitive impairments and neuropathological assessments which may affect gait. In future studies we intend to study both sexes, examine various injury directions, study cognitive impairment, quantify neuropathology, and observe changes in gait impairments for longer periods of time to ensure that there are no further declines after Day Seven post-injury.

6. Conclusions

In conclusion, we observed that sagittal RNR injury can lead to significant acute increase in gait time, decrease in velocity, decrease in cadence, shorter stride length, increase in stance time, and increase in cycle time, much like pediatric TBI patients. Multiple RNR injury was observed to cause worse gait impairments compared to single RNR injury, and multiple RNR injury metrics were significantly outside the healthy reference range. Based on the similarities between our findings and pediatric TBI studies, as well as the anatomy, development, and size of a piglet and a child's brain, these results indicate that a sagittal RNR piglet model can serve as an objective quantitative functional platform in the understanding and treatment of gait impairments due to TBI using novel therapeutics.

Supplementary Materials: The following supporting information can be downloaded at: https://www.mdpi.com/article/10.3390/biomedicines10112976/s1, Table S1: Summary of daily group averages for each parameter; Table S2: Comparison between pediatric studies and single and multiple RNR injury models.

Author Contributions: Conceptualization, S.S.M. and M.H.; methodology, S.S.M., M.M. and O.A.; validation, S.S.M., M.M. and O.A.; formal analysis, M.M. and O.A.; investigation, S.S.M.; resources, S.S.M.; data curation, M.M. and O.A.; writing—original draft preparation, M.M., O.A. and R.A.O.; writing—review and editing, S.S.M., M.H., M.M., O.A. and R.A.O.; visualization, M.M.; supervision,

S.S.M. and R.A.O.; project administration, S.S.M.; funding acquisition, S.S.M. All authors have read and agreed to the published version of the manuscript.

Funding: This study was supported by the Georgia Research Alliance and the National Institutes of Health (R01NS097549).

Institutional Review Board Statement: The animal study protocol was approved by the Institutional Animal Care and Use Committee of Emory University (PROTO201800149, approved 07/16/2020; PROTO201800163, approved 20 May 2019).

Informed Consent Statement: Not applicable.

Data Availability Statement: Not applicable.

Acknowledgments: We acknowledge the valuable collaborative discussion with Kristy Arbogast, Christina Master, and Declan Patton, the early-stage collaboration with Morteza Seidi, and the valuable data collection assistance of Akshara Thakore and William "Harry" Torp during animal studies. We also acknowledge Holly Kinder and Franklin West from the University of Georgia for sharing their porcine gait data for detailed comparisons across injury models.

Conflicts of Interest: The funders had no role in the design of the study; in the collection, analyses, or interpretation of data; in the writing of the manuscript; or in the decision to publish the results.

References

1. Cheng, P.; Li, R.; Schwebel, D.C.; Zhu, M.; Hu, G. Traumatic brain injury mortality among U.S. children and adolescents ages 0–19 years, 1999–2017. *J. Saf. Res.* **2020**, *72*, 93–100. [CrossRef]
2. Dewan, M.C.; Mummareddy, N.; Wellons, J.C., 3rd; Bonfield, C.M. Epidemiology of Global Pediatric Traumatic Brain Injury: Qualitative Review. *World Neurosurg.* **2016**, *91*, 497–509.e1. [CrossRef]
3. Schneier, A.J.; Shields, B.J.; Hostetler, S.G.; Xiang, H.; Smith, G.A. Incidence of Pediatric Traumatic Brain Injury and Associated Hospital Resource Utilization in the United States. *Pediatrics* **2006**, *118*, 483–492. [CrossRef] [PubMed]
4. Faul, M.; Xu, L.; Wald, M.M.; Coronado, V.; Dellinger, A.M. Traumatic brain injury in the United States: National estimates of prevalence and incidence, 2002–2006. *Inj. Prev.* **2010**, *16*, A268. [CrossRef]
5. Prince, C.; Bruhns, M.E. Evaluation and Treatment of Mild Traumatic Brain Injury: The Role of Neuropsychology. *Brain Sci.* **2017**, *7*, 105. [CrossRef]
6. Popernack, M.L.; Gray, N.; Reuter-Rice, K. Moderate-to-Severe Traumatic Brain Injury in Children: Complications and Rehabilitation Strategies. *J. Pediatr. Health Care* **2015**, *29*, e1–e7. [CrossRef]
7. Kuhtz-Buschbeck, J.P.; Hoppe, B.; Gölge, M.; Dreesmann, M.; Damm-Stünitz, U.; Ritz, A. Sensorimotor recovery in children after traumatic brain injury: Analyses of gait, gross motor, and fine motor skills. *Dev. Med. Child Neurol.* **2003**, *45*, 821–828. [CrossRef]
8. Schaaf, P.J.V.; Kriel, R.L.; Krach, L.; Luxenberg, M.G. Late improvements in mobility after acquired brain injuries in children. *Pediatr. Neurol.* **1997**, *16*, 306–310. [CrossRef]
9. Kuhtz-Buschbeck, J.P.; Stolze, H.; Ritz, A. Analyses of gait, reaching, and grasping in children after traumatic brain injury. *Arch. Phys. Med. Rehabil.* **2003**, *84*, 424–430. [CrossRef] [PubMed]
10. Katz-Leurer, M.; Rotem, H.; Keren, O.; Meyer, S. The effect of variable gait modes on walking parameters among children post severe traumatic brain injury and typically developed controls. *NeuroRehabilitation* **2011**, *29*, 45–51. [CrossRef] [PubMed]
11. Howell, D.; Osternig, L.R.; Koester, M.C.; Chou, L.-S. The effect of cognitive task complexity on gait stability in adolescents following concussion. *Exp. Brain Res.* **2014**, *232*, 1773–1782. [CrossRef]
12. Sambasivan, K.; Grilli, L.; Gagnon, I. Balance and mobility in clinically recovered children and adolescents after a mild traumatic brain injury. *J. Pediatr. Rehabil. Med.* **2015**, *8*, 335–344. [CrossRef]
13. Berkner, J.; Meehan, W.P.; Master, C.L.; Howell, D.R. Gait and Quiet-Stance Performance Among Adolescents After Concussion-Symptom Resolution. *J. Athl. Train.* **2017**, *52*, 1089–1095. [CrossRef]
14. Howell, D.R.; Kirkwood, M.W.; Provance, A.; Iverson, G.L.; Meehan, W.P. Using concurrent gait and cognitive assessments to identify impairments after concussion: A narrative review. *Concussion* **2018**, *3*, CNC54. [CrossRef]
15. Stephens, J.; Salorio, C.; Denckla, M.; Mostofsky, S.; Suskauer, S. Subtle Motor Findings During Recovery from Pediatric Traumatic Brain Injury: A Preliminary Report. *J. Mot. Behav.* **2016**, *49*, 20–26. [CrossRef] [PubMed]
16. Lasry, O.; Liu, E.Y.; Powell, G.A.; Ruel-Laliberté, J.; Marcoux, J.; Buckeridge, D.L. Epidemiology of recurrent traumatic brain injury in the general population. *Neurology* **2017**, *89*, 2198–2209. [CrossRef]
17. Sariaslan, A.; Sharp, D.J.; D'Onofrio, B.M.; Larsson, H.; Fazel, S. Long-Term Outcomes Associated with Traumatic Brain Injury in Childhood and Adolescence: A Nationwide Swedish Cohort Study of a Wide Range of Medical and Social Outcomes. *PLoS Med.* **2016**, *13*, e1002103. [CrossRef] [PubMed]
18. Swaine, B.R.; Tremblay, C.; Platt, R.W.; Grimard, G.; Zhang, X.; Pless, I.B. Previous Head Injury Is a Risk Factor for Subsequent Head Injury in Children: A Longitudinal Cohort Study. *Pediatrics* **2007**, *119*, 749–758. [CrossRef] [PubMed]

19. Prins, M.; Giza, C. Repeat traumatic brain injury in the developing brain. *Int. J. Dev. Neurosci.* **2012**, *30*, 185–190. [CrossRef]
20. Joyce, T.; Gossman, W.; Huecker, M.R. Pediatric Abusive Head Trauma. In *Statpearls*; StatPearls Publishing: Tampa, FL, USA, 2022.
21. Hung, K.-L. Pediatric abusive head trauma. *Biomed. J.* **2020**, *43*, 240–250. [CrossRef]
22. Slobounov, S.; Slobounov, E.; Sebastianelli, W.; Cao, C.; Newell, K. Differential rate of recovery in athletes after first and second concussion episodes. *Neurosurgery* **2007**, *61*, 338–344. [CrossRef] [PubMed]
23. Gaetz, D.G.M.; Goodman, D.; Weinberg, H. Electrophysiological evidence for the cumulative effects of concussion. *Brain Inj.* **2000**, *14*, 1077–1088. [CrossRef] [PubMed]
24. Wall, S.E.; Williams, W.H.; Cartwright-Hatton, S.; Kelly, T.P.; Murray, J.; Murray, M.; Owen, A.; Turner, M. Neuropsychological dysfunction following repeat concussions in jockeys. *J. Neurol. Neurosurg. Psychiatry* **2006**, *77*, 518–520. [CrossRef] [PubMed]
25. Hall, A.N.B.; Joseph, B.; Brelsford, J.M.; Saatman, K.E. Repeated Closed Head Injury in Mice Results in Sustained Motor and Memory Deficits and Chronic Cellular Changes. *PLoS ONE* **2016**, *11*, e0159442. [CrossRef]
26. Mountney, A.; Boutté, A.M.; Cartagena, C.M.; Flerlage, W.F.; Johnson, W.D.; Rho, C.; Lu, X.-C.; Yarnell, A.; Marcsisin, S.; Sousa, J.; et al. Functional and Molecular Correlates after Single and Repeated Rat Closed-Head Concussion: Indices of Vulnerability after Brain Injury. *J. Neurotrauma* **2017**, *34*, 2768–2789. [CrossRef]
27. Araki, T.; Yokota, H.; Morita, A. Pediatric Traumatic Brain Injury: Characteristic Features, Diagnosis, and Management. *Neurol. Med. Chir.* **2017**, *57*, 82–93. [CrossRef]
28. Coats, B.; Margulies, S.S. Material Properties of Human Infant Skull and Suture at High Rates. *J. Neurotrauma* **2006**, *23*, 1222–1232. [CrossRef]
29. Duhaime, A.-C.; Margulies, S.S.; Durham, S.R.; O'Rourke, M.M.; Golden, J.A.; Marwaha, S.; Raghupathi, R. Maturation-dependent response of the piglet brain to scaled cortical impact. *J. Neurosurg.* **2000**, *93*, 455–462. [CrossRef] [PubMed]
30. Duberstein, K.J.; Platt, S.R.; Holmes, S.P.; Dove, C.R.; Howerth, E.W.; Kent, M.; Stice, S.L.; Hill, W.D.; Hess, D.C.; West, F.D. Gait analysis in a pre- and post-ischemic stroke biomedical pig model. *Physiol. Behav.* **2014**, *125*, 8–16. [CrossRef]
31. Xiong, Y.; Mahmood, A.; Chopp, M. Animal models of traumatic brain injury. *Nat. Rev. Neurosci.* **2013**, *14*, 128–142. [CrossRef]
32. Povlishock, J.T.; Hayes, R.L.; Michel, M.E.; McIntosh, T.K. Workshop on Animal Models of Traumatic Brain Injury. *J. Neurotrauma* **1994**, *11*, 723–732. [CrossRef]
33. Grate, L.L.; A Golden, J.; Hoopes, P.; Hunter, J.V.; Duhaime, A.-C. Traumatic brain injury in piglets of different ages: Techniques for lesion analysis using histology and magnetic resonance imaging. *J. Neurosci. Methods* **2003**, *123*, 201–206. [CrossRef]
34. Lind, N.M.; Moustgaard, A.; Jelsing, J.; Vajta, G.; Cumming, P.; Hansen, A.K. The use of pigs in neuroscience: Modeling brain disorders. *Neurosci. Biobehav. Rev.* **2007**, *31*, 728–751. [CrossRef] [PubMed]
35. Duhaime, A.-C. Large Animal Models of Traumatic Injury to the Immature Brain. *Dev. Neurosci.* **2006**, *28*, 380–387. [CrossRef] [PubMed]
36. Kinder, H.; Baker, E.W.; Wang, S.; Fleischer, C.C.; Howerth, E.W.; Duberstein, K.J.; Mao, H.; Platt, S.R.; West, F.D. Traumatic Brain Injury Results in Dynamic Brain Structure Changes Leading to Acute and Chronic Motor Function Deficits in a Pediatric Piglet Model. *J. Neurotrauma* **2019**, *36*, 2930–2942. [CrossRef]
37. Kinder, H.A.; Baker, E.W.; West, F.D. The pig as a preclinical traumatic brain injury model: Current models, functional outcome measures, and translational detection strategies. *Neural Regen. Res.* **2019**, *14*, 413–424. [CrossRef] [PubMed]
38. Wang, H.; Baker, E.W.; Mandal, A.; Pidaparti, R.M.; West, F.D.; Kinder, H.A. Identification of predictive MRI and functional biomarkers in a pediatric piglet traumatic brain injury model. *Neural Regen. Res.* **2021**, *16*, 338–344. [CrossRef]
39. Dickerson, J.W.T.; Dobbing, J. Prenatal and postnatal growth and development of the central nervous system of the pig. *Proc. R. Soc. London. Ser. B Boil. Sci.* **1967**, *166*, 384–395. [CrossRef]
40. Ryan, M.C.; Sherman, P.; Rowland, L.; Wijtenburg, S.A.; Acheson, A.; Fieremans, E.; Veraart, J.; Novikov, D.; Hong, L.E.; Sladky, J.; et al. Miniature pig model of human adolescent brain white matter development. *J. Neurosci. Methods* **2017**, *296*, 99–108. [CrossRef] [PubMed]
41. Margulies, S.S.; Coats, B. Biomechanics of pediatric TBI. In *Pediatric Traumatic Brain Injury: New Frontiers in Clinical and Translational Research*; Yeates, K.O., Anderson, V., Eds.; Cambridge University Press: Cambridge, UK, 2010; pp. 7–17.
42. Burrows, P.; Trefan, L.; Houston, R.; Hughes, J.; Pearson, G.; Edwards, R.J.; Hyde, P.; Maconochie, I.; Parslow, R.C.; Kemp, A.M. Head injury from falls in children younger than 6 years of age. *Arch. Dis. Child.* **2015**, *100*, 1032–1037. [CrossRef]
43. Cullen, D.K.; Harris, J.P.; Browne, K.D.; Wolf, J.A.; Duda, J.E.; Meaney, D.F.; Margulies, S.S.; Smith, D.H. A Porcine Model of Traumatic Brain Injury via Head Rotational Acceleration. *Methods Mol. Biol.* **2016**, *1462*, 289–324. [CrossRef] [PubMed]
44. Neumann, M.; Wang, Y.; Kim, S.; Hong, S.M.; Jeng, L.; Bilgen, M.; Liu, J. Assessing gait impairment following experimental traumatic brain injury in mice. *J. Neurosci. Methods* **2009**, *176*, 34–44. [CrossRef]
45. Luo, J.; Nguyen, A.; Villeda, S.; Zhang, H.; Ding, Z.; Lindsey, D.; Bieri, G.; Castellano, J.M.; Beaupre, G.S.; Ewyss-Coray, T. Long-Term Cognitive Impairments and Pathological Alterations in a Mouse Model of Repetitive Mild Traumatic Brain Injury. *Front. Neurol.* **2014**, *5*, 12. [CrossRef] [PubMed]
46. Reed, J.; Grillakis, A.; Kline, A.; Ahmed, A.E.; Byrnes, K.R. Gait analysis in a rat model of traumatic brain injury. *Behav. Brain Res.* **2021**, *405*, 113210. [CrossRef] [PubMed]
47. Abstracts from The 38th Annual National Neurotrauma Symposium July 11–14, 2021 Virtual Conference. *J. Neurotrauma* **2021**, *38*, A125. [CrossRef]

48. Hajiaghamemar, M.; Seidi, M.; Margulies, S.S. Head Rotational Kinematics, Tissue Deformations, and Their Relationships to the Acute Traumatic Axonal Injury. *J. Biomech. Eng.* **2020**, *142*, 0310061–03100613. [CrossRef]
49. Hajiaghamemar, M.; Seidi, M.; Patton, D.; Huber, C.; Arbogast, K.B.; Master, C.; Margulies, S.S. Using On-Field Human Head Kinematics to Guide Study Design for Animal-Model Based Traumatic Brain Injury. In Proceedings of the Biomedical Engineering Society Annual Meeting, Philadelphia, PA, USA, 16–19 October 2020.
50. Gennarelli, T.A.; Thibault, L.E.; Graham, D.I. Diffuse Axonal Injury: An Important Form of Traumatic Brain Damage. *Neurosci.* **1998**, *4*, 202–215. [CrossRef]
51. Ommaya, A.K.; Gennarelli, T.A. Cerebral Concussion and Traumatic Unconsciousness: Correlation of Experimental and Clinical Observations on Blunt Head Injuries. *Brain* **1974**, *97*, 633–654. [CrossRef]
52. Wu, T.; Antona-Makoshi, J.; Alshareef, A.; Giudice, S.; Panzer, M.B. Investigation of Cross-Species Scaling Methods for Traumatic Brain Injury Using Finite Element Analysis. *J. Neurotrauma* **2020**, *37*, 410–422. [CrossRef]
53. Hajiaghamemar, M.; Margulies, S.S. *Traumatic Brain Injury: Translating Head Kinematics Outcomes between Pig and Human*; International Research Council on Biomechanics of Injury: Zurich, Switzerland, 2020; pp. 605–607.
54. Patton, D.A.; Huber, C.M.; Douglas, E.C.; Seacrist, T.; Arbogast, K.B. Laboratory assessment of a head impact sensor for youth soccer ball heading impacts using an anthropomorphic test device. *Proc. Inst. Mech. Eng. Part P J. Sports Eng. Technol.* **2021**, 17543371211063124. [CrossRef]
55. Pusparum, M.; Ertaylan, G.; Thas, O. From Population to Subject-Specific Reference Intervals. In Proceedings of the International Conference on Computational Science—ICCS 2020, Amsterdam, The Netherlands, 3–5 June 2020; Krzhizhanovskaya, V.V., Závodszky, G., Lees, M.H., Dongarra, J.J., Sloot, P.M.A., Brissos, S., Teixeira, J., Eds.; Springer: New York, NY, USA, 2020; pp. 468–482. [CrossRef]
56. Karunakaran, K.K.; Ehrenberg, N.; Cheng, J.; Bentley, K.; Nolan, K.J. Kinetic Gait Changes after Robotic Exoskeleton Training in Adolescents and Young Adults with Acquired Brain Injury. *Appl. Bionics Biomech.* **2020**, *2020*, e8845772. [CrossRef]
57. Gagnon, I.; Swaine, B.; Friedman, D.; Forget, R. Children show decreased dynamic balance after mild traumatic brain injury. *Arch. Phys. Med. Rehabil.* **2004**, *85*, 444–452. [CrossRef] [PubMed]
58. Katz-Leurer, M.; Rotem, H.; Keren, O.; Meyer, S. Effect of concurrent cognitive tasks on gait features among children post-severe traumatic brain injury and typically-developed controls. *Brain Inj.* **2011**, *25*, 581–586. [CrossRef]
59. Katz-Leurer, M.; Rotem, H.; Lewitus, H.; Keren, O.; Meyer, S. Relationship between balance abilities and gait characteristics in children with post-traumatic brain injury. *Brain Inj.* **2008**, *22*, 153–159. [CrossRef]
60. Maki, B.E. Gait Changes in Older Adults: Predictors of Falls or Indicators of Fear? *J. Am. Geriatr. Soc.* **1997**, *45*, 313–320. [CrossRef]
61. Fukuchi, C.A.; Fukuchi, R.K.; Duarte, M. Effects of walking speed on gait biomechanics in healthy participants: A systematic review and meta-analysis. *Syst. Rev.* **2019**, *8*, 153. [CrossRef]
62. Fettrow, T.; Reimann, H.; Grenet, D.; Crenshaw, J.; Higginson, J.; Jeka, J. Walking Cadence Affects the Recruitment of the Medial-Lateral Balance Mechanisms. *Front. Sports Act. Living* **2019**, *1*, 40. [CrossRef] [PubMed]
63. Katz-Leurer, M.; Rotem, H.; Keren, O.; Meyer, S. Balance abilities and gait characteristics in post-traumatic brain injury, cerebral palsy and typically developed children. *Dev. Neurorehabil.* **2009**, *12*, 100–105. [CrossRef] [PubMed]
64. Katz-Leurer, M.; Rotem, H.; Keren, O.; Meyer, S. The relationship between step variability, muscle strength and functional walking performance in children with post-traumatic brain injury. *Gait Posture* **2009**, *29*, 154–157. [CrossRef] [PubMed]
65. Rahman, R.A.A.; Hanapiah, F.A.; Nikmat, A.W.; Ismail, N.A.; Manaf, H. Effects of Concurrent Tasks on Gait Performance in Children with Traumatic Brain Injury Versus Children With Typical Development. *Ann. Rehabil. Med.* **2021**, *45*, 186–196. [CrossRef] [PubMed]
66. Beretta, E.; Cimolin, V.; Picciinini, L.; Turconi, A.C.; Galbiati, S.; Crivellini, M.; Galli, M.; Strazzer, S. Assessment of gait recovery in children after traumatic brain injury. *Brain Inj.* **2009**, *23*, 751–759. [CrossRef]
67. Howell, D.R.; Beasley, M.; Vopat, L.; Meehan, W.P. The Effect of Prior Concussion History on Dual-Task Gait following a Concussion. *J. Neurotrauma* **2017**, *34*, 838–844. [CrossRef] [PubMed]
68. Rahman, R.A.A.; Rafi, F.; Hanapiah, F.A.; Nikmat, A.W.; Ismail, N.A.; Manaf, H. Effect of Dual-Task Conditions on Gait Performance during Timed Up and Go Test in Children with Traumatic Brain Injury. *Rehabil. Res. Pr.* **2018**, *2018*, e2071726. [CrossRef]
69. Coats, B.; Binenbaum, G.; Smith, C.; Peiffer, R.L.; Christian, C.W.; Duhaime, A.-C.; Margulies, S.S. Cyclic Head Rotations Produce Modest Brain Injury in Infant Piglets. *J. Neurotrauma* **2017**, *34*, 235–247. [CrossRef] [PubMed]
70. Prange, M.T.; Coats, B.; Duhaime, A.-C.; Margulies, S.S. Anthropomorphic simulations of falls, shakes, and inflicted impacts in infants. *J. Neurosurg.* **2003**, *99*, 143–150. [CrossRef]
71. Ommaya, A.; Hirsch, A. Tolerances for cerebral concussion from head impact and whiplash in primates. *J. Biomech.* **1971**, *4*, 13–21. [CrossRef]
72. Gupte, R.P.; Brooks, W.; Vukas, R.; Pierce, J.D.; Harris, J.L. Sex Differences in Traumatic Brain Injury: What We Know and What We Should Know. *J. Neurotrauma* **2019**, *36*, 3063–3091. [CrossRef]

Article

Altered Auditory and Visual Evoked Potentials following Single and Repeated Low-Velocity Head Rotations in 4-Week-Old Swine

Anna Oeur [1], William H. Torp [1], Kristy B. Arbogast [2,3], Christina L. Master [2,3,4] and Susan S. Margulies [1,*]

[1] Wallace H. Coulter Department of Biomedical Engineering, Emory University and Georgia Institute of Technology, Atlanta, GA 30332, USA; anna.oeur@emory.edu (A.O.); wtorp3@gatech.edu (W.H.T.)
[2] Center for Injury Research and Prevention, Children's Hospital of Philadelphia, Philadelphia, PA 19146, USA; arbogast@chop.edu (K.B.A.); masterc@chop.edu (C.L.M.)
[3] Perelman School of Medicine, the University of Pennsylvania, Philadelphia, PA 19104, USA
[4] Sports Medicine and Performance Center, Children's Hospital of Philadelphia, Philadelphia, PA 19104, USA
* Correspondence: susan.margulies@emory.edu; Tel.: +1-404-727-9827; Fax: +1-404-727-9873

Abstract: Auditory and visually evoked potentials (EP) have the ability to monitor cognitive changes after concussion. In the literature, decreases in EP are commonly reported; however, a subset of studies shows increased cortical activity after injury. We studied auditory and visual EP in 4-week-old female Yorkshire piglets (N = 35) divided into anesthetized sham, and animals subject to single (sRNR) and repeated (rRNR) rapid non-impact head rotations (RNR) in the sagittal direction. Two-tone auditory oddball tasks and a simple white-light visual stimulus were evaluated in piglets pre-injury, and at days 1, 4- and 7 post injury using a 32-electrode net. Traditional EP indices (N1, P2 amplitudes and latencies) were extracted, and a piglet model was used to source-localize the data to estimate brain regions related to auditory and visual processing. In comparison to each group's pre-injury baselines, auditory Eps and brain activity (but not visual activity) were decreased in sham. In contrast, sRNR had increases in N1 and P2 amplitudes from both stimuli. The rRNR group had decreased visual N1 amplitudes but faster visual P2 latencies. Auditory and visual EPs have different change trajectories after sRNR and rRNR, suggesting that injury biomechanics are an important factor to delineate neurofunctional deficits after concussion.

Keywords: brain concussion; auditory evoked potentials; visually evoked potentials; auditory perception; visual perception; traumatic brain injuries; electrodes; swine

1. Introduction

Sports-related traumatic brain injuries (TBI) are one of the leading causes of emergency department visits [1], with an estimated 1.6–3.8 million sports-related TBI occurring in the United States every year [2]. The International Concussion in Sport Group (CISG) defines a sport-related concussion as a TBI induced by a direct or indirect blow that transmits force to the head, resulting in short-duration impairments that can evolve into longer-lasting signs and symptoms [3]. Forces incurred from a TBI affect the brain at the cellular level, with perturbations disrupting membranes and proteins that inhibit transport of ions across energy channels necessary for homeostasis [4]. This, in turn, impairs neural function and can initiate neuroinflammatory responses that are secondary to the initial insult [5]. Typical cognitive impairments associated with concussion are attention, memory, and information-processing, as well as somatic symptoms such as headaches and dizziness [6]. Auditory and visual dysfunctions are a common finding after closed head injury [7] including concussion [8–10] and may disproportionately disadvantage children, as problems with learning, reading, and speech may impede developmental social and scholastic success [11–13].

Electroencephalography (EEG) is a promising tool for concussion assessment, as it permits millisecond (ms) measurement of synchronous postsynaptic potentials of cortical pyramidal neurons, with the possibility of detecting neurologic dysfunction [14]. EEG assessments for TBI include continuous measurements of brain activity or those evoked by the presentation of a stimulus, defined as evoked potentials (EP) [15]. The EEG waveform resulting from EPs are a series of positive (P) and negative (N) peaks and troughs that are denoted by approximate time of presentation post-stimuli: P1 (50 ms), N1 (75–140 ms), P2 (150–230 ms), N2 (150–250 ms) and P3 (250–350 ms) [16].

Auditory EPs permit an integrative assessment of the auditory pathway from the cochlea, auditory nerve and brainstem pathways, as well as auditory cortical functions that reflect sound detection and early stimulus-processing in the primary auditory cortex [17]. Visual EPs holistically assess pathways from light stimuli on the retina to the optic nerves, the optic chiasm, thalamus and occipital cortex [18]. Auditory and visual EPs have the potential to provide an indication of functional brain activity and related changes in information processing post-TBI that are specific to the brain structures involved with each respective pathway.

The injured brain has been hypothesized to have lower-amplitude and longer-latency responses compared to a healthy brain, reflecting a decreased capacity for information-processing and slower transmission speeds [16,19]. In the literature, there have been studies that support this hypothesis, those that partially support (in amplitude but not latency or vice versa), and even studies that report no differences between concussed and healthy groups. It is also noteworthy that there seems to be a pattern in the literature showing groups with a history of concussion having alternative cognitive processes compared to a no-injury group. To highlight some examples, symptomatic and asymptomatic concussed adults had reduced N1 amplitudes in comparison to a control group; however, there were no differences in latency [20]. Vander Werff and Rieger [8] reported no differences in P1, N1, and P2 between controls and adults with long-term concussion (up to 18 months); however, P3 amplitudes were reduced for the injured group.

In a study involving junior ice hockey players, athletes with a history of concussion (3+) had significantly longer P3 latencies than a no concussion group [21]. More recently, Bennys et al. [22] showed that athletes with a history of concussion (1+ in the last 3 years) had decreased P300 amplitudes and a trend towards longer latencies (in comparison to no injury) from auditory oddball tasks.

Interestingly, some studies report contrary findings where amplitudes are greater and latencies are increased. For example, concussed athletes at 4 years post-TBI had increased N2 and P3 amplitudes, in addition to longer latencies, in comparison to healthy controls [23]. In a separate study, male ice hockey players were subject to a two tone-auditory oddball test within 24 h of concussion and similarly reported increased amplitudes (N100, P300, N400) and decreased latencies in comparison to baseline values [24]. One hypothesis for the increase in cortical activity after TBI posits that recovery from injury could result in a compensatory mechanism by which an increase in neural activity is required to meet the same executive functioning demands as the non-injured brain [23]. It is thought that, after injury, new neural networks (and combinations of neural networks) can engage in different temporal and spatial patterns affecting EP attributes recorded from the scalp [25].

The variability in the timepoints studied post-TBI (acute versus chronic), and the history of concussion per subject (total number and time between injuries) across studies are contribute to the mixed EP findings in the literature. Other factors include heterogeneity in the mechanisms of injury [26], as the mechanical parameters governing the loading conditions in a head injury event, such as velocity and direction of motion, are key causal factors in the observed patterns of neural trauma [27]. In addition, age and cortical maturation have an effect on EP amplitudes and latencies throughout development in childhood [28,29] and adulthood [30], contributing to varying effects of concussion on these measures across the lifespan [31].

Pre-clinical animal models of TBI provide an idealized platform to allow for biomechanical control over the head trauma load and direction; the timepoints studied post-TBI, total number and timing of multiple traumas, as well as animal age and brain maturation to better isolate their effects on brain injury [32,33]. Animal models provide an opportunity to systematically assess the subtleties of TBI and improve our understanding of the auditory and visual impairments related to structurual and functional deficits post-concussion [34,35]. In comparison to other animal models (monkey, dog, rodents), the rapid post-natal development of pigs after birth makes this species a suitable model across a number of different fields, including the skeletal and neuromuscular, pulmonary and cardiovascular, central nervous system and gastrointestinal system [36]. A 4–14-week-old piglet approximates a young child of roughly 2–12 years old [37]. A 4-week-old swine is an established model of pediatric TBI, with neuroanatomical structures (gyri and sulci) and gray and white matter distributions that are similar to the developing brain [38,39] and are important biomechanical characteristics to model brain movement within the skull [40]. The spectrum of diffuse axonal injuries, a form of TBI including concussion, was achieved in our large animal model using a rapid non-impact head rotation (RNR) device and employed in studies from our laboratory examining neurobehavioral deficits, histopathology, and drug efficacy in piglets of different ages [32,41–45].

This study utilizes methods established in our laboratory for measuring evoked potentials in healthy 4-week-old swine, in addition to studying the large-animal model under single or repeated head loads using an RNR device. Sixteen healthy animals presenting with a passive two-tone auditory oddball test were used; infrequent target tones produced greater N1 amplitudes for frontal electrodes and produced consistent day-to-day responses [46]. In a separate set of healthy animals (N = 11), cognitive activity from auditory and visual stimuli were compared using traditional evoked potential measures and cortical activations estimated from source localization techniques. In healthy animals, N1 amplitudes were greater from auditory stimuli in comparison to visual stimuli. P2 amplitudes were greater from visual stimuli and latencies (N1 and P2) were faster for visual stimuli than auditory stimuli. Patterns of cortical activation showed that visual stimulation had greater levels of early (50 ms) activity than auditory stimulation; however, at 85 ms, auditory had greater left-temporal activations. At 110 ms, visual stimulation had greater activity in the left and right occipital regions [47]. The objective of this study was to examine the role of single and repeated head rotations on auditory and visually evoked potentials to better describe the effects of head biomechanics and loading patterns on neurocognitive deficits. Derived from this prior work, our hypotheses for the current study are as follows: (1) there will be a 'between experimental group' effect where RNR will significantly reduce EP indices and cortical activation compared to sham; (2) within each experimental group, there will be a 'day effect' where cognitive processing is decreased at various timepoints after RNR or anesthesia compared to pre-injury baselines; and (3) there will be a 'stimulus effect' where the patterns of reduced cognitive activity are unique to auditory stimuli and are different from visual stimuli.

2. Methods

To study the effects of concussion on auditory and visual processing, we employed the 4-week-old swine pediatric TBI model subject to prescribed controlled mild head rotations. Auditory and visually evoked potentials were captured using methods published previously in healthy piglets [46,47] on each animal before head rotations, to establish a pre-injury baseline, and again at 1, 4, and 7 days after to examine the acute time-course of mild TBI on stimulus processing. Other neurofunctional measures were also collected using this experimental injury paradigm and include piglet gait and pupillary light reflexes, using previously published methods [48,49].

2.1. Animal Subjects

Thirty-five 4-week-old female Yorkshire piglets were allocated into three experimental groups: sham (N = 10), single RNR (N = 12), and repeated RNR (N = 13). Awake subjects were fitted with a 32-electrode EEG net and evaluated using two-tone auditory oddball tests at baseline and multiple days after rapid non-impact head rotation (RNR) or an anesthesia-only event (sham). A subset of these animals was also evaluated using a simple white-light visually evoked potential: sham (N = 5), single RNR (N = 6), multiple RNR (N = 9). Animals were socially housed in cages on a 12 h light and 12 h dark cycle, and were freely permitted food (LabDiet 5080, St. Louis, MO, USA) and water. EEG measurements were taken in a separate behaviour test room where auditory and visual EP measurements were taken in the awake animal while gently restrained in a sling. Animals were acclimated to the sling and head gear (EEG net or stretchable nylon) for at least two 30 min sessions prior to the first study day. All subjects that were included for analysis survived to the end of study. All animal procedures were approved by Emory University's Institutional Animal Care and Use Committee (IACUC).

2.2. Rapid Non-Impact Rotational Injury (RNR)

On the injury day (day 0), each injured and sham piglet was sedated with an intramuscular injection (Ketamine:4 mg/kg, Xylazine: 2 mg/kg, and Midazolam: 0.2 mg/kg) and underwent anesthesia via inhalation of 1.5% isoflurane using a fitted snout mask. While vitals were monitored and body temperature was maintained, a lack of response to a mild toe pinch confirmed the appropriate depth of anesthesia; then, the RNR was delivered. Sham animals received anesthesia only, without RNR. Animals receiving a head rotation were secured to a HYGE device (HYGE Inc., Kittaning, PA, USA) via a bite plate. The bite plate was attached to a linkage system that transfers the linear motion of the pneumatic actuator to a rapid head rotation of the snout and head [32,41]. Angular transducers were mounted onto the linkage system to measure the angular velocity of the system (ARS-06, ATA Engineering, Inc., Herndon, VA or ARS Pro, DTS Inc., Seal Beach, CA, USA). The HYGE device is capable of RNR at levels consistent with the spectrum of diffuse axonal injury pathologies [50–52]. Piglets in the single and repeated RNR groups experienced sagittal rotation on day 0 (Figure 1). The levels of head rotations prescribed to each experimental animal group was computationally scaled from soccer participants instrumented with head-impact sensors [53]. The load levels were scaled from 267 headers in high-school soccer players, primarily causing sagittal motions of the head [53]. The human kinematic data were input into a finite element model of the human brain and maximum axonal strain (MAS) was estimated for each of the 267 impacts [53,54]. The 50th (medium) and 90th (high) percentile MAS values were extracted from the human header data and scaled using a piglet finite-element model to determine the corresponding peak angular velocity and angular accelerations associated with the 50th (medium) and 90th (high) levels. The single RNR group received a single 'high header' and the repeated RNR group experienced one 'high header' load followed by four 'medium header' loads. The target 'high header' rotations scaled to the pig were 104 ± 2.36 rad/s and 37.8 ± 6.06 krad/s^2 and 'medium header' rotations were 61.2 ± 2.02 rad/s and 15.0 ± 1.72 krad/s^2. For the repeated-RNR group, rotations were delivered 8.4 min apart (±1.1 min). The 8 min interval between impacts for the repeated RNR group was determined from the same high-school soccer heading data, where girls and boys received 4 and 6 impacts per h, respectively, where impacts were spaced 8 min apart for both sexes. Further details on the determination of piglet head rotation magnitudes are described in [48]. EEG measurements were taken on a pre-injury day (D-1), one (D1), four (D4) and seven days (D7) post-injury (Figure 1). The Institutional Animal Care and Use Committees (IACUC) at Emory University approved all animal procedures conducted in this study.

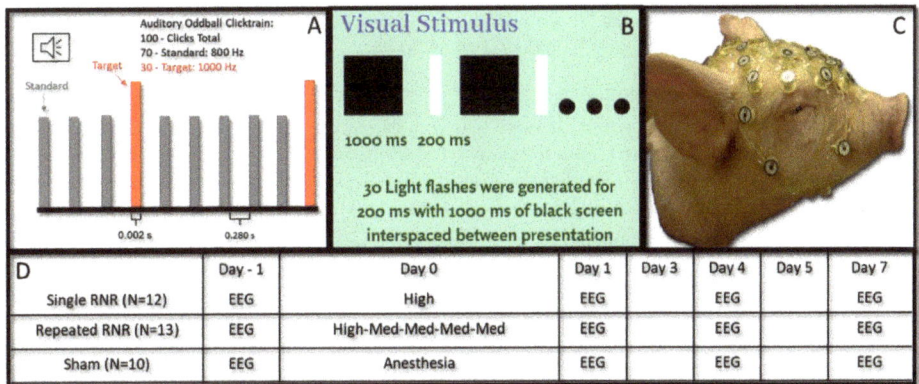

Figure 1. Summary auditory oddball paradigm with 30 target (1000 Hz) tones (0.002 s) randomly played in between standard (800 Hz) tones with an interstimulus interval of 0.280 s (**A**). Simple white light visual stimulus (**B**). Custom 32-electrode EEG net on piglet (**C**). Experimental timeline for sRNR, rRNR, and sham animals for EEG data collection and injury/anesthesia (**D**) at pre-injury (pre), injury (D0), day 1 (D1), day 4 (D4) and day 7 (D7). RNR = rapid nonimpact rotation.

2.3. Electroencephalography Measurements

Non-invasive EEG data were collected using scalp electrodes embedded in a custom 32-channel EGI HyrdoCel piglet electrode net at 1000 Hz using a Net Amps 400 amplifier (Electric Geodesics Inc., EGI, Eugene, OR, USA). Prior to application on the piglets, the net was soaked in baby shampoo (5 mL), potassium chloride (10 mL) and water (1 L) solution. The net was then placed on the piglet's head and electrical impedance was checked and maintained below 1 kΩ [55]. EEG data acquisition was accomplished using a MacBook Pro laptop with Netstation (Version 5.0, Electric Geodesics Inc., EGI, Eugene, OR, USA) and synchronized to auditory and visual stimuli using E'Prime on a PC computer (Version 2.0, Psychology Software Tools, Inc., Pittsburgh, PA, USA). The auditory oddball train consisted of a 100-tone clicktrain comprising 70 standard tones (800 Hz) and 30 target tones (1000 Hz) played in random order (Figure 1A). For visual trains, 30 white-light flashes were presented using a 7-inch LCD screen (Figure 1B). Auditory sounds were presented to the centre of the piglets' head, attempting to stimulate both ears equally; however, for visual stimuli, the light stimuli was presented to the left eye only. Stimuli were presented in the same manner as in our study with healthy piglets [47]. Each animal received 6 trials of auditory trains, and the animals who were studied for visually evoked potentials also received 6 trials of visual trains [47].

EEG data were pre-processed in Netstation Tools (Electric Geodesics Inc., EGI, Eugene, OR, USA) and included band-pass filtering from 0.1 to 30 Hz, segmentation of evoked potential into 300 ms epochs that incorporate a 50 ms pre-stimulus baseline and a 250 ms post-stimulus response period. Bad channels were replaced if greater than 200 uV via interpolations of two nearby electrodes [47]. Data were post-processed in EEGlab Version 14.12 [56] and Matlab Version R2018b (The Mathworks, Inc., Natick, MA, USA) [57] and included baseline correction, waveform averaging and independent component analysis to remove noise artefacts and eye movements [58].

EEG waveforms were averaged per animal, per day and per stimulus type (auditory-target, -standard, and visual). N1 and P2 peak amplitudes and latencies were extracted from the averaged waveforms from each electrode. N1 and P2 pertain to the first negative and second positive peak following stimulus presentation and are the most consistent attributes of the evoked potential for piglets [47]. Peak data from electrodes 1, 2, 3, 4, 17, and 27 were averaged together to represent the response at the front of the head. Similarly, electrodes were averaged together to represent activity in the left (5, 11, 13, 15, 23) and right

(6, 12, 14, 16, 24) temporal regions. Values were removed if the peak was non-negative and non-positive for N1 and P2, respectively.

2.4. Source Localization

EEG waveforms were source-localized using a finite-element model (FEM) of the piglet head and brain to estimate the electrical activity distribution in the brain in sham and RNR piglets. Details regarding the development of this model for source localization in healthy piglets are described elsewhere [47]. Briefly, the model was derived from magnetic resonance images (MRI) of an infant piglet and scaled to the 4-week-old pig [59]. The model comprises 1.6 million tetrahedral elements making up the scalp, skull, brain, ventricles, and eyes [59,60]. The 32-electrode array was imported on the scalp of the model [61]; however, eight electrodes (9, 10, 19, 20, 21, 22, 25, 26) were not included as they lay in a region behind the ears. Conductivity values were assigned for the scalp (0.4 S/m), skull (0.03 S/m), cerebral spinal fluid (1.79 S/m), brain (0.5 S/m) and eyes (1.5 S/m) [59]. Standardized low-resolution brain electromagnetic tomography algorithms (sLORETA) were employed to calculate the inverse solution [62,63]. Mean current density was extracted from five regions of interest (frontal, left and right temporal, and left and right occipital) at three timepoints post-stimuli (50 ms, 85 ms, and 110 ms). These timepoints were selected to capture the cortical activity surrounding the N1 amplitude, which was found in our previous study to have the largest amplitude response from auditory and visual stimuli [47].

2.5. Statistics

Three-way repeated-measure ANOVAs were run per electrode region (frontal, left and right temporal) to evaluate the effect of day (repeated measures), experimental group (sham, sRNR, rRNR), and stimulus (standard, target, and visual) on N1 and P2 amplitudes and latencies. Statistical analyses for source localization were run in a similar manner, employing 3-way repeated measures ANOVAs to evaluate the effect of day (repeated measures), experimental group (sham, sRNR, rRNR), and stimulus (standard, target, and visual) on current density. This analysis was repeated 15 times, to stratify according to timepoint (50 ms, 85 ms, 110 ms) and brain region (frontal, left and right temporal, left and right occipital regions). Post hoc analyses employed one-way ANOVAs with Bonferroni corrections. All statistics were conducted using IBM SPSS (Version 25, Armonk, NY, USA: IBM Corp.) and significance was accepted at $p < 0.05$.

3. Results

3.1. Overview

As an overview, the mean and standard deviation of the angular velocities and angular accelerations corresponding to medium and high load levels for the single-RNR and repeated-RNR groups are reported in Table 1. The measured head kinematics were consistent with our target load levels, as determined from scaling. Exemplar current density patterns for each group at 85 and 110 ms are illustrated in Figures 2 and 3. The results for source localization are shown in Figures 2 and 3. Evoked potential results are shown in Figure 4, presented by stimulus type because our findings from healthy animals showed that visual stimuli produced larger responses in the occipital areas and more auditory stimuli in the temporal regions [47]. Figures 5–7 illustrate 50, 85, and 110 ms current density results, respectively for an exemplar sRNR animal across study day and stimulus type.

Comparisons of EP and current density measurements that were significantly different from each experimental group's pre-injury values and contrary to those found in healthy animals from a previous study [47] are reported in this section. Table 2 presents a summary of the main findings for each experimental group, highlighting changes from pre-injury baselines.

Table 1. Mean ± standard error of angular velocity and acceleration loading levels for single- and repeated-RNR groups.

	Load Level	Angular Velocity (rad/s)	Angular Acceleration (rad/s^2)
sRNR	High	104 ± 0.495	36,900 ± 1120
rRNR	Medium	61.2 ± 0.182	14,900 ± 175
	High	104 ± 0.414	38,300 ± 533

Table 2. Summary table of significant changes in auditory and visual EP findings for sham and RNR piglets.

	Region	Stimulus	Significant Change from Pre-Injury
Sham	Frontal	Auditory–Standard	Decrease (85 ms, current density)
	Frontal	Auditory–Target	Decrease (N1 and P2 amplitudes)
	Left Temporal	Auditory–Standard	Decrease (85 ms, current density)
	Left Temporal	Auditory –Target	Decrease (85 ms, current density)
	Right Temporal	Auditory–Standard	Decrease (85 ms, current density)
sRNR	Right Temporal	Visual	Increase (N1 amplitude)
	Right Temporal	Auditory–Target	Increase (P2 amplitude)
	Right Temporal	Visual	Increase (50 ms, current density)
	Right Occipital	Visual	Increase (50 ms, current density)
	Left Occipital	Auditory–Target	Decrease (110 ms, current density)
	Right Occipital	Auditory–Target	Decrease (110 ms, current density)
rRNR	Right Temporal	Visual	Decrease (N1 amplitude)
	Right Temporal	Auditory–Standard	Decrease (P2 amplitude)
	Frontal	Visual	Decrease (P2 latency) *
	Left Temporal	Auditory–Standard	Decrease (85 ms, current density)

NB: '*' denoting a decrease in latency means P2 latencies were faster rRNR.

3.2. Overall Group Differences

The findings examining experimental group comparisons are presented next. For reliability, if there were significant differences between experimental groups at pre-injury for an extracted EP parameter or current density for a region, further group comparisons on subsequent days were not examined. Figure 2 presents example patterns of cortical activity from source localization analysis at day 4 for a single sham, sRNR, and rRNR animal presented with a standard tone and a visual stimulus depicted at 85 ms. We will discuss the results in each region. In the frontal region, statistically significant results were found, where auditory processing was decreased for RNR groups in comparison to sham. Specifically, on day 4, rRNR (1.117 ± 0.203 µV) had decreased target P2 amplitudes compared to sham (2.672 ± 0.321 µV) and sRNR (0.035 ± 0.005 µV/mm^2) had lower activations from standard tones compared to sham (0.046 ± 0.009 µV/mm^2) at 85 ms.

In the left temporal region, there were pre-injury differences for current density between sRNR and sham (110 ms) for visual stimuli N1 latencies; however, these were not different at pre-injury and, therefore, sRNR had significantly longer visual N1 latencies (62.067 ± 2.811 ms) than sham (49.400 ± 3.443 ms) at day 1. No pre-injury differences were observed between sRNR and rRNR; therefore, on day 4, sRNR had greater left temporal activations than rRNR at 85 ms (sRNR: 0.071 ± 0.012 µV/mm^2, rRNR: 0.031 ± 0.006 µV/mm^2) and 110 ms (sRNR: 0.070 ± 0.012 µV/mm^2, rRNR: 0.035 ± 0.007 µV/mm^2) post visual stimulus (Figure 3).

Regarding pre-injury, in the right temporal regions, standard P2 amplitudes were significantly different between rRNR and sRNR and visual P2 latencies were different between sRNR and sham. P2 amplitudes were not significantly different between rRNR and sham; therefore, contrary trends were found between groups in this region on day 1, where rRNR (2.695 ± 0.360 µV) had greater standard P2 amplitudes than sham (0.830 ± 0.360 µV). In addition, sRNR had greater visual activations (85 ms) than sham (sRNR: 0.050 ± 0.008, sham: 0.022 ± 0.005 µV/mm^2). Further, sRNR (2.824 ± 0.265 µV) and rRNR (2.915 ± 0.342 µV) had greater target P2 amplitudes than sham (1.364 ± 0.42 µV) on day 4, with sRNR (2.538 ± 0.277 µV) remaining significantly greater than sham (0.684 ± 0.358 µV) on day 7.

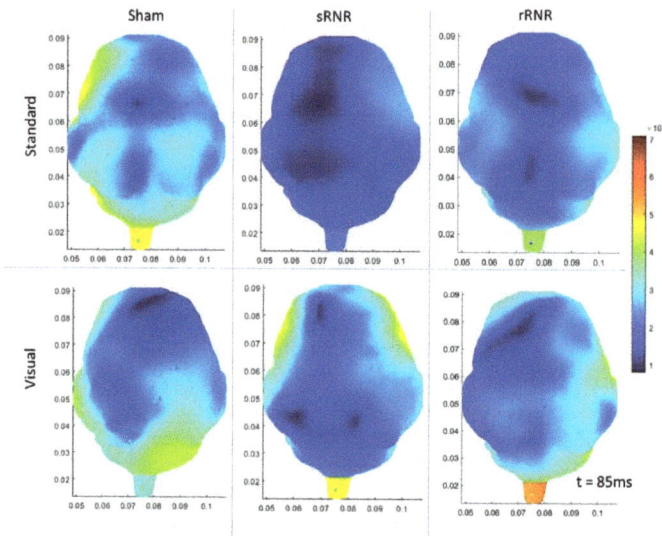

Figure 2. Exemplar current density distributions (85 ms) for sham (**left**), sRNR (**middle**) and rRNR (**right**) for standard tones (**top** row) and visual stimuli (**bottom** row) taken at day 4, illustrating unique cortical activity patterns across experimental group and stimulus type. Between-group comparisons found that sRNR showed significantly decreased frontal activity compared to sham for standard tones. rRNR had significantly decreased left-temporal activity than sRNR for visual stimuli, but was not different than sham.

In the left occipital region, pre-injury current densities at 110 ms were significantly different between sRNR and rRNR for standard tones only. Early cortical activity in the left occipital region (50 ms) showed greater activations from target tones for sRNR (0.027 ± 0.004 µV/mm^2) than rRNR (0.024 ± 0.007 µV/mm^2) on day 1. Interestingly on day 4, sham (0.027 ± 0.003 µV/mm^2) had greater activations than rRNR (0.022 ± 0.004 µV/mm^2) from standard tones; however, this was reversed on day 7, where rRNR (0.024 ± 0.004 µV/mm^2) was greater than sham (0.015 ± 0.003 µV/mm^2). Additionally, sRNR (0.102 ± 0.024 µV/mm^2) had greater left occipital activations (85 ms) from visual stimuli than sham on day 7 (0.025 ± 0.006 µV/mm^2). In the right occipital region, there were pre-injury differences between sRNR and sham, and sRNR and rRNR, with no other significant group comparisons on other days or stimuli.

3.3. Within Group Differences

Differences found within each experimental group (sham, sRNR, and rRNR) are presented below.

3.3.1. Sham–Anaesthesia Only

In sham animals, frontal N1 (−4.848 ± 0.848 µV) and P2 (1.388 ± 0.678 µV) amplitudes from target tones were significantly decreased at day 1 (N1: −2.711 ± 0.598 µV) and at day 7 (P2: 0.476 ± 0.915 µV) compared to baseline pre-anesthesia levels (Figure 4, top panel). In our previous study, examining the auditory oddball paradigm in healthy piglets, there was no effect of day, where standard and target tones produced similar responses on subsequent days of testing [46]. Similarly, frontal activations determined from source localization were decreased at day 1 (0.042 ± 0.008 µV/mm^2) and 7 (0.033 ± 0.004 µV/mm^2) in comparison to pre-anesthesia (0.057 ± 0.009 µV/mm^2) as a result of standard tones (85 ms time point).

At the 85 ms timepoint, the left temporal region showed decreased activations at day 1 for both standard (pre: 0.073 ± 0.014, day 1: 0.059 ± 0.010 μV/mm^2) and target tones (pre: 0.084 ± 0.011, day 1: 0.065 ± 0.010 μV/mm^2). We conclude that anesthesia influenced auditory responses for many metrics on day 1, with some persisting to day 7. No significant changes (from pre-anesthesia) were found for visual stimuli.

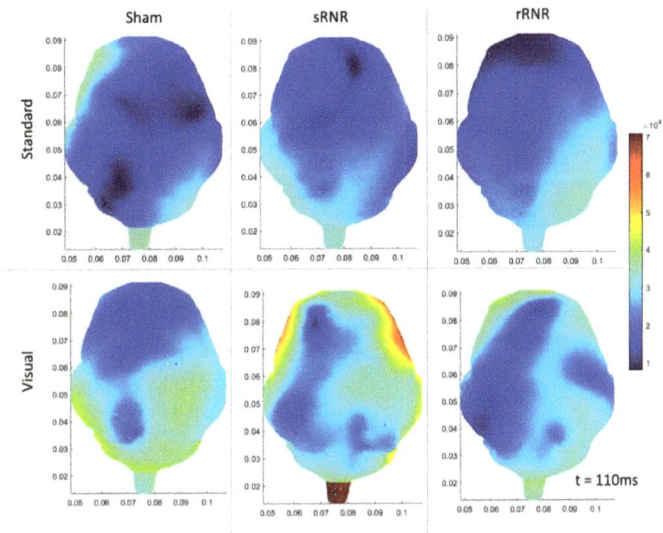

Figure 3. Exemplar current density distributions (110 ms) for sham (**left**), sRNR (**middle**) and rRNR (**right**) for standard tones (**top** row) and visual stimuli (**bottom** row) taken at day 4. Between-group comparisons found that sRNR had significantly increased left-temporal activity compared to rRNR for visual stimuli.

3.3.2. sRNR-Single

For the sRNR group, auditory target tones produced a significant increase in P2 amplitudes in the right temporal region from pre-injury (1.878 ± 0.245 μV) to day 7 (2.538 ± 0.277 μV). Similarly, visual stimuli (Figure 4) produced an increase in right temporal N1 amplitudes from pre-injury (-0.622 ± 0.429 μV) to day 1 (-1.827 ± 0.368 μV). Source localization results from visual stimuli show that, at 50 ms (Figure 5), the right temporal had significantly greater activations at day 7 (0.101 ± 0.011 μV/mm^2) than both pre-injury (0.050 ± 0.009 μV/mm^2) and day 1 (0.056 ± 0.007 μV/mm^2) timepoints. At 50 ms, the right occipital regions also showed a significant increase in visual stimuli activations at day 7 (0.109 ± 0.017 μV/mm^2) compared to pre-injury levels (0.069 ± 0.011 μV/mm^2). This region also increased at day 4 (0.100 ± 0.014 μV/mm^2), above the pre-injury baseline. At 50 ms, for auditory stimuli, there were no injury-associated effects in any region. Figure 6 shows current density patterns at 85 ms; however, no significant changes were found between days and stimuli for sRNR at this timepoint. At 110 ms, auditory target tones produced significantly decreased activations at day 1 in the left occipital (pre: 0.069 ± 0.011, day 1: 0.043 ± 0.004 μV/mm^2) and right occipital (pre: 0.064 ± 0.005, day 1: 0.049 ± 0.006 μV/mm^2) regions (Figure 7). At 110 msec, for auditory standard tones and visual stimuli, there were no injury-associated effects in any region.

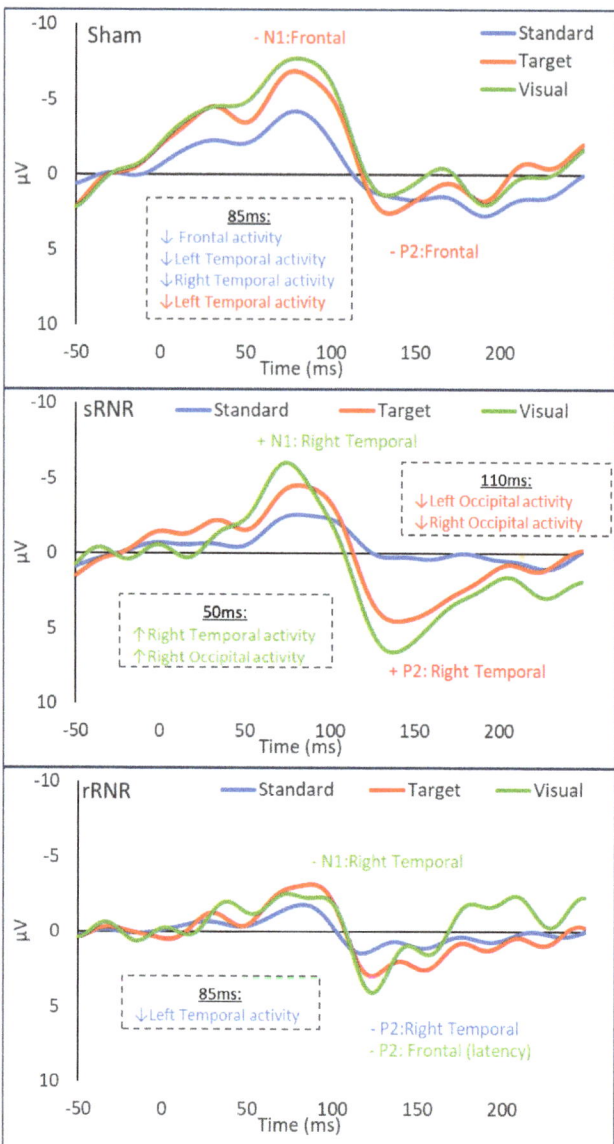

Figure 4. Single-subject exemplar evoked potential (EP) waveforms from channel 17 taken on day 1, summarizing group findings for sham (**top**), sRNR (**middle**), and rRNR (**bottom**), where blue represents auditory standard tones, red target tones, and green visual. Significant EP findings are denoted by '−' and '+', indicating decreases and increases from pre-injury for each brain region. All EP findings pertain to significant amplitude changes, except for P2 frontal latency in rRNR, as indicated in parentheses. For each plot, significant current density findings are summarized in square (dotted) boxes for each specified timepoint (50, 85, and 110 ms) and brain region.

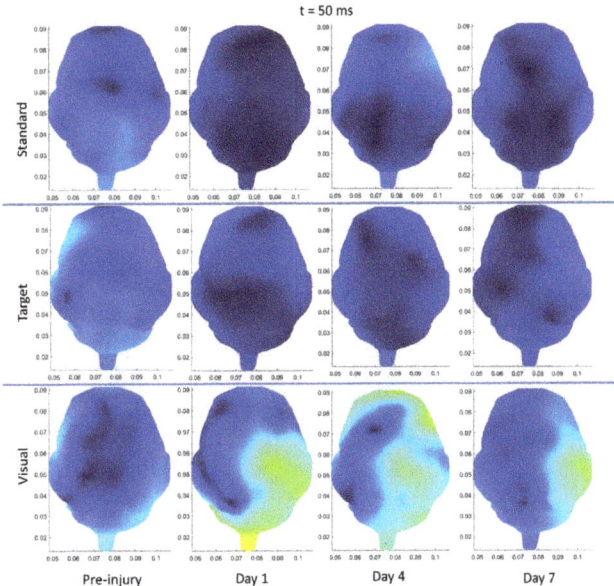

Figure 5. Depiction of source localization analysis at 50 ms post-standard (**top** row), target (**middle** row), and (**bottom** row) visual stimuli for one exemplar animal from the sRNR group across days studied (preinjury, day 1, 4, and 7).

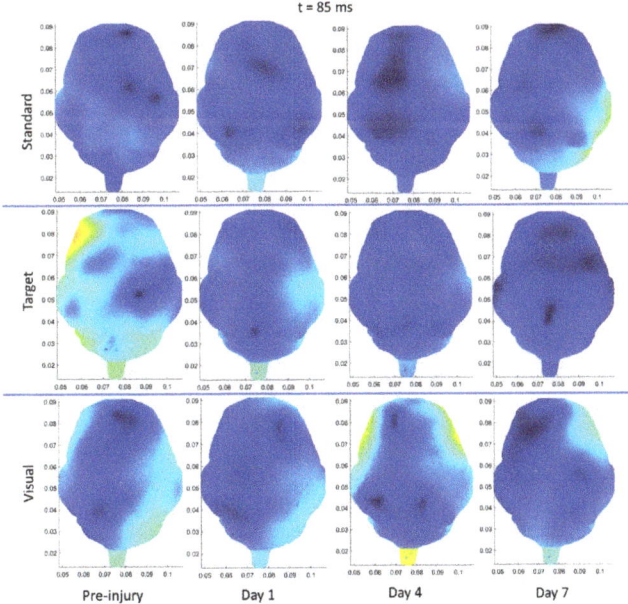

Figure 6. Depiction of source localization analysis at 85 ms post-standard (**top** row), target (**middle** row), and (**bottom** row) visual stimuli for one exemplar animal from the sRNR group across days studied (preinjury, day 1, 4, and 7).

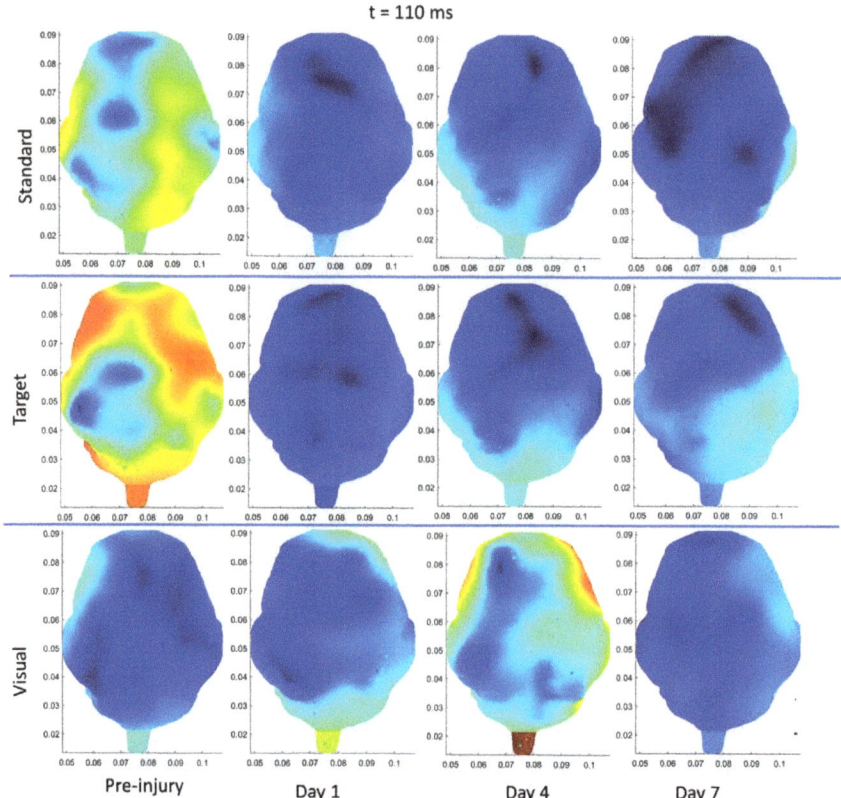

Figure 7. Depiction of source localization analysis at 110 ms post-standard (**top** row), target (**middle** row), and (**bottom** row) visual stimuli for one exemplar animal from the sRNR group across days studied (preinjury, day 1, 4, and 7).

3.3.3. rRNR–Repeated

In rRNR animals, right-temporal N1 and P2 amplitudes were found to be significantly decreased on day 1 for visual (pre: -2.121 ± 0.350, day 1: -1.240 ± 0.350 μV) and day 4 for auditory standard tone (pre: 3.593 ± 0.515 μV, day 4: 1.626 ± 0.431 μV) stimuli, respectively. In addition, frontal P2 latencies were faster at day 4 (111.4 ± 5.497 ms), in comparison to pre-injury (144.0 ± 5.747 ms) from visual stimuli. Source localization for auditory stimuli results yielded significant findings at 85 ms, where standard tones had decreased activations from pre-injury (0.072 ± 0.006 μV/mm^2) compared to day 7 (0.048 ± 0.004 μV/mm^2) in the left temporal region (Figure 3). However, no differences between pre- and post-injury were found at 50 or 110 ms for any stimuli on any day.

Different trajectories of auditory and visual processing changes were noted for sham, sRNR and rRNR groups. In anesthesia-only sham, auditory processing was affected, but not visual processing. In the sRNR group, there were mixed findings, where visual and auditory stimuli resulted in increased and decreased responses, respectively. In the rRNR group, the magnitude of the response (amplitude or current density) had a tendency to decrease. Across all groups, the frontal region emerges as the most vulnerable to auditory deficits, as evidenced by decreased cortical activity; however, visual deficits were observed in the left temporal regions. Interestingly, we found increased activity in the right temporal and right occipital regions that were not specific to any stimulus modality.

4. Discussion

This study examined acute changes (within 7 days) in auditory and visually evoked potentials in a large 4-week-old swine model of TBI under single or repeated head-loads using an RNR device. This current study is an extension of previous work that established methods for measuring and modelling auditory and visually evoked potentials in healthy swine and applied these methods to an experimental cohort subject to RNR. A table summarizing significant EP changes in auditory and visual processing for sham (anesthesia-only) and RNR is displayed in Table 2. In the acute post-injury phase, anaesthesia and RNR were found to have an effect on 4-week-old swines' EP amplitudes and latencies and cortical activations in comparison to pre-injury. In sham, only auditory stimuli were found to be significantly decreased post-anaesthesia on days 1 and 7 in comparison to pre-injury values (Figure 4, top panel). This finding is consistent with the human literature, where auditory processing was suppressed after anesthesia [64]. Visually evoked potential remained unaffected after anaesthesia in piglets, as no significant differences were found on any post-anaesthesia days (1, 4 or 7) following pre-injury.

Interesting findings were observed for the sRNR group, where increased activity (N1 and current density) was observed for visual stimuli in the right temporal and right occipital regions, however auditory-target processing yielded the opposite results, with an increase in P2 amplitudes in the right temporal region but decreased current densities in the left and right occipital regions (Figure 4, middle panel). In healthy piglets, auditory processing localized to the temporal region and visual processing to the occipital regions [47]. In this study, these regions show alternative activity patterns after experimental RNR. The rRNR group yielded different patterns to the sRNR group, where visual stimuli produced decreased N1 amplitudes but faster visual P2 latencies (Figure 4, bottom panel). The different patterns of findings between RNR groups suggest that rRNR is not necessarily a more severe form of sRNR, as this group did not simply reflect greater magnitude deficits within the same parameters or within the same region. sRNR was the only experimental group that showed increased EP responses, while rRNR only decreased responses, in addition to there being fewer significant comparisons than were found in the sRNR group (Table 2). Differences in activation patterns observed between RNR groups further highlight the importance of distinguishing between single and repeated head rotations, as the human literature demonstrates a trend toward the worst outcomes for those suffering from a history of concussion versus no injury or a single concussion [19]. The 'within group' findings for each RNR combination were unique to the mode of stimuli and suggest that the visual and auditory pathways are unequivocally affected by each head-loading paradigm. Taken together, these findings demonstrate the importance of injury biomechanics, particularly head-loading patterns, that create differing trajectories in neurofunctional deficits. A mixture of head injury mechanisms is one factor that likely contributes to the incongruent findings in the human literature, and should be considered when interpreting trends in neurocognitive deficits and symptoms.

Comparisons between groups revealed that both sRNR (current density) and rRNR (P2 amplitude) were decreased in comparison to sham; a finding consistent with the human literature, where auditory information processing is decreased after mild TBI in clinical and athletic populations [26,65,66]. Furthermore, the sRNR group was found to have slowed visual processing (left temporal) in comparison to sham (N1 latency), and rRNR had lower visual activity than sRNR in this region. In previous studies, visual processing in response to a working memory task requiring an active button-press had smaller N350 and P300 amplitudes for patients with mTBI in comparison to healthy controls [67], as well as decreased activity in the dorsolateral prefrontal cortex from functional magnetic resonance imaging (fMRI) [68]. Furthermore, patients with a history of concussion reported decreased P3 amplitudes, as it has been hypothesized that the brain has an increased vulnerability to subsequent concussions resulting in worse outcomes [19,22].

Contrary trends were observed in the right temporal and left occipital regions in our animals, as there were many instances in which increased activity (P2 amplitudes and current densities) were observed for either the sRNR or rRNR groups in comparison to sham. These findings were not specific to a type of stimulus. Increased activity in these regions after RNR suggests that the brain may involve more areas (and at a greater magnitude) to process the same amount of information. Enhanced auditory processing (increased amplitude and decreased latency) was observed in an evaluation of N100, P300, and N400 from a two tone-auditory oddball task in junior ice hockey players taken at pre-season, after a concussion, and multiple timepoints within- and post-season [24]. In a separate study examining varsity athletes, enhanced auditory N2 and P3 and increased electrical activity (from source localization modelling) were reported in athletes with a history of concussion in comparison with a non-concussed group [23]. Similarly, in a visual Go/No-Go task employing an emotional cue (a spider for a threat-related condition or a flower for a neutral condition), enhanced N2-P3 was observed for individuals with a concussion in comparison to control group. The authors postulated that increased attention and cortical activity in response to threat conditions may be related to inefficient control of the emotional response in visual processing in the concussed group [69]. It has been theorized that increased cortical activity after TBI (in contrasted to suppressed activity) can resilt from a breakdown of neural networks, resulting in the inefficient processing of information, requiring more areas to become active to compensate [23,25].

This study examined the effects of single and repeated mild head rotations on auditory and visually evoked potentials in the 4-week-old swine model permitting systematic control of the loading conditions causing transient neurofunctional deficits common to concussion. A limitation of human studies is that the biomechanics causing head injury are often diverse, while factors such as direction, velocity, surface stiffness, are not controlled for in the analysis [27]. These biomechanical factors govern the conditions of energy transfer from an event to cause injury to the head and brain tissues. Despite careful control of the loading conditions in this study, we observed trends where auditory and visual processing not only decreased, but also increased after head rotations. While head rotations were completed in the sagittal plane by applying loads evenly across both hemispheres, it is possible that the dysfunctions were unequal for the auditory and visual pathways because each involves different brain structures. We acknowledge that the piglet brain is much smaller in size than the pediatric brain and the neural axis of the pig is parallel to the ground whereas humans' is perpendicular [40], further limiting the direct translation of findings this study to humans. Furthermore, we noted that patterns of increases or decreases in auditory and visual processing were specific to the examined brain region, thus supporting the notion of whole-brain analysis to capture the full picture of deficits or over-compensation mechanisms after TBI. We did observe an effect of anaesthesia on decreased auditory processing in sham animals; however, we conducted within-group comparisons in our attempt to delineate the effects of anaesthesia versus RNR on outcomes to better isolate and understand the effects of RNR. Further limitations in this study include employing passive auditory and visual tasks that were suitable for piglets, while the majority of tasks in the human literature employ active tasks requiring a response or more complex and cognitively demanding tasks. We studied RNR in a single direction and only in female pigs. We hypothesize that the magnitude and direction (increase or decrease) of auditory and visual processing would change if a different direction, i.e., coronal loading, was employed, as the biomechanical loads would disproportionally affect these pathways as the head would moved from medial to lateral instead of from anterior to posterior directions. Male pigs may have different auditory and visual processing trends after RNR as it has been demonstrated in the human literature that males with concussion have a greater N1 suppression after the visual presentation of human faces of different emotions in comparison to females with concussions [70]. Lastly, our findings were limited to the 35 animal subjects studied, and it is possible that, if more animals were included, comparisons that were found to be not significant may become significant.

5. Conclusions

In summary, the trajectory of alterations in auditory and visually evoked potentials were increased after single RNR but decreased after repeated RNR. This suggests that the injury processes affecting cortical activity for the rRNR group may be different than those for the sRNR group since rRNR did not simply reflect greater changes in the same direction (increased activity). The frontal region seems to be most vulnerable to auditory deficits, as reflected by decreased cortical activity, and visual deficits were found in the left temporal regions. Interestingly, we found increased activity in the right temporal and right occipital regions that were not specific to any stimulus modality. Auditory and visual EPs have different change trajectories after sRNR and rRNR, suggesting that injury biomechanics are important factors when delineating the patterns of neurofunctional deficits after concussion. Auditory and visual stimuli evaluate separate and specific neural pathways of the brain; however, future work should include other stimulus modalities, such as motor or sensory pathways, to examine the integrative brain functionality after TBI and provide a complete understanding of the functional changes after concussion.

Author Contributions: Conceptualization, A.O. and S.S.M.; Methodology, A.O. and S.S.M.; Software, A.O. and W.H.T.; Validation, A.O. and W.H.T.; Formal Analysis, A.O. and W.H.T.; Investigation, A.O., W.H.T. and S.S.M.; Resources, S.S.M.; Data Curation, A.O. and W.H.T.; Writing—Original Draft Preparation, A.O., W.H.T., K.B.A., C.L.M. and S.S.M.; Writing—Review and Editing, A.O., W.H.T., K.B.A., C.L.M. and S.S.M.; Visualization, A.O. and W.H.T.; Supervision, A.O. and S.S.M.; Project Administration, A.O. and S.S.M.; Funding Acquisition, K.B.A., C.L.M. and S.S.M. All authors have read and agreed to the published version of the manuscript.

Funding: Supported by NIH R01NS097549 and the Georgia Research Alliance.

Institutional Review Board Statement: The Institutional Animal Care and Use Committee at Emory University approved all procedures in this study under PROTO201800149 (approved 16 July 2020) and PROTO201800163 (approved 20 May 2019).

Informed Consent Statement: Not applicable.

Data Availability Statement: Not applicable.

Acknowledgments: We acknowledge the assistance of Mackenzie Mull, Akshara Thakore, Gbemi Aderibigbe, Melissa Crowe and Ethan Karstedt during animal studies and we acknowledge Mariano Fernandez-Corrazza for his expertise in piglet model analysis.

Conflicts of Interest: The authors declare no conflict of interest. The sponsors had no role in the design, execution, interpretation, or writing of the study.

References

1. Coronado, V.G.; Haileyesus, T.; Cheng, T.A.; Bell, J.M.; Haarbauer-Krupa, J.; Lionbarger, M.R.; Flores-Herrera, J.; McGuire, L.C.; Gilchrist, J. Trends in Sports- and Recreation-Related Traumatic Brain Injuries Treated in US Emergency Departments: The National Electronic Injury Surveillance System-All Injury Program (NEISS-AIP) 2001–2012. *J. Head Trauma Rehabil.* **2015**, *30*, 185–197. [CrossRef] [PubMed]
2. Langlois, J.A.; Rutland-Brown, W.; Wald, M.M. The epidemiology and impact of traumatic brain injury: A brief overview. *J. Head Trauma Rehabil.* **2006**, *21*, 375–378. [CrossRef]
3. McCrory, P.; Meeuwisse, W.; Dvorak, J.; Aubry, M.; Bailes, J.; Broglio, S.; Cantu, R.C.; Cassidy, D.; Echemendia, R.J.; Castellani, R.J.; et al. Consensus statement on concussion in sport—The 5th international conference on concussion in sport held in Berlin, October 2016. *Br. J. Sport. Med.* **2017**, *51*, 838–847. [CrossRef]
4. Giza, C.C.; Prins, M.L.; Hovda, D.A. It's Not All Fun and Games: Sports, Concussions, and Neuroscience. *Neuron* **2017**, *94*, 1051–1055. [CrossRef]
5. Giza, C.C.; Hovda, D.A. The new neurometabolic cascade of concussion. *Neurosurgery* **2014**, *75* (Suppl. S4), S24–S33. [CrossRef] [PubMed]
6. Lumba-Brown, A.; Yeates, K.O.; Sarmiento, K.; Breiding, M.J.; Haegerich, T.M.; Gioia, G.A.; Turner, M.; Benzel, E.C.; Suskauer, S.J.; Giza, C.C.; et al. Diagnosis and Management of Mild Traumatic Brain Injury in Children: A Systematic Review. *JAMA Pediatr.* **2018**, *172*, e182847. [CrossRef]
7. Munjal, S.K.; Panda, N.K.; Pathak, A. Audiological Deficits After Closed Head Injury. *J. Trauma* **2010**, *68*, 13–18. [CrossRef]

8. Vander Werff, K.R.; Rieger, B. Impaired auditory processing and neural representation of speech in noise among symptomatic post-concussion adults. *Brain Inj.* **2019**, *33*, 1320–1331. [CrossRef]
9. Nandrajog, P.; Idris, Z.; Azlen, W.N.; Liyana, A.; Abdullah, J.M. The use of event-related potential (P300) and neuropsychological testing to evaluate cognitive impairment in mild traumatic brain injury patients. *Asian J. Neurosurg.* **2017**, *12*, 447–453. [CrossRef]
10. Master, C.L.; Bacal, D.; Grady, M.F.; Hertle, R.; Shah, A.S.; Strominger, M.; Whitecross, S.; Bradford, G.E.; Lum, F.; Donahue, S.P. Vision and Concussion: Symptoms, Signs, Evaluation, and Treatment. *Pediatrics* **2022**, *150*, e2021056047. [CrossRef]
11. Rose, S.C.; Weber, K.D.; Collen, J.B.; Heyer, G.L. The Diagnosis and Management of Concussion in Children and Adolescents. *Pediatr. Neurol.* **2015**, *53*, 108–118. [CrossRef] [PubMed]
12. Kraus, N.; Thompson, E.C.; Krizman, J.; Cook, K.; White-Schwoch, T.; Labella, C.R. Auditory biological marker of concussion in children. *Sci. Rep.* **2016**, *6*, 39009. [CrossRef] [PubMed]
13. Master, C.L.; Podolak, O.E.; Ciuffreda, K.J.; Metzger, K.B.; Joshi, N.R.; McDonald, C.C.; Margulies, S.S.; Grady, M.F.; Arbogast, K.B. Utility of Pupillary Light Reflex Metrics as a Physiologic Biomarker for Adolescent Sport-Related Concussion. *JAMA Ophthalmol.* **2020**, *138*, 1135–1141. [CrossRef]
14. Luck, S.J.; Mathalon, D.H.; O'Donnell, B.F.; Hmlinen, M.S.; Spencer, K.M.; Javitt, D.C.; Uhlhaas, P.J. A roadmap for the development and validation of event-related potential biomarkers in schizophrenia research. *Biol. Psychiatry* **2011**, *70*, 28–34. [CrossRef] [PubMed]
15. Rapp, P.E.; Keyser, D.O.; Albano, A.; Hernandez, R.; Gibson, D.B.; Zambon, R.A.; David Hairston, W.; Hughes, J.D.; Krystal, A.; Nichols, A.S. Traumatic brain injury detection using electrophysiological methods. *Front. Hum. Neurosci.* **2015**, *9*, 11. [CrossRef] [PubMed]
16. Washnik, N.J.; Anjum, J.; Lundgren, K.; Phillips, S. A review of the role of auditory evoked potentials in mild traumatic brain injury assessment. *Trends Hear.* **2019**, *23*, 2331216519840094. [CrossRef]
17. Näätänen, R.; Picton, T. The N1 wave of the human electric and magnetic response to sound: A review and an analysis of the component structure. *Psychophysiology* **1987**, *24*, 375–425. [CrossRef]
18. Lascano, A.M.; Lalive, P.H.; Hardmeier, M.; Fuhr, P.; Seeck, M. Clinical evoked potentials in neurology: A review of techniques and indications. *J. Neurol. Neurosurg. Psychiatry* **2017**, *88*, 688–696. [CrossRef]
19. Broglio, S.P.; Moore, R.D.; Hillman, C.H. A history of sport-related concussion on event-related brain potential correlates of cognition. *Int. J. Psychophysiol.* **2011**, *82*, 16–23. [CrossRef]
20. Gosselin, N.; Thériault, M.; Leclerc, S.; Montplaisir, J.; Lassonde, M. Neurophysiological anomalies in symptomatic and asymptomatic concussed athletes. *Neurosurgery* **2006**, *58*, 1151–1160. [CrossRef]
21. Gaetz, M.; Goodman, D.; Weinberg, H. Electrophysiological evidence for the cumulative effects of concussion. *Brain Inj.* **2000**, *14*, 1077–1088. [CrossRef]
22. Bennys, K.; Busto, G.U.; Touchon, J. Cumulative effects of subsequent concussions on the neural patterns of young rugby athletes: Data from event-related potentials. *Res. Sport. Med.* **2023**, 1–12. [CrossRef] [PubMed]
23. Ledwidge, P.S.; Molfese, D.L. Long-Term Effects of Concussion on Electrophysiological Indices of Attention in Varsity College Athletes: An Event-Related Potential and Standardized Low-Resolution Brain Electromagnetic Tomography Approach. *J. Neurotrauma* **2016**, *33*, 2081–2090. [CrossRef]
24. Fickling, S.D.; Smith, A.M.; Pawlowski, G.; Ghosh Hajra, S.; Liu, C.C.; Farrell, K.; Jorgensen, J.; Song, X.; Stuart, M.J.; D'Arcy, R.C.N. Brain vital signs detect concussion-related neurophysiological impairments in ice hockey. *Brain J. Neurol.* **2019**, *142*, 255–262. [CrossRef]
25. Molfese, D.L. The Need for Theory to Guide Concussion Research. *Dev. Neuropsychol.* **2015**, *40*, 1–6. [CrossRef] [PubMed]
26. Gomes, J.; Damborská, A. Event-Related Potentials as Biomarkers of Mild Traumatic Brain Injury. *Act. Nerv. Super.* **2017**, *59*, 87–90. [CrossRef]
27. Ommaya, A.K.; Goldsmith, W.; Thibault, L. Biomechanics and neuropathology of adult and paediatric head injury. *Br. J. Neurosurg.* **2002**, *16*, 220–242. [CrossRef] [PubMed]
28. Silva, L.A.F.; Magliaro, F.C.L.; Carvalho, A.C.M.; Matas, C.G. Cortical maturation of long latency auditory evoked potentials in hearing children: The complex P1-N1-P2-N2. *CoDAS* **2017**, *29*, e20160216. [CrossRef] [PubMed]
29. Fitzroy, A.B.; Krizman, J.; Tierney, A.; Agouridou, M.; Kraus, N. Longitudinal maturation of auditory cortical function during adolescence. *Front. Hum. Neurosci.* **2015**, *9*, 530. [CrossRef]
30. Lightfoot, G. Summary of the N1-P2 Cortical Auditory Evoked Potential to Estimate the Auditory Threshold in Adults. *Semin. Hear.* **2016**, *37*, 1–8. [CrossRef] [PubMed]
31. Baillargeon, A.; Lassonde, M.; Leclerc, S.; Ellemberg, D. Neuropsychological and neurophysiological assessment of sport concussion in children, adolescents and adults. *Brain Inj.* **2012**, *26*, 211–220. [CrossRef] [PubMed]
32. Margulies, S.S.; Kilbaugh, T.; Sullivan, S.; Smith, C.; Propert, K.; Byro, M.; Saliga, K.; Costine, B.A.; Duhaime, A.C. Establishing a clinically relevant large animal model platform for TBI therapy development: Using cyclosporin a as a case study. *Brain Pathol.* **2015**, *25*, 289–303. [CrossRef] [PubMed]
33. Sullivan, S.; Friess, S.H.; Ralston, J.; Smith, C.; Propert, K.J.; Rapp, P.E.; Margulies, S.S. Behavioral Deficits and Axonal Injury Persistence after Rotational Head Injury Are Direction Dependent. *J. Neurotrauma* **2013**, *30*, 538–545. [CrossRef]
34. Atlan, L.S.; Lan, I.S.; Smith, C.; Margulies, S.S. Changes in event-related potential functional networks predict traumatic brain injury in piglets. *Clin. Biomech.* **2018**, *64*, 14–21. [CrossRef]

35. Fang, M.; Li, J.; Rudd, J.A.; Wai, S.M.; Yew, J.C.C.; Yew, D.T. fMRI Mapping of cortical centers following visual stimulation in postnatal pigs of different ages. *Life Sci.* **2006**, *78*, 1197–1201. [CrossRef]
36. Barrow, P. Use of the swine pediatric model. In *Pediatric Non-Clinical Drug Testing: Principles, Requirements, and Practice*; Hoberman, A.M., Lewis, E.M., Eds.; John Wiley & Sons, Inc.: Hoboken, NJ, USA, 2012; pp. 213–229.
37. Barrow, P. Toxicology testing for products intended for pediatric populations. In *Nonclinical Drug Safety Assessment: Practical Considerations for Successful Registration*; Sietsema, W.K., Schwen, R., Eds.; FDA News: Washington, DC, USA, 2007; pp. 411–440.
38. Weeks, D.; Sullivan, S.; Kilbaugh, T.; Smith, C.; Margulies, S.S. Influences of developmental age on the resolution of diffuse traumatic intracranial hemorrhage and axonal injury. *J. Neurotrauma* **2014**, *31*, 206–214. [CrossRef] [PubMed]
39. Ryan, M.C.; Kochunov, P.; Sherman, P.M.; Rowland, L.M.; Wijtenburg, S.A.; Acheson, A.; Hong, L.E.; Sladky, J.; McGuire, S. Miniature pig magnetic resonance spectroscopy model of normal adolescent brain development. *J. Neurosci. Methods* **2018**, *308*, 173–182. [CrossRef] [PubMed]
40. Finnie, J. Comparative approach to understanding traumatic injury in the immature, postnatal brain of domestic animals. *Aust. Vet. J.* **2012**, *90*, 301–307. [CrossRef] [PubMed]
41. Cullen, D.K.; Harris, J.P.; Browne, K.D.; Wolf, J.A.; Duda, J.E.; Meaney, D.F.; Margulies, S.S.; Smith, D.H. A porcine model of traumatic brain injury via head rotational acceleration. *Methods Mol. Biol.* **2016**, *1462*, 289–324. [CrossRef]
42. Raghupathi, R.; Margulies, S.S. Traumatic axonal injury after closed head injury in the neonatal pig. *J. Neurotrauma* **2002**, *19*, 843–853. [CrossRef]
43. Friess, S.H.; Ichord, R.N.; Owens, K.; Ralston, J.; Rizol, R.; Overall, K.L.; Smith, C.; Helfaer, M.A.; Margulies, S.S. Neurobehavioral functional deficits following closed head injury in the neonatal pig. *Exp. Neurol.* **2007**, *204*, 234–243. [CrossRef] [PubMed]
44. Friess, S.H.; Ichord, R.N.; Ralston, J.; Ryall, K.; Helfaer, M.A.; Smith, C.; Margulies, S.S. Repeated traumatic brain injury affects composite cognitive function in piglets. *J. Neurotrauma* **2009**, *26*, 1111–1121. [CrossRef]
45. Coats, B.; Binenbaum, G.; Smith, C.; Peiffer, R.L.; Christian, C.W.; Duhaime, A.C.; Margulies, S.S. Cyclic head rotations produce modest brain injury in infant piglets. *J. Neurotrauma* **2017**, *34*, 235–247. [CrossRef] [PubMed]
46. Oeur, R.A.; Margulies, S.S. Target detection in healthy 4-week old piglets from a passive two-tone auditory oddball paradigm. *BMC Neurosci.* **2020**, *21*, 52. [CrossRef]
47. Oeur, R.A.; Palaniswamy, M.; Ha, M.; Fernandez-Corazza, M.; Margulies, S.S. Regional variations distinguish auditory from visual evoked potentials in healthy 4 week old piglets. *Physiol. Meas.* **2023**, *44*, 025006. [CrossRef]
48. Mull, M.; Aderibigbe, O.; Hajiaghamemar, M.; Oeur, R.A.; Margulies, S.S. Multiple Head Rotations Result in Persistent Gait Alterations in Piglets. *Biomedicines* **2022**, *10*, 2976. [CrossRef]
49. Oeur, A.; Mull, M.; Riccobono, G.; Arbogast, K.B.; Ciuffreda, K.J.; Joshi, N.; Fedonni, D.; Master, C.L.; Margulies, S.S. Pupillary Light Response Deficits in 4-Week-Old Piglets and Adolescent Children after Low-Velocity Head Rotations and Sports-Related Concussions. *Biomedicines* **2023**, *11*, 587. [CrossRef] [PubMed]
50. Hajiaghamemar, M.; Seidi, M.; Oeur, R.A.; Margulies, S.S. Toward development of clinically translatable diagnostic and prognostic metrics of traumatic brain injury using animal models: A review and a look forward. *Exp. Neurol.* **2019**, *318*, 101–123. [CrossRef]
51. Eucker, S.A.; Smith, C.; Ralston, J.; Friess, S.H.; Margulies, S.S. Physiological and histopathological responses following closed rotational head injury depend on direction of head motion. *Exp. Neurol.* **2011**, *227*, 79–88. [CrossRef]
52. Raghupathi, R.; Mehr, M.F.; Helfaer, M.A.; Margulies, S.S. Traumatic Axonal Injury is Exacerbated following Repetitive Closed Head Injury in the Neonatal Pig. *J. Neurotrauma* **2004**, *21*, 307–316. [CrossRef]
53. Hajiaghamemar, M.; Seidi, M.; Patton, D.; Huber, C.; Arbogast, K.B.; Master, C.L.; Margulies, S.S. Using On-Field Human Head Kinematics to Guide Study Design for Animal-Model Based Traumatic Brain Injury. In Proceedings of the National Biomedical Engineering Society (BMES) Annual Meeting, Virtual, 14–17 October 2020.
54. Wu, T.; Hajiaghamemar, M.; Giudice, J.S.; Alshareef, A.; Margulies, S.S.; Panzer, M.B. Evaluation of Tissue-Level Brain Injury Metrics Using Species-Specific Simulations. *J. Neurotrauma* **2021**, *38*, 1879–1888. [CrossRef]
55. Atlan, L.S.; Margulies, S. Frequency-Dependent Changes in Resting State EEG Functional Networks After Traumatic Brain Injury in Piglets. *J. Neurotrauma* **2019**, *36*, 2558–2578. [CrossRef]
56. Delorme, A.; Makeig, S. EEGLAB: An open source toolbox for analysis of single-trial EEG dynamics including independent component analysis. *J. Neurosci. Methods* **2004**, *134*, 9–21. [CrossRef] [PubMed]
57. Onton, J.; Westerfield, M.; Townsend, J.; Makeig, S. Imaging human EEG dynamics using independent component analysis. *Neurosci. Biobehav. Rev.* **2006**, *30*, 808–822. [CrossRef] [PubMed]
58. Makeig, S.; Onton, J. ERP features and EEG dynamics: An ICA perspective. In *The Oxford Handbook of ERP Components*; Luck, S.J., Kappenman, E.S., Eds.; Oxford University Press: New York, NY, USA, 2012; pp. 51–86.
59. Fernandez-Corazza, M.; Sadleir, R.; Turovets, S.; Tucker, D. MRI Piglet Head Model for EIT and IVH Simulation. In Proceedings of the 17th Conference on Electrical Impedance Tomography, Stockholm, Sweden, 19–23 June 2016.
60. Fang, Q.; Boas, D.A. Tetrahedral mesh generation from volumetric binary and grayscale images. In Proceedings of the 2009 IEEE International Symposium on Biomedical Imaging: From Nano to Macro, Boston, MA, USA, 28 June–1 July 2009.
61. Von Ellenrieder, N.; Beltrachini, L.; Blenkman, A.; Fernandez-Corrazza, M.; Kochen, S.; Muravchik, C.H. A Robust Photogrammetry Method to Measure Electrode Positions. 2013; *Unpublished work*.
62. Pascual-Marqui, R.D. Standardized low-resolution brain electromagnetic tomography (sLORETA): Technical details. *Methods Find. Exp. Clin. Pharmacol.* **2002**, *24* (Suppl. D), 5–12.

63. Sekihara, K.; Nagarajan, S.S. Adaptive spatial filters. In *Adaptive Spatial Filters for Electromagnetic Brain Imaging*; Sekihara, K., Nagarajan, S.S., Eds.; Springer: Berlin/Heidelberg, Germany, 2008; pp. 37–63.
64. Van Hooff, J.C.; de Beer, N.A.M.; Brunia, C.H.M.; Cluitmans, P.J.M.; Korsten, H.H.M. Event-related potential measures of information processing during general anesthesia. *Electroencephalogr. Clin. Neurophysiol.* **1997**, *103*, 268–281. [CrossRef] [PubMed]
65. Mortazavi, M.; Lucini, F.A.; Joffe, D.; Oakley, D.S. Electrophysiological trajectories of concussion recovery: From acute to prolonged stages in late teenagers. *J. Pediatr. Rehabil. Med.* **2023**, *16*, 287–299. [CrossRef]
66. Clayton, G.; Davis, N.; Holliday, A.; Joffe, D.; Oakley, D.S.; Palermo, F.X.; Poddar, S.; Rueda, M. In-clinic event related potentials after sports concussion: A 4-year study. *J. Pediatr. Rehabil. Med.* **2020**, *13*, 81–92. [CrossRef]
67. Gosselin, N.; Erg, C.B.; Chen, J.K.; Huntgeburth, S.C.; Beaumont, L.D.; Petrides, M.; Cheung, B.; Ptito, A. Evaluating the cognitive consequences of mild traumatic brain injury and concussion by using electrophysiology. *Neurosurg. Focus* **2012**, *33*, E7. [CrossRef]
68. Gosselin, N.; Bottari, C.; Chen, J.-K.; Petrides, M.; Tinawi, S.; De Guise, É.; Ptito, A. Electrophysiology and Functional MRI in Post-Acute Mild Traumatic Brain Injury. *J. Neurotrauma* **2011**, *28*, 329–341. [CrossRef]
69. Mäki-Marttunen, V.; Kuusinen, V.; Brause, M.; Peräkylä, J.; Polvivaara, M.; dos Santos Ribeiro, R.; Öhman, J.; Hartikainen, K.M. Enhanced attention capture by emotional stimuli in mild traumatic brain injury. *J. Neurotrauma* **2015**, *32*, 272–279. [CrossRef] [PubMed]
70. Carrier-Toutant, F.; Guay, S.; Beaulieu, C.; Léveillé, E.; Turcotte-Giroux, A.; Papineau, S.D.; Brisson, B.; D'Hondt, F.; De Beaumont, L. Effects of Repeated Concussions and Sex on Early Processing of Emotional Facial Expressions as Revealed by Electrophysiology. *J. Int. Neuropsychol. Soc.* **2018**, *24*, 673–683. [CrossRef] [PubMed]

Disclaimer/Publisher's Note: The statements, opinions and data contained in all publications are solely those of the individual author(s) and contributor(s) and not of MDPI and/or the editor(s). MDPI and/or the editor(s) disclaim responsibility for any injury to people or property resulting from any ideas, methods, instructions or products referred to in the content.

Article

Multimodal Neuromonitoring and Neurocritical Care in Swine to Enhance Translational Relevance in Brain Trauma Research

John C. O'Donnell [1,2,*], Kevin D. Browne [1,2], Svetlana Kvint [2], Leah Makaron [3], Michael R. Grovola [1,2], Saarang Karandikar [1,4], Todd J. Kilbaugh [2,5], D. Kacy Cullen [1,2,4] and Dmitriy Petrov [2,*]

[1] Center for Neurotrauma, Neurodegeneration & Restoration, Corporal Michael J. Crescenz Veterans Affairs Medical Center, Philadelphia, PA 19104, USA
[2] Center for Brain Injury & Repair, Department of Neurosurgery, Perelman School of Medicine, University of Pennsylvania, Philadelphia, PA 19104, USA
[3] University Laboratory Animal Resources, Department of Pathobiology, School of Veterinary Medicine, University of Pennsylvania, Philadelphia, PA 19104, USA
[4] Department of Bioengineering, School of Engineering and Applied Science, University of Pennsylvania, Philadelphia, PA 19104, USA
[5] Department of Anesthesiology and Critical Care Medicine, Perelman School of Medicine, University of Pennsylvania, The Children's Hospital of Philadelphia, Philadelphia, PA 19104, USA
* Correspondence: odj@pennmedicine.upenn.edu (J.C.O.); dmitriy.petrov@pennmedicine.upenn.edu (D.P.); Tel.: +1-215-294-9494 (D.P.)

Abstract: Neurocritical care significantly impacts outcomes after moderate-to-severe acquired brain injury, but it is rarely applied in preclinical studies. We created a comprehensive neurointensive care unit (neuroICU) for use in swine to account for the influence of neurocritical care, collect clinically relevant monitoring data, and create a paradigm that is capable of validating therapeutics/diagnostics in the unique neurocritical care space. Our multidisciplinary team of neuroscientists, neurointensivists, and veterinarians adapted/optimized the clinical neuroICU (e.g., multimodal neuromonitoring) and critical care pathways (e.g., managing cerebral perfusion pressure with sedation, ventilation, and hypertonic saline) for use in swine. Moreover, this neurocritical care paradigm enabled the first demonstration of an extended preclinical study period for moderate-to-severe traumatic brain injury with coma beyond 8 h. There are many similarities with humans that make swine an ideal model species for brain injury studies, including a large brain mass, gyrencephalic cortex, high white matter volume, and topography of basal cisterns, amongst other critical factors. Here we describe the neurocritical care techniques we developed and the medical management of swine following subarachnoid hemorrhage and traumatic brain injury with coma. Incorporating neurocritical care in swine studies will reduce the translational gap for therapeutics and diagnostics specifically tailored for moderate-to-severe acquired brain injury.

Keywords: swine; acquired brain injury; traumatic brain injury; coma; disorders of consciousness; subarachnoid hemorrhage; neurocritical care; neurointensive care unit; multimodal neuromonitoring; translational neurotrauma

1. Introduction

Acquired brain injury—event-related brain damage, such as traumatic brain injury (TBI) or stroke—is frequently debilitating when not outright fatal, and outcomes are often dependent on neurocritical care. TBI is a leading cause of death and disability, with global incidence in approximately 69 million people per year [1]. The Institute for Health Metrics and Evaluation's Global Burden of Diseases, Injuries, and Risk Factors study found that, in 2016, there were approximately 27 million new TBIs that required hospital care, which likely skews toward the "moderate-to-severe" due to the oversimplified nature of the injury severity spectrum [2]. That study also found that there were approximately 12 million new stroke cases in 2019, half of which were fatal, making it the second leading cause of death worldwide [3]. Ischemic stroke accounts for the majority of cases, while subarachnoid hemorrhage (SAH) accounted for approximately 10% [3]. Acute brain injury is also associated with increased risk for developing dementia and neurodegenerative diseases such as Alzheimer's and chronic traumatic encephalopathy [4–8].

Due to low prognostic accuracy and a paucity of treatment options for acquired brain injury, there can be a tendency to give in to nihilism when making care decisions in the neurointensive care unit (neuroICU). Indeed, the leading cause of death in the neuroICU is withdrawal of care [9–11]. However, recent clinical studies have shown that the potential for recovery is greater than expected, revealing that the prevalent negative prognostic bias is unwarranted [11–15]. In addition, an exhaustive meta-analysis recently found that neurocritical care significantly improves outcome for adults following brain injury [16]. As the neurotrauma field dismisses nihilism and moves forward with renewed determination, the preclinical study of neurocritical care would be an invaluable paradigm to improve clinical prognostic accuracy and offer a viable path for discovery and translation of effective treatments. Indeed, the Neurocritical Care Society's Curing Coma Campaign has repeatedly called for bidirectional translation in preclinical modeling [10,17,18]. Furthermore, the mounting evidence of the essential role neurocritical care plays in improving neurological outcomes after moderate-to-severe TBI dictates that any preclinical translational work should strive to incorporate neurocritical care techniques and paradigms.

Swine are ideal subjects for the preclinical study of acquired brain injury and neurocritical care. Since swine are large mammals, human neuromonitoring equipment is directly compatible with them, greatly increasing clinical relevance. Compared to ubiquitous small animal models, swine have large gyrencephalic brains with high white:gray matter ratios, similar to what is found in humans (60:40 in swine and humans versus 14:86 in rats and 10:90 in mice), and these physical properties have major implications for injury mechanisms and pathological manifestations [19–22]. Indeed, swine enable the study of white matter damage and effects on connectivity due to stroke and TBI. Furthermore, the meningeal subarachnoid space around the swine brain is similar to that of humans and, as such, allows for blood and clot accumulation similar to what is observed clinically with SAH [23–25]. These and other advantages of the use of swine in stroke research have been explored extensively in a comprehensive review from Melià-Sorolla and colleagues [26].

In addition to the factors that make swine an ideal translational model for stroke, modeling human closed-head TBI presents challenges that are uniquely addressed by swine. In humans, TBI begins with an intense instant in which mechanical forces are exerted on the brain, followed by days, weeks, or even years of secondary injury mechanisms. Different mechanical forces can result in dramatically different injuries and injury manifestations. The loading mechanisms that generate the mechanical forces of human TBI include impact-loading, which can result in focal lesion (usually cortical) with a gradient of pathology emanating from it; and rotational acceleration/deceleration-induced inertial loading, which results in diffuse injury to neurons, glia, and vasculature throughout the brain [22,27,28]. These loading mechanisms can occur in combination (e.g., impact causing acceleration or vice versa), and inertial loading due to acceleration often occurs without any significant impact loading, but it is exceedingly rare for impact loading to occur in humans without any resultant head acceleration and inertial loading. Inertial loading is unique to humans and other large animals because the injurious forces caused by rotational acceleration are dependent on the mass of the brain, and therefore even extremely high accelerations generate very little force within small brains [28–30]. Given their relatively large brain mass, we can scale up acceleration in pigs to achieve the same forces experienced by humans and even tease apart kinematic elements of the injury (e.g., max acceleration, max deceleration, and jerk) to test their influence on recovery and pathology [31]. The aforementioned white:gray matter ratio and gyrencephalic cortex also influence the distribution and effects of injurious forces generated by inertial loading [22,27].

Among the manifestations of human TBI that are specifically due to rotational acceleration, the most obvious is traumatic loss of consciousness (coma) [32–34], which is the primary diagnostic for guiding neurocritical triage following TBI in humans. Indeed, in the absence of other major polytrauma, coma duration and severity are the primary criteria for determining whether a TBI patient will enter intensive care, as a score of ≤ 8 on the Glasgow Coma Scale (GCS) typically mandates intubation. Following the brief mechanical injury, secondary injury mechanisms include ischemia due to increased intracranial pressure (ICP) impairing brain perfusion, as well as a variety of interwoven cell/molecular injury cascades, such as inflammation, excitotoxicity, oxidative stress, and others. Because the ubiquitous impact-loaded rodent models share similar cell/molecular mechanisms of secondary injury with human TBI, these cell/molecular mechanisms have historically been the primary focus of preclinical TBI research. Unfortunately, rodent models cannot recreate the mechanisms (inertial loading) or manifestations of human TBI that are due to rotational loading (e.g., loss of consciousness) [28–30]. In addition to an inability to produce traumatic loss of consciousness, without mass lesion or hypoxia, rodent TBI models also do not reproduce the secondary increase in ICP that guides most treatment decisions in the neuroICU [35]. Thus, due to a variety of reasons, in addition to an overreliance on impact-only small animal models, our field has yet to translate any of the therapeutics found to be effective for treating impact-only brain injuries in rodents. Rodent models will always be the foundation of preclinical brain injury research, but for reliable, fail-early, pre-IND/IDE (investigational new drug/investigational device exemption) studies, we must employ large animal models that better replicate the mechanisms and manifestations of the human injury.

The swine model of rotational-acceleration-induced TBI is currently the only preclinical model that scales rotational-acceleration-induced inertial loading to recreate the forces of mild-to-severe human TBI [22,27]. Importantly for our purposes, this model can reliably produce a prolonged coma that would lead to admission to the neuroICU if presented in a human patient [36,37]. Swine models of SAH also provide high-fidelity modeling of the human injury due to the similarities of the subarachnoid space and their high white matter content [23–26]. However, if we are to recreate the mechanisms and manifestations of the human injury, we must go beyond the inciting incident. The course of both injuries typically takes patients through a neuroICU, where they receive neurocritical care and monitoring, introducing influential variables that can affect the injury course and provide extensive neuromonitoring data for which we are striving to improve prognostic value.

Generally, the only existing treatments for acquired brain injury in the acute/subacute phase involve managing endophenotypes in the neuroICU guided by multimodal neuromonitoring (MMNM). There are medical options for clot clearance within a tight window following ischemic stroke (tissue plasminogen activator or mechanical thrombectomy), as well as limited surgical options for clot clearance following hemorrhagic stroke. Decompressive craniectomy offers a surgical approach to reduce damage from intracranial pressure (ICP) after brain injury, though results of clinical trials that defy straightforward explanation leave some questions surrounding its efficacy unresolved [38–41]. Beyond that, monitoring and responding to secondary injury processes—including altered ICP and partial brain tissue oxygen ($PbtO_2$)—following moderate-to-severe acquired brain injury are central to achieving a positive outcome [16]. It is therefore important to recreate the intensive care environment in a preclinical model to (a) account for and study the primary variables experienced by humans after injury (including the interventions encountered in the ICU); (b) provide clinically relevant neuromonitoring data in preclinical studies; (c) enable translational development of new treatments specific to the neurocritical care space; and (d) provide the human-level care necessary to extend the study period for preclinical moderate-to-severe TBI with coma, which has historically been limited to 8 h due to the need for sophisticated neurocritical care to survive animals beyond this point [36,37].

Others have made valuable progress recreating elements of the clinical neurocritical care environment in swine models of brain injury, as displayed in Table 1. These researchers have employed a variety of injury mechanisms, study durations, monitoring modalities, and other data collection techniques [42–49]. Notably, Friess and colleagues utilized a pediatric swine model of rotational acceleration TBI and measured ICP, PbtO$_2$, cerebral blood flow (CBF), and microdialysis for lactate:pyruvate ratio (LPR) over a period of 6 h [42–44]. Those studies established a correlation between neurocritical care monitoring and pathology and investigated the interplay of vasopressors with CBF and cerebral perfusion pressure (CPP). The pediatric pig model of rotational acceleration TBI is essential to understanding the unique pathophysiology of pediatric TBI and developing treatments that can prevent long-term consequences [50]. In a swine model of SAH, Nyberg and colleagues collected ICP and microdialysis for glucose and LPR. They validated their model by confirming that cerebral ischemia and metabolic changes after SAH were consistent with the clinical condition [45]. To our knowledge, the longest study duration to date was in a swine model of acute subdural hematoma, in which Datzman and colleagues collected ICP, PbtO$_2$, and microdialysis for LPR over a period of 52 h [46]. More recent studies include a model of blast TBI in swine that demonstrated increased coagulopathy and ICP after injury [51], as well as a fluid percussion injury model in swine that monitored ICP, along with transesophageal echo, to study left ventricular function [52], and a swine model directly controlling increases in ICP to test the effects on CBF [53]. We sought to integrate the strengths of these innovative studies while drawing on state-of-the-art clinical practice to maximize the translational relevance of this platform.

Here, we present the development of a comprehensive multimodal neuromonitoring and neurocritical care suite for use with swine—an accepted large animal species commonly used for IND/IDE-enabling studies. We describe in detail the protocols we developed and present individual examples of monitoring and medical management in subjects administered a sham injury, subarachnoid hemorrhage, or moderate-to-severe TBI with coma under these protocols. Through an iterative development approach to systematically optimize medical management and improve survival, developing this paradigm allowed for data collection from swine experiencing TBI with coma that exceeded 8 h for the first time, yielding notable technical advancements and observations.

Table 1. Examples of neuromonitoring in swine neurotrauma research.

Authors, Year	Title	Injury	Time	Anesthesia	Neuromonitoring Modalities	EEG	Other	A Line	Central Line	Lumbar Drain
Friess et al., 2011 [42]	"Neurocritical Care Monitoring Correlates with Neuropathology in a Swine Model of Pediatric Traumatic Brain Injury"	rotational TBI	6 h	isoflurane and CRI fentanyl	ICP, PbtO$_2$, microdialysis for LPR	N	IHC	Y	Y	N
Friess et al., 2012 [43]	"Early cerebral perfusion pressure augmentation with phenylephrine after traumatic brain injury may be neuroprotective in a pediatric swine model"	rotational TBI	6 h	isoflurane and CRI fentanyl	ICP, PbtO$_2$, CBF, microdialysis for LPR	N	IHC	Y	Y	N
Weenink et al., 2012 [47]	"Quantitative electroencephalography in a swine model of cerebral arterial gas embolism"	arterial gas embolism	4 h	IV ketamine, sufentanil, midazolam, and pancuronium bromide	ICP, PbtO$_2$, microdialysis for lactate and glucose	Y (surface)	n/a	Y	Y	N
Nyberg et al., 2014 [45]	"Metabolic Pattern of the Acute Phase of Subarachnoid Hemorrhage in a Novel Porcine Model: Studies with Cerebral Microdialysis with High Temporal Resolution"	SAH	135 min	CRI ketamine, morphine, and rocuronium bromide	ICP, microdialysis for glucose and LPR	N	CT scan after experiment	Y	Y	N
Friess et al., 2015 [44]	"Differing effects when using phenylephrine and norepinephrine to augment cerebral blood flow after traumatic brain injury in the immature brain"	rotational TBI	6 h	CRI midazolam and fentanyl	ICP, PbtO$_2$, CBF, microdialysis for LPR	N	IHC	Y	Y	N
Chen et al., 2017 [48]	"Quantitative electroencephalography in a swine model of blast induced brain injury"	blast TBI	2 h	IV propofol	none	Y (surface)	n/a	N	N	N
Mader et al., 2018 [49]	"Evaluation of a New Multiparameter Brain Probe for Simultaneous Measurement of Brain Tissue Oxygenation, Cerebral Blood Flow, Intracranial Pressure, and Brain Temperature in a Porcine Model"	CCI, physiological challenges	~5 h	CRI thiopental and piritramide	testing single probe for ICP, PbtO$_2$, CBF	N	n/a	Y	Y	N
Datzman et al., 2019 [46]	"In-depth characterization of a long-term, resuscitated model of acute subdural hematoma–induced brain injury"	ASDH	54 h	CRI propofol and fentanyl	ICP, PbtO$_2$, microdialysis for lactate and glucose	N	mGCS; IHC; brain tissue mitochondrial respiration (Oroboros); plasma GFAP and NSE	Y	Y	N
Cralley et al., 2022 [51]	"Zone 1 REBOA in a combat DCBI swine model does not worsen brain injury"	dismounted complex blast injury (DCBI)	6 h	CRI propofol and fentanyl	ICP	N	brain water content, MAP	Y	Y	N

Table 1. Cont.

Authors, Year	Title	Injury	Time	Anesthesia	Neuromonitoring Modalities	EEG	Other	A Line	Central Line	Lumbar Drain
Adedipe et al., 2022 [52]	"Left Ventricular Function in the Initial Period After Severe Traumatic Brain Injury in Swine"	fluid percussion injury	8 h	isoflurane	ICP	N	transesophageal echocardiography, coagulation, blood flow	Y	Y	N
Abdou et al., 2022 [53]	"Characterizing Brain Perfusion in a Swine Model of Raised Intracranial Pressure"	raised ICP via intracranial Fogarty balloon	2 h	isoflurane	ICP	N	computed tomography perfusion for CBF	Y	Y	N

2. Materials and Methods

All procedures were approved by the University of Pennsylvania's Institutional Animal Care and Use Committee (IACUC) and the Corporal Michael J. Crescenz VA Medical Center in Philadelphia (both AAALAC accredited) and were completed in accordance with the Guide for the Care and Use of Laboratory Animals [54]. All studies were performed in the Porcine Neurointensive Care and Assessment Facility (NCAF) at the University of Pennsylvania, a state-of-the-art swine injury and behavioral assessment facility created in January 2015 to study acute and long-term responses in porcine models of brain injury.

An alphabetical list of abbreviations can be found in the Abbreviations section after the Discussion.

2.1. Iterative Technique Development

This study was designed to adapt and optimize clinical multimodal neuromonitoring and neurocritical care techniques for use in swine. As such, iterative changes were made between each subject as part of the development process. Examples of these changes include sampling frequency, processing procedure, and types of biological samples collected; adjustments to sedation and/or ventilation strategies; streamlining surgical procedures to optimize timing; and altering the animal protocol to add drugs for medical management. Therefore, although our monitoring and care was optimized with 3 sham injury animals, 3 experimental SAH animals, and 2 severe TBI animals with coma administered via rapid angular rotational acceleration of the head, the iterative nature of this study makes it difficult to group subjects together in a way that allows for clear summarizations/comparisons or valid statistical testing beyond the intended purposes of creating a comprehensive neurointensive care unit for studying severe brain injury in swine. Instead, we describe in this Methods section the techniques that we developed, and in the Results section, we present general observations for each condition and medical management in individual subjects.

2.2. Workflow Summary

Our workflow included pre-injury surgeries and monitoring, TBI or SAH injuries, and post-injury monitoring and neurocritical care (Figure 1). Female Yorkshire swine (25–30 kg) were induced via ketamine/midazolam (Hospira, Lake Forest, IL, USA), intubated, and maintained under isoflurane anesthesia. Femoral artery and internal jugular vein catheterizations were performed to allow for continuous blood pressure (BP) monitoring and serial blood draws from an arterial line (A-line), as well as drug administration via triple lumen central line. In some cases, a lumbar drain was placed to facilitate cerebrospinal fluid (CSF) sampling.

For all subjects in the swine neuroICU, a quad-lumen bolt (Hemedex, MA, USA) was secured 1 cm rostral to bregma for placement of a parenchymal ICP probe (Natus, WI, USA), $PbtO_2$/temperature sensors (Integra Licox, Princeton, NJ), Spencer depth electroencephalography (EEG) electrode (SD08R-AP58X-000; Ad-Tech, WI, USA), and microdialysis catheter (mDialysis, MA, USA) per institutional clinical paradigm at the University of Pennsylvania. In some cases, a Bowman CBF monitor (Hemedex, MA, USA) was placed in the quad bolt, and a separate burr hole was used for the depth EEG electrode. Likewise, a surface EEG array was placed on the scalp. In the TBI group, the invasive cranial monitors were placed after injury induction to prevent shearing during rapid acceleration/deceleration of the head. In the SAH animals, cranial monitors were placed prior to injury. Animals in the neuroICU were switched from isoflurane to total intravenous anesthesia, utilizing titrations of propofol and fentanyl, delivered via the central line.

TBI was administered in two subjects via the HYGE pneumatic actuator to achieve inertial loading via rapid rotational acceleration of the head in the sagittal plane (113–114 rad/s). Following head rotational TBI, these subjects were moved to the swine neuroICU. SAH was administered in three subjects via injection of autologous blood into the subarachnoid space at the skull base, using a contralateral external ventricular drain catheter placed down to the skull base. Three sham subjects underwent all ICU-related procedures but received neither TBI nor SAH.

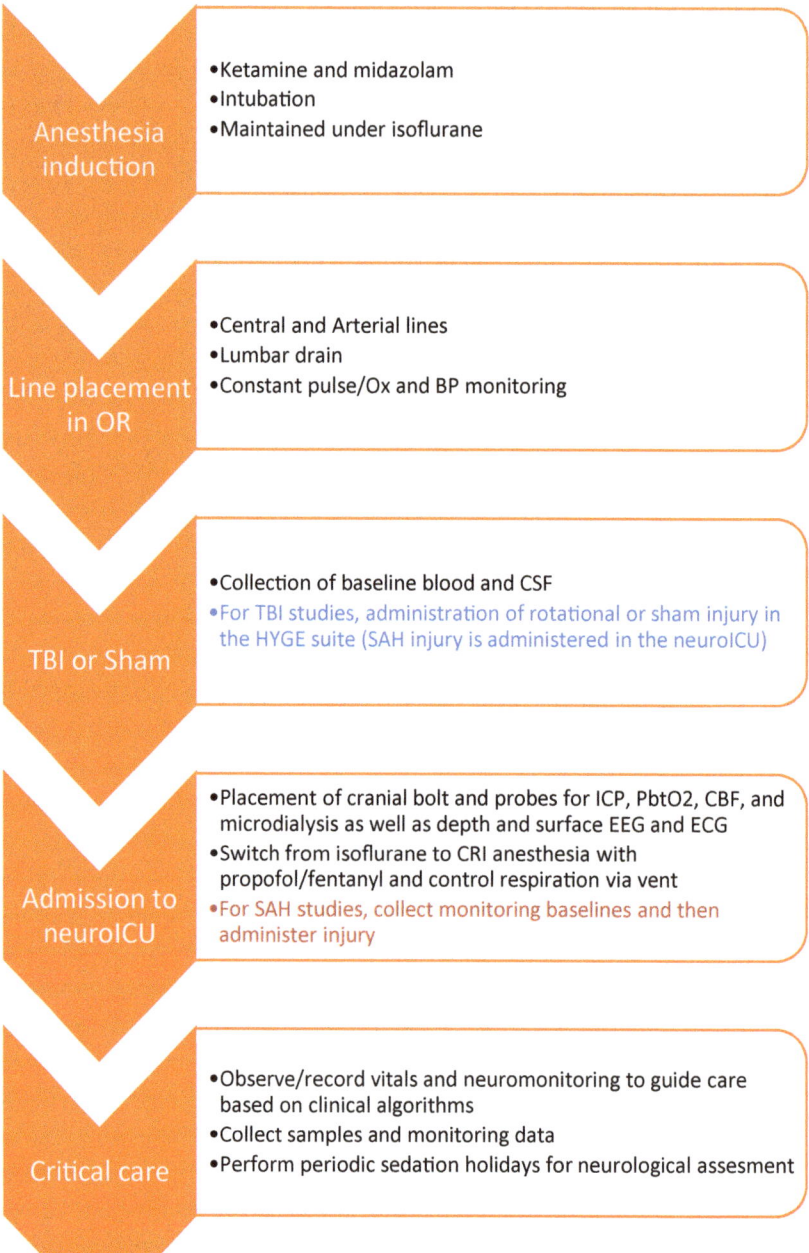

Figure 1. Workflow for SAH and TBI in the swine neuroICU. This flowchart details the order of events for SAH and TBI experiments utilizing neurocritical care and monitoring in the swine neuroICU.

Animals were monitored continuously up to 36 h. Neurological assessments were performed during sedation holidays (propofol off, fentanyl low). Electrocardiogram (EKG), blood oxygen saturation (SpO$_2$), capnography, BP, ICP, PbtO$_2$/temperature, and EEG (depth + scalp) were time-synchronized and continuously recorded with waveform resolution on a Moberg CNS-200 (Moberg Research Inc., Ambler, PA, USA). Arterial blood, CSF, microdialysate, and urine were collected, processed, and frozen for analysis and biobanking. Plasma biomarker assays were run using the Neurology 4-plex B assay on the ultrasensitive Simoa-HDX bead-based immunoassay platform (Quanterix, Billerca, MA, USA). At the end of the experiment, subjects were deeply anesthetized and exsanguinated via cardiac perfusion with normal saline, followed by 10% neutral buffered formalin for tissue fixation. Formalin-fixed, paraffin-embedded brain sections were stained with hematoxylin (Modified Mayer's Hematoxylin; 22-110-639; Fisher Scientific, Hampton, NH, USA) and eosin (Eosin Y—aqueous; 6766009; Fisher Scientific, Hampton, NH, USA) (H&E) and immunostained for amyloid precursor protein (APP; mouse; MAB348; 1:80,000; Millipore Sigma, St. Louis, MO, USA) and ionized calcium binding adaptor molecule 1 (IBA1; rabbit; 019-19741, 1:4000; Wako, Richmond, VA, USA) with DAB (3,3'-Diaminobenzidine; SK-4100; Vector Labs, Newark, CA, USA) secondary staining for colorimetric microscopy. Staining for APP reveals white matter pathology, as APP accumulates in damaged or degenerating axons. Staining for IBA1 reveals the morphology, location, and number of microglia in a given brain area, which provides information on the nature and severity of the neuroinflammatory response to injury. Entire sections were scanned using an Aperio CS2 digital slide scanner. Detailed staining and pathological scoring methods can be found in Grovola et al. (2021) [55].

2.3. Induction and Line Placement

Prior to injury, animals were anesthetized, and indwelling catheters were placed for repeated blood draws and mean arterial pressure monitoring. Pigs were induced with an intramuscular injection of midazolam (0.4–0.6 mg/kg) and ketamine (20–30 mg/kg), intubated, and maintained on isoflurane (1–5%). Glycopyrrolate (0.01–0.02 mg/kg) or atropine (0.02–0.05 mg/kg) was used to mitigate excessive secretions during intubation. Animals were continuously monitored, and anesthetic levels were adjusted as needed throughout the procedure to maintain a plane of anesthesia, ensuring that the SpO$_2$, heart rate, and respiration rate were within acceptable ranges. Thermal support was provided using blankets, a BairHugger warmer, or a HotDogger unit. Temperature was monitored continuously throughout the procedure. For all surgeries, hair at the site was clipped prior to aseptic skin prep. The surgeon wore a cap, booties, a mask, sterile gloves, and a sterile gown. The surgery was performed aseptically, following the IACUC Guidelines for USDA species' survival surgery.

For the jugular/cephalic vein catheterization (central line), the pig was placed in dorsal recumbency, and Bupivacaine (1–2 mg/kg) was injected subcutaneously over the identified incision site prior to incision. An incision was made, and then the subcutaneous tissue and cutaneous colli muscle were dissected, identifying the vein. The vein was retracted and then catheterized. The catheter was secured in the vessel with silk ties. The catheter was then passed via a sterilized trocar tunneled through the subcutaneous tissues exiting the skin on the dorsum of the neck; alternatively, the exteriorized catheter could be exited through the incision and taped down. The original incision was closed in multiple layers (muscle, subcutaneous, and buried skin), using absorbable or nonabsorbable sutures. If unsuccessful at achieving access, the procedure was performed on the contralateral side. This access was used for drug delivery, including continuous rate infusion (CRI) anesthesia in the neuroICU.

For the femoral artery catheterization (A-line), the pig was placed in dorsal recumbency, and the rear leg was retracted caudally. The medial aspect of the leg (starting at the stifle and extending inguinally) was draped and prepped using chlorhexidine scrub. Under aseptic conditions, the femoral artery was catheterized with an arterial catheterization kit, using the Seldinger technique. This access was used to record the mean arterial pressure and draw samples. The catheter was flushed with saline at the time of each blood draw.

For the placement of the lumbar drain, the pig was placed in a lateral recumbent position. The hair over the lower lumbar spinous processes was clipped, and the skin was prepped aseptically with repeated surgical scrubs of chlorhexidine and betadine. Under aseptic conditions, a Touhy needle was inserted in the midline and through the interspinous space at the L5–L6 region. Once the thecal sac was entered—confirmed by CSF flow—the Tuohy needle stylet was removed, and the lumbar drain catheter was threaded through the Tuohy needle. The drain was then secured in place and attached to an external collection system.

2.4. Rotational Traumatic Brain Injury (TBI)

Following induction, intubation, and line placement, the subjects—still maintained under isoflurane via a portable anesthesia cart—were covered with a blanket and/or Bair-Hugger unit for thermal control and moved to an adjacent procedure room in the NCAF, where brain trauma could be induced via head rotational acceleration. The anesthesia level, heart rate, respiration rate, blood oxygen saturation, and temperature were continuously monitored. The level of anesthesia was gauged by assessing the animals' jaw tone, as well as a strong pinch of the ear or the webbing between the toes.

This model produces pure inertial non-impact head rotation in different planes at controlled rotational acceleration levels (thus controlling severity). The powerful HYGE pneumatic actuator device can generate 20,000 kg of thrust in less than 6 ms, resulting in peak angular accelerations of over 300,000 rad/s^2. Of note, the loading conditions generated by this device closely approximate the conditions of inertial brain injury in humans based on brain mass scaling. While this model is labor-, resource-, and skill-intensive, it is currently the only model that can recreate the inertial loading and head rotational acceleration central to human TBI and is therefore the most clinically relevant animal model of closed-head TBI. The subject's head was secured to a padded snout clamp, which, in turn, was mounted to the linkage assembly of the HYGE pneumatic actuator device. The padded snout clamp is specially designed to convert the linear motion of the HYGE to an angular (rotational) motion in the sagittal plane. To produce closed-head, diffuse, moderate-to-severe TBI, pure head rotation (approximately 55 degrees in <12 ms) was induced. For each experiment, head angular velocity/acceleration traces were recorded. The peak rotational velocities of the two subjects included in this study were 113 and 114 rad/s. Immediately following rotation, animals were detached from the device, assessed for apnea, examined for any injury to the mouth/snout (none was found), and transferred to the swine neuroICU in an adjacent procedure room in the NCAF. Buprenorphine SR was provided when recovery was sufficient to graduate from the neuroICU.

2.5. Multimodal Neuromonitoring (MMNM)

After induction of anesthesia and placement of lines—and in the case of the TBI group, following induction of injury—animals were transferred to the swine neuroICU for placement of cranial probes, monitoring, and critical care. A summary illustration of the lines and probes we utilize in the swine neuroICU can be found in Figure 2.

Subjects were connected to a clinical ICU vital sign monitor (Phillips, Netherlands) for continuous monitoring of the heart rate, arterial and non-invasive BP, oxygen saturation, and EKG tracing. The temperature was continuously monitored via a rectal thermometer probe. End-tidal CO_2 (EtCO_2) was monitored at the junction of the breathing tube and endotracheal tube. Cranial monitor placement was then performed aseptically, following the IACUC Guidelines for USDA species' survival surgery. After a left paramedian incision, a small burr hole was made above the frontal cortex, 5–15 mm paramedian and 5–10 mm cranial of the coronal suture. A quad lumen bolt to house several neuromonitoring modalities was placed into the burr hole. Specifically, an ICP probe (Natus, WI, USA) was placed in the brain parenchyma approximately 1–1.5 cm deep to reach the junction of cortex and subcortical white matter. A PbO_2 and temperature probe (Integra Licox, Princeton, NJ, USA) was placed in the subcortical white matter, in combination with a Bowman CBF monitor (Hemedex, MA, USA). A microdialysis bolt catheter with either a standard 20 kD pore size (8002823E; mDialysis, MA, USA) or a high-pass 100 kD pore size (8050194A; mDialysis, MA, USA) was placed 10 mm deep to obtain extracellular fluid solutes for metabolic analysis. The microdialysis catheter was perfused with lactated ringers at low flow (1 µL/min), with dextran added in the case of the high pass membrane. A 1 mm burr hole was made posterior to the bolt for implantation of an 8-lead Spencer depth electrode for continuous EEG monitoring across 17 mm of cortical layers (SD08R-AP58X-000; Ad-Tech, WI, USA). The incision was sutured closed around the bolt housing. Once the bolt and sensors were placed and the animals were stable, they were transitioned to CRI intravenous propofol (5–20 mg/kg/h) anesthesia and fentanyl (0.1–100 µg/kg/h) analgesia for the duration of their time in the neuroICU. A basic EEG surface array was placed over the frontal lobes comprising two electrodes over each hemisphere with grounding electrode. EEG electrodes were routed through a standard Moberg headbox EEG array. All monitors were connected to a Moberg ICU monitor for time-synced recording of monitoring data, vital sign data (Phillips ICU monitor), and EEG (Figure 3).

2.6. Administration of Subarachnoid Hemorrhage (SAH)

SAH injury was administered following admission to the neuroICU and placement of lines and cranial monitors. An incision was made over the left frontal lobe, mirroring the placement of the quad lumen bolt. A contralateral burr hole was made at the same landmarks. An external ventricular drain (EVD) catheter was augmented to have a single opening at the distal end. The drain was inserted perpendicular to the outer table of the skull until it reached the skull base. The inner stylet was withdrawn. CSF was carefully withdrawn, confirming placement in a basal cistern. Arterial blood was collected from the arterial line and injected under pressure through the external drain into the basal cisterns. The injection was stopped when cerebral perfusion pressure, as measured by the difference between mean arterial pressure (MAP) and ICP, was 0 mmHg (requiring injection of approximately 10 mL of autologous blood). After the injection was halted, the catheter was withdrawn, the wound was irrigated, and the skin was closed with running nylon suture.

2.7. Neurocritical Care

Neurocritical care in the pigs was modeled after clinical practice, and therefore it was primarily focused on maintaining brain perfusion. This is achieved by monitoring cerebral perfusion pressure (CPP), which is the difference between MAP and ICP. MAP must be high enough to overcome ICP in order to provide sufficient brain perfusion. For adult swine and humans, we seek to maintain CPP above 60 mmHg. The focus on CPP is primarily due to the prevalence of increased ICP observed after brain injury. For that reason, treatment in humans and in our pig model usually involves maintaining ICP below 20 mmHg through a combination of ventilation management, sedation, and hyperosmotic therapy. While increased ICP after brain injury is common, it is also possible to encounter MAP below the optimal range of 80–100 mmHg (also consistent with target values for humans), in which case application of a vasopressor such as norepinephrine may be needed. The focus on maintaining brain perfusion is primarily to prevent hypoxia/ischemia, but hypoxia may sometimes be present even when ICP and CPP are normal [56,57]. Observational clinical studies have reported that low PbO_2 is common after severe TBI and associated with poor outcome [58–62], while PbO_2-directed care is associated with improved outcome [63–66]. The BOOST-II clinical trial was a randomized, controlled study conducted in 119 severe TBI patients from 10 ICUs designed to test the efficacy of including PbO_2, along with ICP, in guiding treatment [67]. They found that monitoring and treating PbO_2 significantly reduced total hypoxia burden, and a phase III trial is currently underway to assess the impact on neurological outcome [68]. Therefore, we also sought to maintain PbO_2 above 20 mmHg in the swine neuroICU.

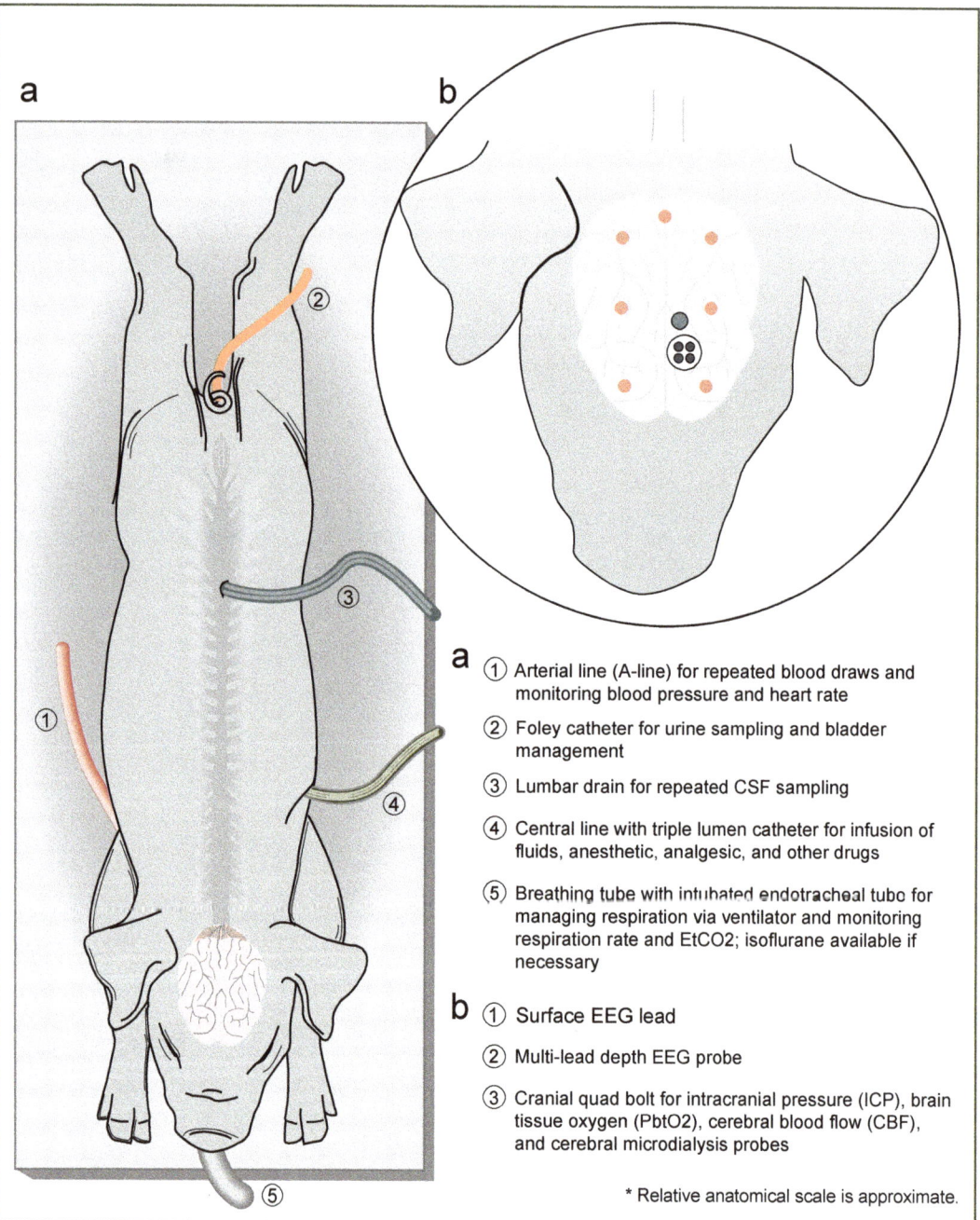

Figure 2. **Instrumentation and line placement in the swine neuroICU.** (**a**) Illustration of a pig in the neuroICU with lines labeled; (**b**) a close-up drawing of cranial monitoring in the swine neuroICU, with probes labeled. Artwork by Paul Schiffmacher.

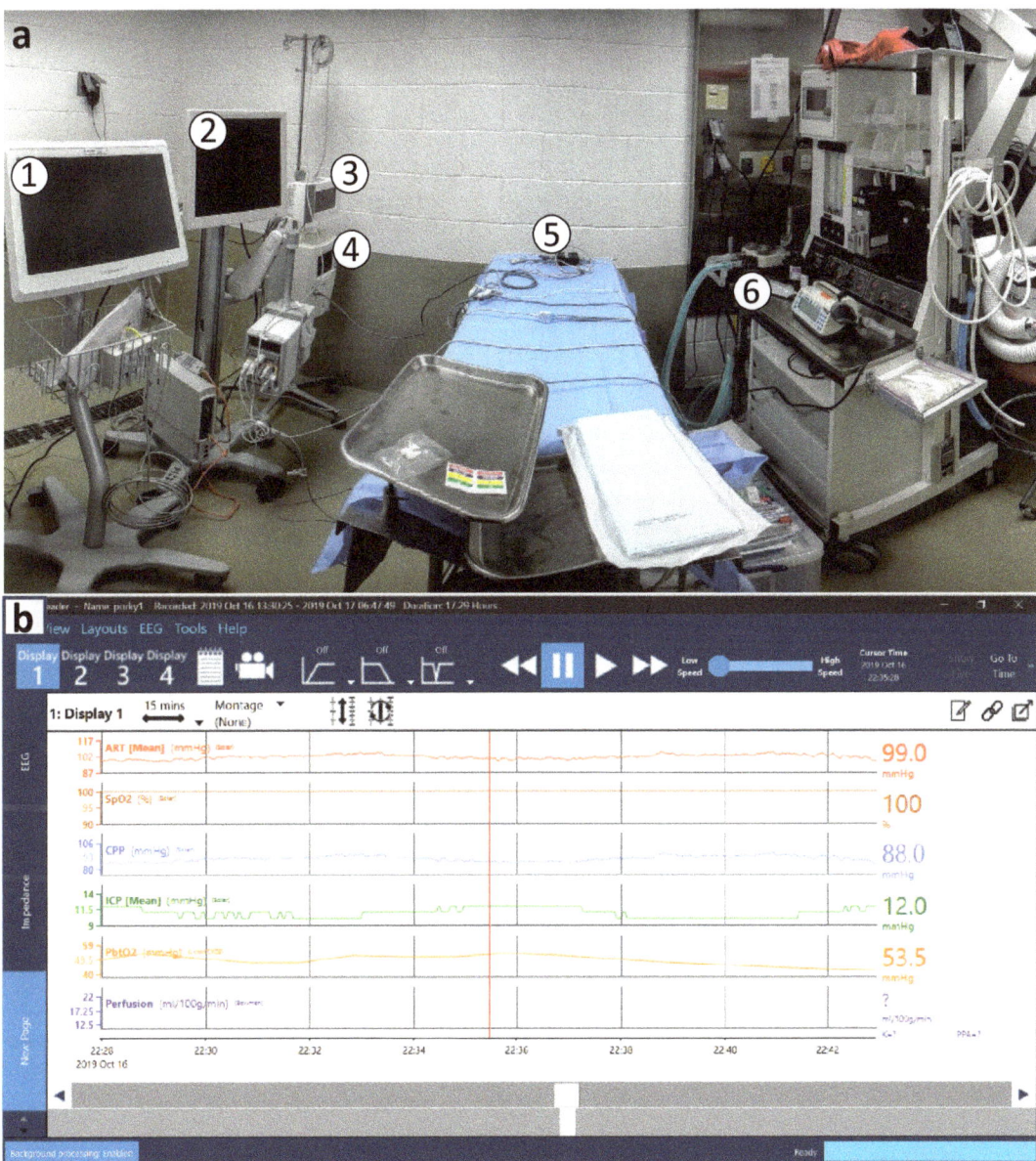

Figure 3. Panoramic view of the swine neuroICU. (**a**) Numbers indicate a few key pieces of equipment pictured in this panoramic view of the swine neuroICU. (1) Moberg CNS Monitor for time-synchronizing continuous EEG with systemic physiology from over 30 ICU monitoring devices. (2) GE Solar monitor for core vital signs. (3) Licox $PbtO_2$ and temperature sensor. (4) Bowman Hemedex thermal diffusion cerebral perfusion monitor for CBF. (5) Various probes atop a variable pressure pad on the patient table. (6) Anesthesia and respiratory support. (**b**) A screen-capture image of example traces from the Moberg CNS Monitor during hour 22 of a swine neuroICU experiment (cerebral perfusion monitor cycles on and off by design).

Surface and depth EEGs provide extensive data for research applications, and they also allow us to detect seizures that develop due to post-traumatic epilepsy (PTE). We monitored for seizures and managed them with midazolam and propofol. Acquired brain injury also frequently results in medical events extending beyond the central nervous system. Apnea is common and must be detected immediately in order to administer manual resuscitation via bag valve mask until the animal can be placed on the ventilator. Cardiac events (e.g., stress myocardia) may also occur after injury. Monitoring vital signs such as heart rate (HR), SpO_2, and respiratory rate provides information for maintaining anesthetic depth, along with providing early—or immediate—signs that intervention is necessary. We also measure $EtCO_2$, which informs our ventilator settings and can indicate potentially dangerous changes in pH. Those pH changes, as well as metabolic and other perturbations, are also detected via blood gas analysis, which we performed via an i-STAT handheld blood analyzer when we drew blood samples from the A-line. Neurological assessments were performed during regular sedation holidays (propofol off, fentanyl low), using parameters from the human GCS [69,70] and the canine-modified GCS [71–73] that we adapted for use in swine in a sternal position, with limited head movement, due to the presence of cranial probes. Our prototype of the swine GCS can be found in the Results section in Table 2, with scoring for a TBI subject. While the animals were fully sedated, we also monitored them for signs of responsiveness, which may include palpebral reflex, anal or jaw tone, EEG activity, or breathing against the ventilator detectable on the $EtCO_2$ trace.

Table 2. Swine Coma Scale scores during TBI-induced coma and prolonged emergence.

To maximize arousal (5 min after stopping propofol): grip and roll between fingers; first cheek, then neck (sternocleidomastoid), then shoulder/back (trapezious)						
Swine Coma Scale - MAXIMIZE AROUSAL PRIOR TO EXAM -		At each timepoint, indicate the score for each category.				
Score	EYE BLINK	Baseline	2 h	4 h	8 h	12 h
3	Blinks spontaneously (wait 3 min)	3			3	3
2	Blinks upon stimulation (e.g. pinch or ear tickle; not near eye)		2	2		
1	No blinking (without directly touching the eye)					
	MOTOR ACTIVITY					
6	Voluntary walking	6				
5	Sitting					
4	Isolated spontaneous movements (e.g. limbs or head)		4	4	4	4
3	Withdraws forepaw and/or hindpaw in response to noxious stimulation					
2	Muscle contractions in response to noxious stimulation of the limbs					
1	Absence of motor response to noxious stimulation					
	AUDITORY RESPONSE					
2	Auditory startle	2			2	2
1	None		1	1		
	BRAIN STEM REFLEXES (score 3 and skip if motor score is 6)					
3	Both palpebral AND pinna reflexes present	3	3	3	3	3
2	Palpebral OR pinna reflex present					
1	Absence of palpebral and pinna reflexes					

Table 2. Cont.

	RESPIRATION					
4	Not on ventilator, breathes with a regular pattern	4			4	4
3	Not on ventilator, breathes with an irregular pattern			3		
2	Breathes above ventilator rate (initiates spontaneous breaths)		2			
1	Breathes at ventilator rate or apnea (no spontaneous breaths)					
	TOTAL	18	12	13	16	16

The tests and measurements detailed here are key indicators of potentially life-threatening events that occur after brain injury, but simply maintaining CPP > 60 mmHg, ICP < 20 mmHg, and $PbtO_2$ > 20 mmHg is not sufficient to provide the best chance of recovery. Unexpected or subtle medical events often emerge. Individuals must be trained with clinical neurointensivists and veterinarians to provide care in the swine neuroICU, and a neurointensivist and veterinarian should always be available in person or via phone or video chat for consultation.

2.8. Sample Collection and Analyses

Animals were monitored continuously up to 36 h, and all monitoring waveforms that informed clinical care decisions (e.g., ICP, MAP, HR, SpO_2, capnography, BP, $PbtO_2$/temperature, EEG, etc.) were also time synched and recorded via the Moberg ICU monitor for subsequent analyses. We continuously collected cerebral microdialysate while sampling arterial blood (via A-line), CSF (via lumbar drain), and urine (via an aseptically placed urinary catheter) every 2–4 h, and these samples were all processed and frozen for analysis and biobanking. At the end of the study period, subjects were perfusion fixed with 10% formalin, after which brains were extracted and submerged in 10% formalin overnight for complete fixation. Fixed brains were then sectioned coronally into blocks for gross pathology (5 mm blocks with an initial cut immediately rostral to the optic chiasm), after which the blocks were run through an automated tissue processor and finally embedded in paraffin wax for long-term archival preservation and ease of sectioning. A microtome was used to cut 8 μm thick slices from the formalin-fixed, paraffin-embedded (FFPE) brain blocks, and then the slices were mounted on slides for histological analyses. We stained with H&E and immunostained for either amyloid precursor protein (APP; mouse; 1:80,000) or ionized calcium binding adaptor molecule 1 (IBA1; rabbit; 1:4000) with DAB secondary staining for colorimetric microscopy. Entire sections were scanned using an automated Aperio CS2 digital slide scanner (Leica Biosystems, Germany) at multiple magnifications to allow for cloud-based examination of multiple different brain regions across the breadth of the large gyrencephalic brain sections. Detailed staining methods and pathology scoring techniques can be found in Grovola et al. (2021) [55]. Cranial dialysate was assayed for lactate, pyruvate, glycerol, and glucose via an Iscus clinical microdialysis analyzer. Plasma samples were analyzed via the Neurology 4-plex B assay on the ultrasensitive Quanterix Simoa-HDX bead-based immunoassay platform, a kit designed for detecting brain injury biomarkers, such as glial fibrillary acidic protein (GFAP) and ubiquitin carboxyl-terminal esterase L1 (UCHL-1), in human plasma.

3. Results

This study was designed for the iterative development of comprehensive neurocritical care and monitoring in pigs with severe brain injury, and therefore the primary research products are the capabilities that were successfully developed, as reported in the Methods section. While animals received similar injuries, data were not pooled across animals because critical care variables were changed between each animal to facilitate model development. In addition to the techniques that were developed in this study, several useful observations were made, and notable medical events were encountered that were successfully mitigated. In this Results section, we present these observations and medical management reports.

Medical management of sham subjects

Maintaining an uninjured control subject under CRI anesthesia in a neuroICU for extended periods of time represents one of the unique advantages of preclinical research in this area, but it also presents unique challenges. We cannot overstate the importance of gaining experimental insights and precision by gathering invasive multimodal neuromonitoring, biological fluid samples, neurological assessments, and brain tissue from control subjects that experience all experimental variables, except for the injury under study. Monitoring modalities in our three sham animals did not typically deviate beyond acceptable parameters, and blinded histopathology analyses found signs of pathology and inflammation only near sites of cranial probe implantation. However, special care must be taken to maintain sedation in animals that have not sustained any awareness-altering injury. While remaining within the same dose ranges for anesthetic agents as injured subjects on the same protocol, administration rates will need to be increased over time to maintain sedation during wake cycles. Furthermore, it is important to maintain only the doses required for sedation, not only for experimental reasons but also because tolerance can build to anesthetic agents even within a 24 h period. Therefore, we found that it is optimal to employ a gradual increase in administration rates as needed, with later reductions during low arousal periods when possible.

3.1. High $EtCO_2$

Even when no injury was administered, the close monitoring of anesthetized subjects remained essential. For example, 12 h after entering the ICU, one sham subject experienced a decreased HR and BP, along with moderately increased $EtCO_2$ and ICP. The ventilation rate was increased, and anesthesia was reduced to compensate. The HR and BP remained low but consistent throughout the night. $EtCO_2$ remained high (55–65 mmHg) until the I:E ratio was adjusted from 1:2 to 1:1 (reduced $EtCO_2$ to <40 mmHg). In the morning, the subject was temporarily removed from the ventilator to gauge autonomous breathing. As expected, the subject breathed autonomously, the $EtCO_2$ fluctuated, the HR increased to >120 bpm, the BP spiked to >120 mmHg, and the ICP spiked to 30 mmHg but decreased to normal levels over 10 min. This example serves to illustrate that there can be medical challenges to keeping an animal anesthetized for long periods of time—whether injured or not—and care must be taken.

Medical management of SAH

The administration of the SAH injuries occurred after the subjects were stabilized in the neuroICU and baseline samples were collected. Autologous blood was injected into the basal cistern via an EVD implanted to the skull base. When the MAP and ICP reached the same value (CPP = 0), the injections were stopped. This technique reliably recreated the presence of a large volume of blood in the subarachnoid space that was clearly evident during brain extraction and gross pathology, resulting in signs of localized ischemia, as evidenced by H&E staining (Figure 4). Our blinded histopathology analyst correctly noted that the localized ventral appearance of blood, pathology, and inflammation that was observed was not consistent with the pathology from closed-head TBI [55,74–76]. As with the sham animals, we sometimes needed to manage rising $EtCO_2$ with sedation and ventilator adjustments. Unlike with the shams, we needed to resort to hyperosmotic therapy to manage the increased ICP. In addition, one animal experienced seizures during the administration of the SAH injury, and another, as described below, experienced a major cardiac event nearly 12 h after experimental SAH.

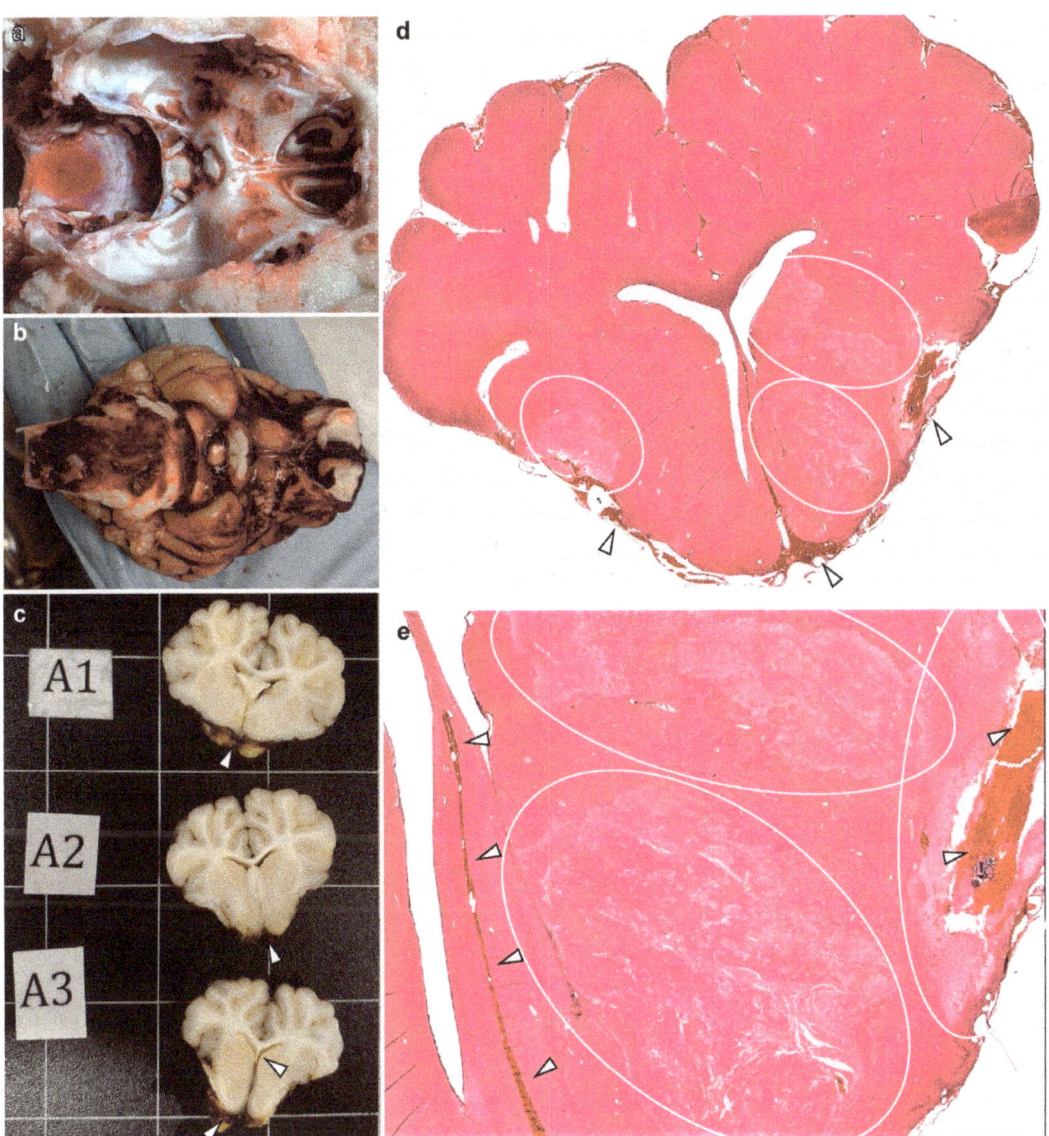

Figure 4. SAH gross and histopathology. Upon brain extraction, clear confirmation of blood deposition at the skull base was observed (**a,b**). Confirmation of blood deposition (white arrows) on the ventral brain surface and in the parenchyma was also confirmed during gross pathology (**c**). With H&E staining, blood (white arrows) can be found in sulci, fissure, and parenchyma, with pale staining indicative of ischemia (white circles) adjacent to the deposited blood (**d,e**).

Approximately 11 h and 40 min after SAH administration, one subject experienced a precipitous drop in MAP and CPP, along with loss of peripheral and brain tissue oxygenation. This cardiac event was followed by a large increase in ICP to 49 mmHg, but brain perfusion would have been of little help at the time because there was little-to-no oxygen in circulation. We started a hypertonic saline infusion, and the ICP gradually came down to acceptable levels over the next 30 min, with restoration of PbtO₂ following the same time course (Figure 5).

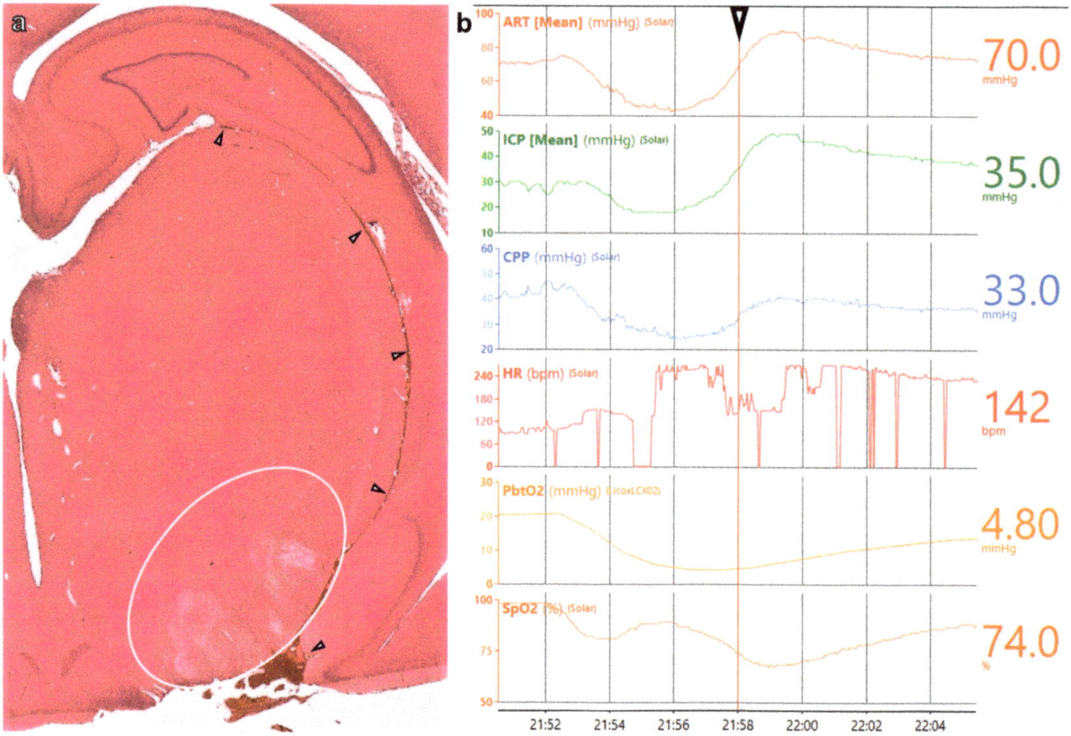

Figure 5. SAH shock myocardia. H&E staining reveals an ischemic area adjacent to blood at the base of the brain (white circle), with white arrows tracking blood traveling through a fissure to the hippocampus (**a**). Multimodal monitoring from the Moberg surrounding the time of the cardiac event (**b**). White arrow and vertical red line indicate the point on the traces for which specific values are provided on the right.

Medical management of TBI with coma

Administration of moderate-to-severe head rotational acceleration TBI occurred after lines were placed, but prior to entry into the neuroICU and placement of cranial probes (which would have been damaged and caused major local trauma if present during the head rotational acceleration). The care team must be fully prepared for ICU admission prior to injury, and critical care must begin immediately following injury, as evidenced by the examples below. As with the sham and SAH animals, underlying ventilator and sedation management was key to achieving stability. However, apnea, cardiovascular events, post-traumatic epilepsy, ICP increases, and other factors made medical management of the TBI animals more challenging overall. Fortunately, the management of ICP with sedation and hyperosmotic therapy was effective (Figure 6a).

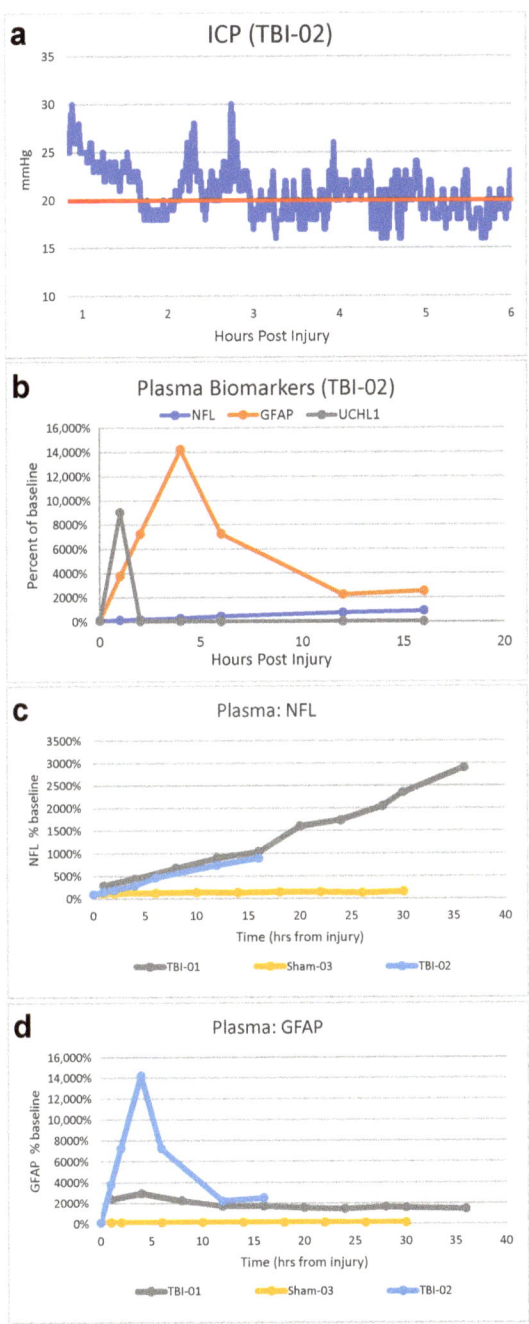

Figure 6. **TBI ICP and biomarkers.** Continuous ICP recording for 6 h following moderate-to-severe TBI with coma, with a red horizontal line to indicate the target of maintaining pressure below 20 mmHg (**a**). Results for the same animal demonstrating compatibility of swine plasma with a standard human TBI biomarker panel with NfL, GFAP, and UCHL1 (**b**). Plasma-biomarker-assay results for two TBI animals and a Sham animal, showing changes in NfL (**c**) and GFAP (**d**) after injury.

The only FDA-approved plasma biomarkers for TBI are glial fibrillary acidic protein (GFAP) and ubiquitin C-terminal hydrolase-L1 (UCHL1) [77,78]. While they have good prognostic value for mild TBI outcome, their value for severe TBI is questionable, and their use in place of a computed tomography scan is very controversial since every bleed that goes undetected could be fatal [79–81]. In general terms, plasma concentrations of GFAP and UCHL1 increase acutely following TBI in humans, while neurofilament light (NfL) increases gradually over time [82]. We found that human assay kits were compatible with pig plasma for detection of GFAP, UCHL1, and NfL, and that in a pair of severe TBI pigs, the time course of these plasma biomarkers after injury was in line with what is observed clinically (Figure 6b).

3.2. Cardiovascular Distress and Post-Traumatic Epilepsy (PTE)

Following injury, one subject experienced repeated periods of apnea and an instance of sudden cardiac arrest, requiring manual resuscitation via bag valve mask until transfer to the ICU, where breathing was controlled by the ventilator; during this time, MAP was monitored externally and found to be low (42 mmHg). Norepinephrine administered via a central line at 0.01 mg/h for 5 min corrected MAP (105 mmHg) and produced a brief spike in the HR (180 bpm). After placement of the cranial bolt and switching to CRI anesthesia, the animal was relatively stable for approximately 15 h (Figure 7a). However, approximately 18 h after injury, the animal began experiencing seizures. Initially, a 5 mg bolus of midazolam was sufficient to terminate seizures, as seen in the EEG trace (Figure 7b). Seizures returned throughout the following morning, requiring propofol boluses in addition to midazolam. CRI midazolam at 10 mg/h effectively terminated seizures. Infusion continued throughout the remainder of the 36 h experiment, during which seizures were infrequent and terminated by propofol bolus.

3.3. Coma and Wakefulness without Awareness

In a different subject, we observed signs of ventilator desynchrony in the neuroICU approximately 90 min after injury. During their first sedation holiday for neurological assessment at 2 h post injury (Table 2), there was no jaw tone, some reflexive head movement, and a spike in ICP that resolved when stimulation was discontinued. The animal was breathing without the ventilator at 3 h post injury. During the second assessment holiday, we observed immediate hind-limb reactivity, and responsiveness was observed sooner than in the previous assessment. In addition, the ICP increase was also less severe after removing propofol sedation. Therefore, the animal appeared to be stable and emerging from coma by 6 h post injury, so we switched to isoflurane anesthesia and removed instrumentation to exit the neuroICU and return to the home cage for recovery.

Over the next 10 h after anesthesia was discontinued, the animal did not fully regain consciousness but was clearly not comatose. The subject was responsive to tactile stimuli and could briefly sit up with assistance. After 4 h, they were responsive to auditory startle and displayed spontaneous eye blinks. The animal swallowed water from a syringe but did not display any other response to the water, indicating that it may have been reflexive behavior. We observed a stable HR (98–111 bpm) and SpO_2 (97–99%) during the 10 h in the home cage, but the temperature was elevated, briefly reaching 104.8 °F. Cool wet towels were applied to the feet, followed by a chilled normosol drip via a central line. In accordance with our protocol at the time, the experiment ended when the animal had not recovered volitional mobility or awareness of environment 10 h after leaving the ICU. During postmortem brain extraction, extensive subdural bleeding was observed across the entire surface of the brain and brainstem (Figure 8a,b). Even at this relatively acute timepoint, histopathological analyses revealed a high burden of APP (scored 3 on a 0–3 scale by a blinded analyst), which is indicative of significant diffuse axonal injury throughout the brain, and elevated reactive microglia with short, thick processes that are indictive of a neuroinflammatory response. This pathology was particularly severe in the periventricular white matter dorsolateral to the lateral ventricles and in the fornix/septum pellucidum

(Figure 8c–e), as compared to the same areas in the brain of a sham animal (Figure 8f–h). In these areas, differences in APP pathology were stark, with severe axonal damage being apparent in the TBI brain and a near absence of any APP signal in the sham brain. Similarly, IBA1 staining in the dorsolateral periventricular white matter revealed overt tissue damage and dense ameboid microglia along the edge of the ventricle in the TBI brain (Figure 8e). Farther from the edge, the regions of interest magnified in Figure 8e show reactive microglia with a short, stubby process in the TBI brain, while the processes in the sham brain (8h) create a fine network throughout the parenchyma passing in and out of plane to give a somewhat speckled appearance. Enlarged lateral ventricles are also evident in the images from the TBI animal when compared to the sham animal.

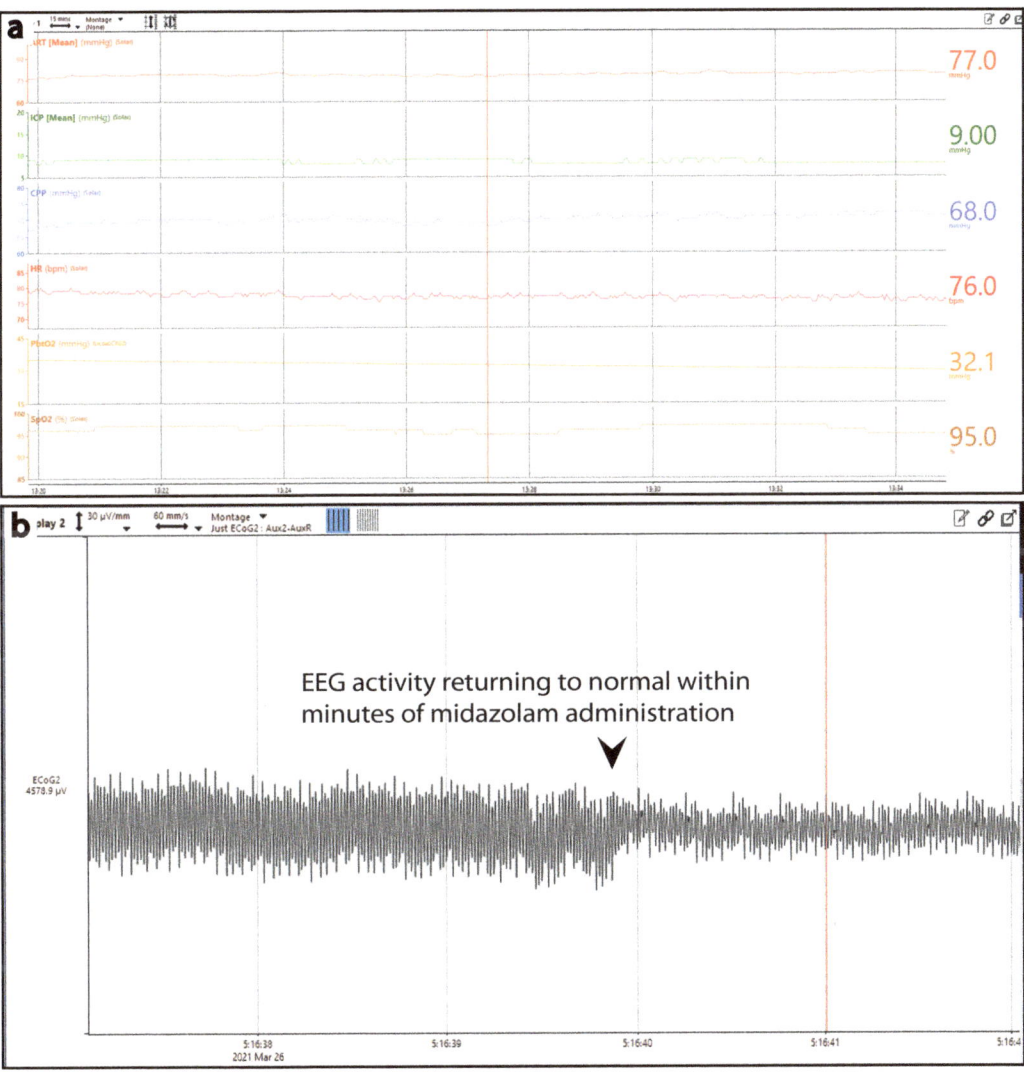

Figure 7. Post-traumatic epilepsy following TBI with coma. Multimodal monitoring during a 15-hour period of relative stability following early apnea and cardiac events (**a**). Single EEG trace showing termination of seizure immediately following midazolam administration (**b**).

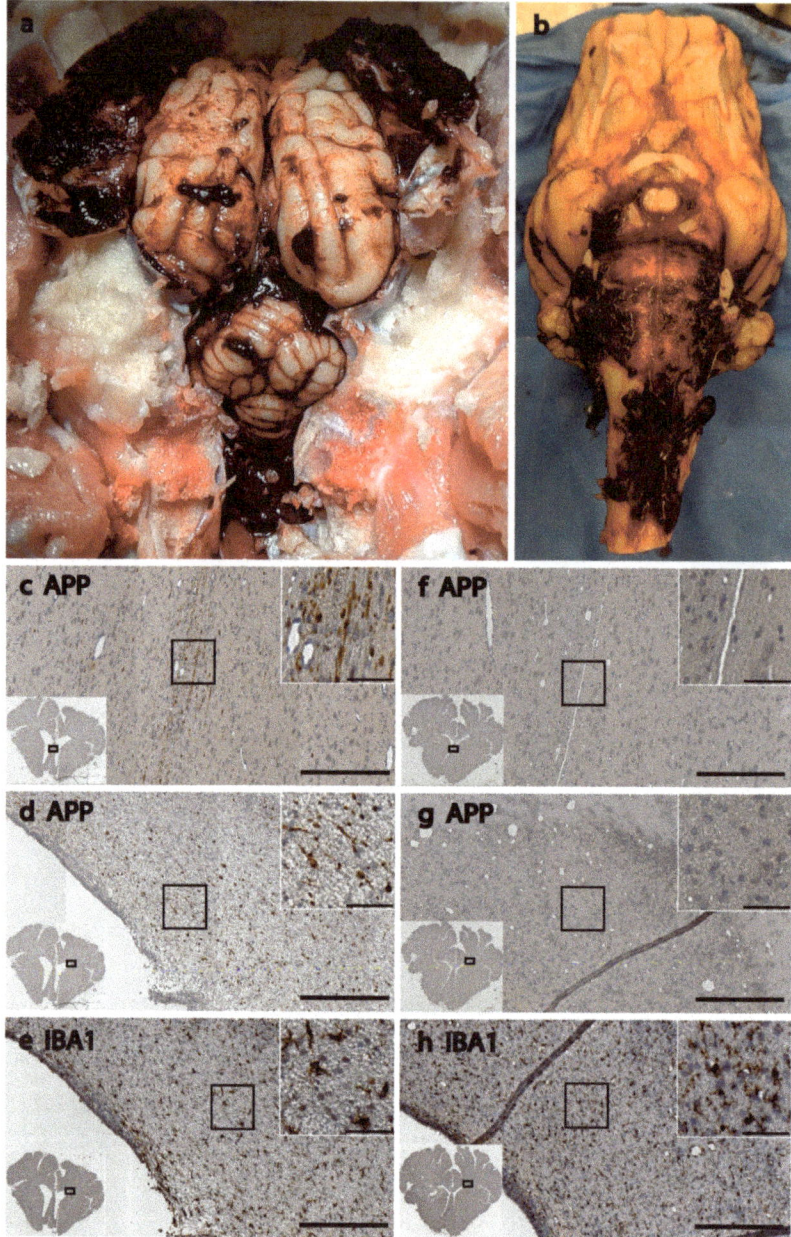

Figure 8. General pathology following TBI with coma and prolonged emergence. Extensive subdural bleeding was present during brain extraction, with clotting evident on the underside of the dura, on the brain's surface, in the tentorium, and surrounding the brainstem (**a,b**). The same post-TBI brain is pictured in panels (**a,b**), and histology from this brain is pictured in panels (**c–e**). (**c,f**) Colorimetric histological images from the medial periventricular white matter comprising the fornix and septum pellucidum. (**d,e,g,h**) Images from the dorsolateral periventricular white matter. Each panel contains a large 15× image with an overlay in the bottom left showing the entire section, with a rectangle to

indicate the region from which the 15× image was acquired, as well as a magnified view of cellular features in a call out box in the top right, also with a square on the 15× image to indicate the region being magnified. Enlarged ventricles are apparent in the TBI brain as compared to the sham brain, as shown in the full section insets. APP burden was scored as 3 out of 3 by a blinded analyst in the TBI brain (c,d), with no detectable APP pathology in similar sections from the sham brain (f,g); IBA1 stain reveals reactive microglia with short, thick processes in the TBI brain (e), while microglia in the sham brain (h) appear to have more ramified networks of processes extending out into the surrounding area, moving in and out of plane to create a speckled staining pattern in the parenchyma; large scale bars = 250 μm, and smaller inset scale bars = 50 μm.

4. Discussion

To develop appropriate diagnostics and treatments in the preclinical space, the field must strive to model the physiologic and clinical factors that collectively dictate patient outcomes. To that effect, the acute period with the most significant change in physiology and the initiation of the injury cascade, and the subacute period presenting with peak brain swelling and deleterious pathophysiological changes, require the highest-fidelity modeling. The best way to accurately predict the translational potential of newly developed tools and treatments is to not only model the mechanisms/manifestations of the human injury but also the unique environment of the neuroICU in which the patient will be treated. We hope that by addressing both needs, the capabilities developed in this study will help to improve the translational pipeline in neurotrauma.

Swine and human brains share many key characteristics that are not found in small mammals which influence both the mechanisms and consequences of stroke and TBI. In addition to their gyrencephalic cortical architecture and high white matter content, pigs are also more similar to humans than smaller mammals in the anatomy of their limbic, subcortical, diencephalic, and brainstem structures [19–22,83,84]. These features, along with similarities between human and pig basal cisterns in the subarachnoid space and the aforementioned high white matter content, make the pig an excellent model for SAH [23–25]. These anatomical similarities also lead to high-fidelity modeling of human TBI, particularly for rotational acceleration/deceleration-mediated inertial loading, which is dependent on brain size [22,27,28]. Indeed, despite rotational acceleration injury attempts made with the rodent "CHIMERA" model, it is not physically possible to scale up the acceleration high enough (8000%) to recreate the forces of human TBI given the small brain mass of rodents [28–30]. Fortunately, some CHIMERA users have dutifully reported failure to reach scaled thresholds for inertial loading, with diffuse pathology clearly emanating from the impact site, but a large swath of the field may still be unaware that the CHIMERA model does not provide any acceleration-induced brain injury in rodents [85,86]. By recreating the mechanism of inertial loading in the pig model, we are able to recreate the manifestations of human TBI—such as loss of consciousness and increased ICP—that require closed-head rotational acceleration acting upon a sufficient brain mass, which cannot be recreated in rodents [27–30,32–35]. Rodent models will always be highly relevant and foundational to neurotrauma research given their low cost, time for iteration, well-characterized and readily altered genetics, and significant overlap with secondary cell/molecular injury mechanisms, but there are insurmountable obstacles to direct translation from rodents to humans that are best addressed by filling the gap with pig models.

Utilizing the monitoring and care capabilities of our reverse-translated neuroICU, we observed and addressed many manifestations of the human injuries being modeled. The collaboration between neurotraumatologists, clinicians, and veterinarians in the administration of this study was vital to guiding the medical management decisions that, in turn, informed algorithms for future work, and this multidisciplinary team will be vital as we further refine our protocols. The inclusion of sham controls emphasized that rising $EtCO_2$ can emerge in anesthetized animals on a ventilator even when they are not injured, and if this is not appropriately addressed via sedation, ventilator adjustments, or by taking the animal off of the ventilator, it can have systemic effects that negatively impact brain perfusion. The stress cardiomyopathy that we observed in one of our SAH animals (Figure 5) also commonly occurs in human patients after SAH and is thought to be associated with excessively high catecholamine signaling [87–92]. During the stress cardiomyopathy that we observed following experimental SAH, there were instances of artifactual interference that made it appear as if the subject's heart stopped for up to 30 s. However, while these were artifact and not cardiac arrest, in some instances, cardiac arrest may follow SAH stress cardiomyopathy and is thought to be connected to larger volume bleeds that damage the hypothalamus or brainstem vasomotor areas, though the mechanisms connecting SAH to these potentially fatal cardiac events are not well understood [90,92–95]. This model seems ideal for testing the dose/response of bleed volume to outcomes after SAH.

Our plasma biomarker results demonstrate compatibility with the human brain biomarker assay kit and show promising trends that are consistent with what is observed in humans following TBI. The enlargement of the lateral ventricles that we observed histologically following TBI is also observed in human TBI patients, and it correlates with outcome; though delayed enlargement is associated with prolonged coma, the acute enlargement that we observed is typically indicative of cerebrospinal fluid obstruction due to hemorrhage and/or edema [96–99]. Periventricular white matter pathology is also common to closed-head moderate-to-severe TBI in humans and pigs, and the extensive axonal injury in the fornix/septum pellucidum that we observed is associated with cognitive deficits in humans following TBI, as well as neurodegenerative diseases such as Alzheimer's, and can even predict cognitive decline in healthy adults [5,100–104]. The post-traumatic epilepsy and coma observed in our pigs are also frequently observed following moderate-to-severe TBI in humans, and importantly, the wakefulness without awareness that we observed in one of our TBI subjects is a human manifestation of TBI that, to our knowledge, has not previously been reported preclinically. While we are not drawing any conclusions based on these observations per se, they suggest that future studies in this model will allow us to test new clinically relevant hypotheses that were previously difficult or impossible to test preclinically.

While this study focused on acute critical care, we are also developing these capabilities to provide a path to recovery to enhance the translational relevance of rehabilitation and regenerative medicine research. Therefore, we monitor for signs of emergence from coma. During this study, we learned that merely emerging from a coma (i.e., responsive to noxious stimuli) may not indicate that the animal will soon be able to be extubated, become ambulatory, and resume self-feeding. Indeed, it appears that after high-rate sagittal angular rotational acceleration/deceleration of the head in pigs, animals may linger in a wakeful/unaware state for some time after emerging from a coma prior to regaining full awareness of their environment. This emphasizes the need for additional reverse translational development to adopt aspects of post-ICU care. We are currently consulting with physiatrists specializing in rehabilitative care for traumatic disorders of consciousness to establish protocols for administering post-ICU care to wakeful/unaware pigs after severe TBI. This extended post-coma, pre-recovery state will require replication, characterization, and unique care considerations, but the initial observation raises the exciting prospect of opening the study of traumatic disorders of consciousness to translational research [36].

This study was focused on technique development and optimization, and future studies will be needed to further characterize these injury manifestations, their similarities to those observed in humans, the mechanisms involved, and potential mitigation strategies. Furthermore, there are several modalities that were not employed in this study due to our focus on optimizing critical care. Most notably absent is neuroimaging due to our reluctance to transport subjects under these conditions during the critical care development phase. In other ongoing studies, we are utilizing sequences for in vivo T1, T2, susceptibility weighted imaging (SWI), and diffusion tensor imaging (DTI), as well as high-resolution ex vivo DTI sequences that have been validated in swine. Neuroimaging offers rich, vital data for brain trauma research and is a highly relevant clinical correlate, and, as such, it will be incorporated into upcoming studies utilizing this swine neurocritical care and monitoring platform.

We recognize the current bottleneck that exists due to the highly specialized equipment and expertise necessary to work in this model, as well as the significant demands placed upon resources and personnel. Therefore, we will make every effort to share our data from future studies by tabulating compatible data for upload to platforms such as the Open Data Commons for TBI (ODC-TBI) [105]; scanning slides via Aperio for upload to eSlide manager; and biobanking FFPE brain blocks, as well as blood, CSF, and other samples, for future analyses with collaborators. We will also continue to advocate for funding agencies to work with research institutions and provide the necessary startup funds and resources to qualified researchers to expand the use of swine models, specifically the swine model of rotational-acceleration-induced TBI. Ultimately, engagement and investment from funding agencies and research institutions will be necessary to relieve this bottleneck in the translational neurotrauma pipeline.

4.1. Future Directions for Translational SAH and TBI Studies

With the development of these techniques and the observations made along the way, there are many new avenues of investigation opened before us. We hope to utilize these capabilities for future studies of SAH, exploring the role of complement activation in vasospasm following injury, along with the pathological mechanisms linking SAH and stress cardiomyopathy. While this is not the only model of SAH, the inclusion of multimodal neuromonitoring and critical care enhances both the translational relevance and the depth of clinically relevant data collected. Future studies of TBI with coma will begin with investigating pathological correlates of coma severity, in particular, the pontine projections of the reticular activating system. The swine model of rotational-acceleration-induced TBI is a well-characterized model that has generated significant advancements in neurotrauma for decades. However, without the sophisticated neurocritical care afforded to human patients, moderate-to-severe TBI is very difficult to manage and often fatal. Therefore, with the humane treatment of research animals as a top priority, chronic studies have been limited to mild TBI [55,75,106], and studies of swine in extended coma have not exceeded 8 h [37,42–44]. Only by providing these subjects with the same neurocritical care afforded to a human patient can we ethically and practically extend beyond the acute period for translational study of rehabilitation [107]. Therefore, in addition to developing new diagnostics, therapeutics, and improving prognostic capabilities in the neurocritical care space, we can now do the same for the recovery/rehabilitation phase following moderate-to-severe TBI with coma.

4.2. Conclusions

Large animal models are both resource- and labor-intensive, and the inclusion of SAH or head rotational acceleration injury and neurocritical care significantly increases the specialized training and expertise required to have success. However, by establishing a multidisciplinary team, we demonstrated that it is feasible to replicate the mechanisms and manifestations of severe human neurotrauma in a large animal model with the inclusion of neurocritical care and monitoring. The preclinical study of SAH or TBI with neurocritical care requires an animal model that recapitulates injury mechanisms (e.g., head rotational loading) and manifestations (e.g., coma and increased ICP) observed in humans to bridge the translational gap between rodent studies and clinical trials. The integration of neuromonitoring and critical care into a model that uniquely recreates the forces of human TBI to create true moderate and severe TBI with coma further increases translational relevance and allows for preclinical study of the unique neurocritical care environment, while also extending the study period for moderate-to-severe TBI with coma and offering a path toward the preclinical study of traumatic disorders of consciousness.

Author Contributions: Conceptualization, J.C.O., K.D.B., D.K.C., and D.P.; Methodology, J.C.O., K.D.B., S.K. (Svetlana Kvint), L.M., T.J.K., and D.P.; Formal Analysis, J.C.O., K.D.B., S.K. (Svetlana Kvint), M.R.G., and D.P.; Investigation, J.C.O., K.D.B., S.K. (Svetlana Kvint), L.M., M.R.G., S.K. (Saarang Karandikar), T.J.K., D.K.C., and D.P.; Resources, J.C.O., K.D.B., D.K.C., and D.P.; Writing—Original Draft Preparation, J.C.O.; Writing—Review and Editing, J.C.O., D.K.C., and D.P.; Visualization, J.C.O.; Supervision, J.C.O., D.K.C., and D.P.; Project Administration, J.C.O., K.D.B., and D.P.; Funding Acquisition, J.C.O., D.K.C., and D.P. All authors have read and agreed to the published version of the manuscript.

Funding: Financial support was provided by the Department of Veterans Affairs (RR&D IK2-RX003376 (O'Donnell); BLR&D I01-BX005017 (Cullen)) and the Department of Neurosurgery, Perelman School of Medicine, University of Pennsylvania (Petrov). Opinions, interpretations, conclusions, and recommendations are those of the authors and are not necessarily endorsed by the Department of Veterans Affairs or the University of Pennsylvania.

Institutional Review Board Statement: All procedures were approved by the University of Pennsylvania's Institutional Animal Care and Use Committee (IACUC) and the Corporal Michael J. Crescenz VA Medical Center in Philadelphia (both AAALAC accredited) and were completed in accordance with the Guide for the Care and Use of Laboratory Animals [54].

Informed Consent Statement: Not applicable.

Data Availability Statement: The data presented in this study are available upon reasonable request.

Acknowledgments: We would like to thank Stella Spears and Philip Latourette for their expert veterinary consultation, along with the veterinary technicians of the University of Pennsylvania. We would also like to thank Cillian Lynch and Ramon Diaz-Arrastia for performing biomarker assays; and John Wolf, Cassidy Fetterman, and Daniel Han for their technical contributions.

Conflicts of Interest: The authors declare no conflict of interest.

Abbreviations

A-line	arterial line
APP	amyloid precursor protein
BP	blood pressure
CBF	cerebral blood flow
CPP	cerebral perfusion pressure
CRI	continuous rate infusion
CSF	cerebrospinal fluid
DAB	3,3′-diaminobenzidine
DTI	diffusion tensor imaging
EEG	electroencephalography
EKG	electrocardiogram
$EtCO_2$	end-tidal CO_2
EVD	external ventricular drain
FFPE	formalin-fixed, paraffin-embedded
GCS	Glasgow Coma Scale
GFAP	glial fibrillary acidic protein
H&E	hematoxylin and eosin
HR	heart rate
IACUC	Institutional Animal Care and Use Committee
IBA1	ionized calcium binding adaptor molecule 1
ICP	intracranial pressure
IND/IDE	investigational new drug/investigational device exemption
LPR	lactate:pyruvate ratio
MAP	mean arterial pressure
MMNM	multimodal neuromonitoring
NCAF	Neurointensive Care and Assessment Facility
neuroICU	neurointensive care unit

ODC-TBI	Open Data Commons for TBI
PbtO$_2$	partial brain tissue oxygen
PTE	post-traumatic epilepsy
rad/sec	radians/second
SAH	subarachnoid hemorrhage
SpO$_2$	blood oxygen saturation
SWI	susceptibility weighted Imaging
TBI	traumatic brain injury
UCHL1	ubiquitin carboxyl-terminal esterase L1

References

1. Dewan, M.C.; Rattani, A.; Gupta, S.; Baticulon, R.E.; Hung, Y.-C.; Punchak, M.; Agrawal, A.; Adeleye, A.O.; Shrime, M.G.; Rubiano, A.M.; et al. Estimating the global incidence of traumatic brain injury. *J. Neurosurg.* **2019**, *130*, 1080–1097. [CrossRef] [PubMed]
2. James, S.L.; Theadom, A.; Ellenbogen, R.G.; Bannick, M.S.; Montjoy-Venning, W.; Lucchesi, L.R.; Abbasi, N.; Abdulkader, R.; Abraha, H.N.; Adsuar, J.C.; et al. Global, regional, and national burden of traumatic brain injury and spinal cord injury, 1990–2016: A systematic analysis for the Global Burden of Disease Study 2016. *Lancet Neurol.* **2019**, *18*, 56–87. [CrossRef] [PubMed]
3. Feigin, V.L.; Stark, B.A.; Johnson, C.O.; Roth, G.A.; Bisignano, C.; Abady, G.G.; Abbasifard, M.; Abbasi-Kangevari, M.; Abd-Allah, F.; Abedi, V.; et al. Global, regional, and national burden of stroke and its risk factors, 1990–2019: A systematic analysis for the Global Burden of Disease Study 2019. *Lancet Neurol.* **2021**, *20*, 795–820. [CrossRef] [PubMed]
4. Turner, R.C.; Lucke-Wold, B.P.; Logsdon, A.F.; Robson, M.; Lee, J.M.; Bailes, J.E.; Dashnaw, M.L.; Huber, J.D.; Petraglia, A.L.; Rosen, C.L. Modeling Chronic Traumatic Encephalopathy: The Way Forward for Future Discovery. *Front. Neurol.* **2015**, *6*, 223. [CrossRef] [PubMed]
5. Lacalle-Aurioles, M.; Iturria-Medina, Y. Fornix degeneration in risk factors of Alzheimer's disease, possible trigger of cognitive decline. *Cereb. Circ. Cogn. Behav.* **2023**, *4*, 100158. [CrossRef]
6. Swanson, R.L.; Acharya, N.K.; Cifu, D.X. Cerebral Microvascular Pathology Is a Common Endophenotype Between Traumatic Brain Injury, Cardiovascular Disease, and Dementia: A Hypothesis and Review. *Cureus* **2022**, *14*, e25318. [CrossRef]
7. Wilson, L.; Stewart, W.; Dams-O'Connor, K.; Diaz-Arrastia, R.; Horton, L.; Menon, D.K.; Polinder, S. The chronic and evolving neurological consequences of traumatic brain injury. *Lancet Neurol.* **2017**, *16*, 813–825. [CrossRef]
8. Johnson, V.E.; Stewart, W.; Arena, J.D.; Smith, D.H. Traumatic Brain Injury as a Trigger of Neurodegeneration. *Adv. Neurobiol.* **2017**, *15*, 383–400. [CrossRef]
9. Izzy, S.; Compton, R.; Carandang, R.; Hall, W.; Muehlschlegel, S. Self-Fulfilling Prophecies Through Withdrawal of Care: Do They Exist in Traumatic Brain Injury, Too? *Neurocrit. Care* **2013**, *19*, 347–363. [CrossRef]
10. Provencio, J.J.; Hemphill, J.C.; Claassen, J.; Edlow, B.L.; Helbok, R.; Vespa, P.M.; Diringer, M.N.; Polizzotto, L.; Shutter, L.; Suarez, J.I.; et al. The Curing Coma Campaign: Framing Initial Scientific Challenges—Proceedings of the First Curing Coma Campaign Scientific Advisory Council Meeting. *Neurocrit. Care* **2020**, *33*, 1–12. [CrossRef]
11. Turgeon, A.F.; Lauzier, F.; Simard, J.-F.; Scales, D.C.; Burns, K.E.; Moore, L.; Zygun, D.A.; Bernard, F.; Meade, M.O.; Dung, T.C.; et al. Mortality associated with withdrawal of life-sustaining therapy for patients with severe traumatic brain injury: A Canadian multicentre cohort study. *Can. Med. Assoc. J.* **2011**, *183*, 1581–1588. [CrossRef]
12. Egbebike, J.; Shen, Q.; Doyle, K.; Der-Nigoghossian, C.A.; Panicker, L.; Gonzales, I.J.; Grobois, L.; Carmona, J.C.; Vrosgou, A.; Kaur, A.; et al. Cognitive-motor dissociation and time to functional recovery in patients with acute brain injury in the USA: A prospective observational cohort study. *Lancet Neurol.* **2022**, *21*, 704–713. [CrossRef]
13. McCrea, M.A.; Giacino, J.T.; Barber, J.; Temkin, N.R.; Nelson, L.D.; Levin, H.S.; Dikmen, S.; Stein, M.; Bodien, Y.G.; Boase, K.; et al. Functional Outcomes Over the First Year After Moderate to Severe Traumatic Brain Injury in the Prospective, Longitudinal TRACK-TBI Study. *JAMA Neurol.* **2021**, *78*, 982. [CrossRef]
14. Kowalski, R.G.; Hammond, F.M.; Weintraub, A.H.; Nakase-Richardson, R.; Zafonte, R.D.; Whyte, J.; Giacino, J.T. Recovery of Consciousness and Functional Outcome in Moderate and Severe Traumatic Brain Injury. *JAMA Neurol.* **2021**, *78*, 548. [CrossRef]
15. Edlow, B.L.; Fins, J.J. Assessment of Covert Consciousness in the Intensive Care Unit: Clinical and Ethical Considerations. *J. Head Trauma Rehabil.* **2018**, *33*, 424–434. [CrossRef]
16. Pham, X.; Ray, J.; Neto, A.S.; Laing, J.; Perucca, P.; Kwan, P.; O'brien, T.J.; Udy, A.A. Association of Neurocritical Care Services With Mortality and Functional Outcomes for Adults With Brain Injury: A Systematic Review and Meta-Analysis. *JAMA Neurol.* **2022**, *79*, 1049–1058. [CrossRef]
17. Claassen, J.; Akbari, Y.; Alexander, S.; Bader, M.K.; Bell, K.; Bleck, T.P.; Boly, M.; Brown, J.; Chou, S.H.-Y.; Diringer, M.N.; et al. Proceedings of the First Curing Coma Campaign NIH Symposium: Challenging the Future of Research for Coma and Disorders of Consciousness. *Neurocrit. Care* **2021**, *35*, 4–23. [CrossRef]
18. Luppi, A.I.; Cain, J.; Spindler, L.R.B.; Górska, U.J.; Toker, D.; Hudson, A.E.; Brown, E.N.; Diringer, M.N.; Stevens, R.D.; Massimini, M.; et al. Mechanisms Underlying Disorders of Consciousness: Bridging Gaps to Move Toward an Integrated Translational Science. *Neurocrit. Care* **2021**, *35*, 37–54. [CrossRef]

19. Bailey, E.L.; McCulloch, J.; Sudlow, C.; Wardlaw, J.M. Potential Animal Models of Lacunar Stroke: A Systematic Review. *Stroke* 2009, *40*, e451–e458. [CrossRef]
20. Howells, D.W.; Porritt, M.J.; Rewell, S.S.J.; O'Collins, V.; Sena, E.S.; Van Der Worp, H.B.; Traystman, R.J.; Macleod, M.R. Different Strokes for Different Folks: The Rich Diversity of Animal Models of Focal Cerebral Ischemia. *J. Cereb. Blood Flow Metab.* 2010, *30*, 1412–1431. [CrossRef]
21. Zhang, K.; Sejnowski, T.J. A universal scaling law between gray matter and white matter of cerebral cortex. *Proc. Natl. Acad. Sci. USA* 2000, *97*, 5621–5626. [CrossRef] [PubMed]
22. Cullen, D.K.; Harris, J.P.; Browne, K.D.; Wolf, J.A.; Duda, J.E.; Meaney, D.F.; Margulies, S.S.; Smith, D.H. A Porcine Model of Traumatic Brain Injury via Head Rotational Acceleration. *Methods Mol. Biol.* 2016, *1462*, 289–324. [CrossRef] [PubMed]
23. Wagner, K.R.; Xi, G.; Hua, Y.; Kleinholz, M.; De Courten-Myers, G.M.; Myers, R.E.; Broderick, J.P.; Brott, T.G. Lobar Intracerebral Hemorrhage Model in Pigs: Rapid Edema Development in Perihematomal White Matter. *Stroke* 1996, *27*, 490–497. [CrossRef] [PubMed]
24. Hartings, J.A.; York, J.; Carroll, C.P.; Hinzman, J.M.; Mahoney, E.; Krueger, B.; Winkler, M.K.L.; Major, S.; Horst, V.; Jahnke, P.; et al. Subarachnoid blood acutely induces spreading depolarizations and early cortical infarction. *Brain* 2017, *140*, 2673–2690. [CrossRef]
25. Ma, Q.; Khatibi, N.H.; Chen, H.; Tang, J.; Zhang, J.H. *History of Preclinical Models of Intracerebral Hemorrhage*; Springer: Vienna, Austria, 2019.
26. Melià-Sorolla, M.; Castaño, C.; DeGregorio-Rocasolano, N.; Rodríguez-Esparragoza, L.; Dávalos, A.; Martí-Sistac, O.; Gasull, T. Relevance of Porcine Stroke Models to Bridge the Gap from Pre-Clinical Findings to Clinical Implementation. *Int. J. Mol. Sci.* 2020, *21*, 6568. [CrossRef]
27. Keating, C.E.; Cullen, D.K. Mechanosensation in traumatic brain injury. *Neurobiol. Dis.* 2020, *148*, 105210. [CrossRef]
28. Meaney, D.F.; Smith, D.H.; Shreiber, D.I.; Bain, A.C.; Miller, R.T.; Ross, D.T.; Gennarelli, T.A. Biomechanical Analysis of Experimental Diffuse Axonal Injury. *J. Neurotrauma* 1995, *12*, 689–694. [CrossRef]
29. Margulies, S.S.; Thibault, L.E.; Gennarelli, T.A. Physical model simulations of brain injury in the primate. *J. Biomech.* 1990, *23*, 823–836. [CrossRef]
30. Meaney, D.F.; Margulies, S.S.; Smith, D.H. Diffuse Axonal Injury. *J. Neurosurg.* 2001, *95*, 1108–1110. [CrossRef]
31. Wofford, K.L.; Grovola, M.R.; Adewole, D.O.; Browne, K.D.; Putt, M.E.; O'Donnell, J.C.; Cullen, D.K. Relationships between Injury Kinematics, Neurological Recovery, and Pathology Following Concussion. *Brain Commun.* 2021, fcab268. [CrossRef]
32. Denny-Brown, D.E.; Russell, W.R. Experimental Concussion: (Section of Neurology). *Proc. R. Soc. Med.* 1941, *34*, 691–692.
33. Langlois, J.A.; Rutland-Brown, W.; Wald, M.M. The Epidemiology and Impact of Traumatic Brain Injury: A Brief Overview. *J. Head Trauma Rehabil* 2006, *21*, 375–378. [CrossRef]
34. Ommaya, A.K.; Gennarelli, T.A. Cerebral Concussion and Traumatic Unconsciousness. Correlation of Experimental and Clinical Observations of Blunt Head Injuries. *Brain* 1974, *97*, 633–654. [CrossRef]
35. Gabrielian, L.; Willshire, L.W.; Helps, S.C.; Heuvel, C.V.D.; Mathias, J.; Vink, R. Intracranial Pressure Changes following Traumatic Brain Injury in Rats: Lack of Significant Change in the Absence of Mass Lesions or Hypoxia. *J. Neurotrauma* 2011, *28*, 2103–2111. [CrossRef]
36. O'Donnell, J.C.; Browne, K.D.; Kilbaugh, T.J.; Chen, H.I.; Whyte, J.; Cullen, D.K. Challenges and demand for modeling disorders of consciousness following traumatic brain injury. *Neurosci. Biobehav. Rev.* 2019, *98*, 336–346. [CrossRef]
37. Smith, D.H.; Nonaka, M.; Miller, R.; Leoni, M.; Chen, X.-H.; Alsop, D.; Meaney, D.F. Immediate coma following inertial brain injury dependent on axonal damage in the brainstem. *J. Neurosurg.* 2000, *93*, 315–322. [CrossRef]
38. Cooper, D.J.; Rosenfeld, J.V.; Murray, L.; Arabi, Y.M.; Davies, A.R.; D'Urso, P.; Kossmann, T.; Ponsford, J.; Seppelt, I.; Reilly, P.; et al. Decompressive Craniectomy in Diffuse Traumatic Brain Injury. *N. Engl. J. Med.* 2011, *364*, 1493–1502. [CrossRef]
39. Hutchinson, P.J.; Kolias, A.G.; Timofeev, I.S.; Corteen, E.A.; Czosnyka, M.; Timothy, J.; Anderson, I.; Bulters, D.O.; Belli, A.; Eynon, C.A.; et al. Trial of Decompressive Craniectomy for Traumatic Intracranial Hypertension. *N. Engl. J. Med.* 2016, *375*, 1119–1130. [CrossRef]
40. Kolias, A.G.; Adams, H.; Timofeev, I.S.; Corteen, E.A.; Hossain, I.; Czosnyka, M.; Timothy, J.; Anderson, I.; Bulters, D.O.; Belli, A.; et al. Evaluation of Outcomes Among Patients With Traumatic Intracranial Hypertension Treated With Decompressive Craniectomy vs Standard Medical Care at 24 Months: A Secondary Analysis of the RESCUEicp Randomized Clinical Trial. *JAMA Neurol.* 2022, *79*, 664–671. [CrossRef]
41. Hawryluk, G.W.J.; Rubiano, A.M.; Totten, A.M.; O'reilly, C.; Ullman, J.S.; Bratton, S.L.; Chesnut, R.; Harris, O.A.; Kissoon, N.; Shutter, L.; et al. Guidelines for the Management of Severe Traumatic Brain Injury: 2020 Update of the Decompressive Craniectomy Recommendations. *Neurosurgery* 2020, *87*, 427–434. [CrossRef]
42. Friess, S.H.; Ralston, J.; Eucker, S.A.; Helfaer, M.A.; Smith, C.; Margulies, S.S. Neurocritical Care Monitoring Correlates With Neuropathology in a Swine Model of Pediatric Traumatic Brain Injury. *Neurosurgery* 2011, *69*, 1139–1147. [CrossRef] [PubMed]
43. Friess, S.H.; Smith, C.; Kilbaugh, T.; Frangos, S.G.; Ralston, J.; Helfaer, M.A.; Margulies, S.S. Early cerebral perfusion pressure augmentation with phenylephrine after traumatic brain injury may be neuroprotective in a pediatric swine model. *Crit. Care Med.* 2012, *40*, 2400–2406. [CrossRef] [PubMed]
44. Friess, S.H.; Bruins, B.; Kilbaugh, T.; Smith, C.; Margulies, S.S. Differing Effects when Using Phenylephrine and Norepinephrine To Augment Cerebral Blood Flow after Traumatic Brain Injury in the Immature Brain. *J. Neurotrauma* 2015, *32*, 237–243. [CrossRef] [PubMed]

45. Nyberg, C.; Karlsson, T.; Hillered, L.; Engström, E.R. Metabolic Pattern of the Acute Phase of Subarachnoid Hemorrhage in a Novel Porcine Model: Studies with Cerebral Microdialysis with High Temporal Resolution. *PLoS ONE* **2014**, *9*, e99904. [CrossRef]
46. Datzmann, T.; Kapapa, T.; Scheuerle, A.; McCook, O.; Merz, T.; Unmuth, S.; Hoffmann, A.; Mathieu, R.; Mayer, S.; Mauer, U.M.; et al. In-depth characterization of a long-term, resuscitated model of acute subdural hematoma–induced brain injury. *J. Neurosurg.* **2019**, *134*, 223–234. [CrossRef]
47. Weenink, R.P.; Vrijdag, X.C.; van Putten, M.J.; Hollmann, M.W.; Stevens, M.F.; van Gulik, T.M.; van Hulst, R.A. Quantitative electroencephalography in a swine model of cerebral arterial gas embolism. *Clin. Neurophysiol.* **2012**, *123*, 411–417. [CrossRef]
48. Chen, C.; Zhou, C.; Cavanaugh, J.M.; Kallakuri, S.; Desai, A.; Zhang, L.; King, A.I. Quantitative electroencephalography in a swine model of blast-induced brain injury. *Brain Inj.* **2017**, *31*, 120–126. [CrossRef]
49. Mader, M.M.; Leidorf, A.; Hecker, A.; Heimann, A.; Mayr, P.S.M.; Kempski, O.; Alessandri, B.; Wöbker, G. Evaluation of a New Multiparameter Brain Probe for Simultaneous Measurement of Brain Tissue Oxygenation, Cerebral Blood Flow, Intracranial Pressure, and Brain Temperature in a Porcine Model. *Neurocrit. Care* **2018**, *29*, 291–301. [CrossRef]
50. Nwafor, D.C.; Brichacek, A.L.; Foster, C.H.; Lucke-Wold, B.P.; Ali, A.; Colantonio, M.A.; Brown, C.M.; Qaiser, R. Pediatric Traumatic Brain Injury: An Update on Preclinical Models, Clinical Biomarkers, and the Implications of Cerebrovascular Dysfunction. *J. Cent. Nerv. Syst. Dis.* **2022**, *14*, 11795735221098124. [CrossRef]
51. Cralley, A.L.; Moore, E.E.; Kissau, D.; Coleman, J.R.M.; Vigneshwar, N.; DeBot, M.; Schaid, T.R.J.; Moore, H.B.M.; Cohen, M.J.; Hansen, K.; et al. A combat casualty relevant dismounted complex blast injury model in swine. *J. Trauma Inj. Infect. Crit. Care* **2022**, *93*, S110–S118. [CrossRef]
52. Adedipe, A.; John, A.S.; Krishnamoorthy, V.; Wang, X.; Steck, D.T.; Ferreira, R.; White, N.; Stern, S. Left Ventricular Function in the Initial Period After Severe Traumatic Brain Injury in Swine. *Neurocrit. Care* **2022**, *37*, 200–208. [CrossRef]
53. Abdou, H.; Edwards, J.; Patel, N.; Stonko, D.P.; Elansary, N.; Lang, E.; Richmond, M.J.; Ptak, T.; White, J.M.; Scalea, T.M.; et al. Characterizing Brain Perfusion in a Swine Model of Raised Intracranial Pressure. *J. Surg. Res.* **2022**, *278*, 64–69. [CrossRef]
54. National Research Council (US) Committee for the Update of the Guide for the Care and Use of Laboratory Animals Guide for the Care and Use of Laboratory Animals. In *The National Academies Collection: Reports Funded by National Institutes of Health*, 8th ed.; National Academies Press (US): Washington, DC, USA, 2011; ISBN 978-0-309-15400-0.
55. Grovola, M.R.; Paleologos, N.; Brown, D.P.; Tran, N.; Wofford, K.L.; Harris, J.P.; Browne, K.D.; Shewokis, P.A.; Wolf, J.A.; Cullen, D.K.; et al. Diverse changes in microglia morphology and axonal pathology during the course of 1 year after mild traumatic brain injury in pigs. *Brain Pathol.* **2021**, *31*, e12953. [CrossRef]
56. Menon, D.K.; Coles, J.P.; Gupta, A.K.; Fryer, T.D.; Smielewski, P.; Chatfield, D.A.; Aigbirhio, F.; Skepper, J.N.; Minhas, P.S.; Hutchinson, P.J.; et al. Diffusion limited oxygen delivery following head injury. *Crit. Care Med.* **2004**, *32*, 1384–1390. [CrossRef]
57. Vespa, P.M.; O'phelan, K.; McArthur, D.; Miller, C.; Eliseo, M.; Hirt, D.; Glenn, T.; Hovda, D.A. Pericontusional brain tissue exhibits persistent elevation of lactate/pyruvate ratio independent of cerebral perfusion pressure. *Crit. Care Med.* **2007**, *35*, 1153–1160. [CrossRef]
58. van Santbrink, H.; Maas, A.I.; Avezaat, C.J. Continuous Monitoring of Partial Pressure of Brain Tissue Oxygen in Patients with Severe Head Injury. *Neurosurgery* **1996**, *38*, 21–31. [CrossRef]
59. Brink, W.A.V.D.; van Santbrink, H.; Steyerberg, E.W.; Avezaat, C.J.J.; Suazo, J.A.C.; Hogesteeger, C.; Jansen, W.J.; Kloos, L.M.H.; Vermeulen, J.; Maas, A.I.R. Brain Oxygen Tension in Severe Head Injury. *Neurosurgery* **2000**, *46*, 868–878. [CrossRef]
60. Artru, F.; Jourdan, C.; Perret-Liaudet, A.; Charlot, M.; Mottolese, C. Low brain tissue oxygen pressure: Incidence and corrective therapies. *Neurol. Res.* **1998**, *20* (Suppl. 1), S48–S51. [CrossRef]
61. Valadka, A.B.; Goodman, J.C.; Gopinath, S.P.; Uzura, M.; Robertson, C.S. Comparison of Brain Tissue Oxygen Tension to Microdialysis-Based Measures of Cerebral Ischemia in Fatally Head-Injured Humans. *J. Neurotrauma* **1998**, *15*, 509–519. [CrossRef]
62. Bardt, T.F.; Unterberg, A.W.; Härtl, R.; Kiening, K.L.; Schneider, G.H.; Lanksch, W.R. Monitoring of Brain Tissue PO2 in Traumatic Brain Injury: Effect of Cerebral Hypoxia on Outcome. *Acta Neurochir. Suppl.* **1998**, *71*, 153–156. [CrossRef]
63. Kiening, K.; Härtl, R.; Unterberg, A.; Schneider, G.-H.; Bardt, T.; Lanksch, W. Brain tissue pO_2-monitoring in comatose patients: Implications for therapy. *Neurol. Res.* **1997**, *19*, 233–240. [CrossRef] [PubMed]
64. Bohman, L.-E.; Heuer, G.G.; Macyszyn, L.; Maloney-Wilensky, E.; Frangos, S.; Le Roux, P.D.; Kofke, A.; Levine, J.M.; Stiefel, M.F. Medical Management of Compromised Brain Oxygen in Patients with Severe Traumatic Brain Injury. *Neurocrit. Care* **2011**, *14*, 361–369. [CrossRef] [PubMed]
65. Pascual, J.L.; Georgoff, P.; Maloney-Wilensky, E.; Sims, C.; Sarani, B.; Stiefel, M.F.; LeRoux, P.D.; Schwab, C.W. Reduced Brain Tissue Oxygen in Traumatic Brain Injury: Are Most Commonly Used Interventions Successful? *J. Trauma Inj. Infect. Crit. Care* **2011**, *70*, 535–546. [CrossRef]
66. Nangunoori, R.; Maloney-Wilensky, E.; Stiefel, M.; Park, S.; Kofke, W.A.; Levine, J.M.; Yang, W.; Le Roux, P.D. Brain Tissue Oxygen-Based Therapy and Outcome After Severe Traumatic Brain Injury: A Systematic Literature Review. *Neurocrit. Care* **2012**, *17*, 131–138. [CrossRef] [PubMed]
67. Okonkwo, D.O.; Shutter, L.; Moore, C.; Temkin, N.R.; Puccio, A.M.; Madden, C.J.; Andaluz, N.; Chesnut, R.; Bullock, M.R.; Grant, G.A.; et al. Brain Oxygen Optimization in Severe Traumatic Brain Injury Phase-II: A Phase II Randomized Trial. *Crit. Care Med.* **2017**, *45*, 1907–1914. [CrossRef]

68. Barsan, W. Brain Oxygen Optimization in Severe TBI (BOOST3): A Comparative Effectiveness Study to Test the Efficacy of a Prescribed Treatment Protocol Based on Monitoring the Partial Pressure of Brain Tissue Oxygen; 2021. Available online: clinicaltrials.gov (accessed on 20 December 2021).
69. Teasdale, G.; Jennett, B. Assessment of coma and impaired consciousness. A practical scale. *Lancet* **1974**, *2*, 81–84. [CrossRef]
70. Teasdale, G.; Maas, A.; Lecky, F.; Manley, G.; Stocchetti, N.; Murray, G. The Glasgow Coma Scale at 40 years: Standing the test of time. *Lancet Neurol.* **2014**, *13*, 844–854. [CrossRef]
71. Platt, S.R.; Radaelli, S.T.; McDonnell, J.J. The Prognostic Value of the Modified Glasgow Coma Scale in Head Trauma in Dogs. *J. Vet. Intern. Med.* **2001**, *15*, 581–584. [CrossRef]
72. Sharma, D.; Holowaychuk, M.K. Retrospective evaluation of prognostic indicators in dogs with head trauma: 72 cases (January-March 2011). *J. Vet. Emerg. Crit. Care* **2015**, *25*, 631–639. [CrossRef]
73. Shores, A. Craniocerebral Trauma. In *Current Veterinary Therapy X*; Kirk, R.W., Ed.; WB Saunders: Philadelphia, PA, USA, 1983; pp. 847–854.
74. Wofford, K.L.; Harris, J.P.; Browne, K.D.; Brown, D.P.; Grovola, M.R.; Mietus, C.J.; Wolf, J.A.; Duda, J.E.; Putt, M.E.; Spiller, K.L.; et al. Rapid neuroinflammatory response localized to injured neurons after diffuse traumatic brain injury in swine. *Exp. Neurol.* **2017**, *290*, 85–94. [CrossRef]
75. Grovola, M.; Paleologos, N.; Wofford, K.L.; Harris, J.P.; Browne, K.D.; Johnson, V.; Duda, J.E.; Wolf, J.A.; Cullen, D.K. Mossy cell hypertrophy and synaptic changes in the hilus following mild diffuse traumatic brain injury in pigs. *J. Neuroinflamm.* **2020**, *17*, 44. [CrossRef]
76. Grovola, M.R.; von Reyn, C.; Loane, D.J.; Cullen, D.K. Understanding microglial responses in large animal models of traumatic brain injury: An underutilized resource for preclinical and translational research. *J. Neuroinflamm.* **2023**, *20*, 67. [CrossRef]
77. Bazarian, J.J.; Biberthaler, P.; Welch, R.D.; Lewis, L.M.; Barzo, P.; Bogner-Flatz, V.; Brolinson, P.G.; Büki, A.; Chen, J.Y.; Christenson, R.H.; et al. Serum GFAP and UCH-L1 for prediction of absence of intracranial injuries on head CT (ALERT-TBI): A multicentre observational study. *Lancet Neurol.* **2018**, *17*, 782–789. [CrossRef]
78. Papa, L.; Brophy, G.M.; Welch, R.D.; Lewis, L.M.; Braga, C.F.; Tan, C.N.; Ameli, N.J.; Lopez, M.A.; Haeussler, C.A.; Giordano, D.I.M.; et al. Time Course and Diagnostic Accuracy of Glial and Neuronal Blood Biomarkers GFAP and UCH-L1 in a Large Cohort of Trauma Patients With and Without Mild Traumatic Brain Injury. *JAMA Neurol.* **2016**, *73*, 551–560. [CrossRef]
79. Korley, F.K.; Jain, S.; Sun, X.; Puccio, A.M.; Yue, J.K.; Gardner, R.C.; Wang, K.K.W.; Okonkwo, D.O.; Yuh, E.L.; Mukherjee, P.; et al. Prognostic value of day-of-injury plasma GFAP and UCH-L1 concentrations for predicting functional recovery after traumatic brain injury in patients from the US TRACK-TBI cohort: An observational cohort study. *Lancet Neurol.* **2022**, *21*, 803–813. [CrossRef]
80. Gan, Z.S.; Stein, S.C.; Swanson, R.; Guan, S.; Garcia, L.; Mehta, D.; Smith, D.H. Blood Biomarkers for Traumatic Brain Injury: A Quantitative Assessment of Diagnostic and Prognostic Accuracy. *Front. Neurol.* **2019**, *10*, 446. [CrossRef]
81. Papa, L.; Ladde, J.G.; O'brien, J.F.; Thundiyil, J.G.; Tesar, J.; Leech, S.; Cassidy, D.D.; Roa, J.; Hunter, C.; Miller, S.; et al. Evaluation of Glial and Neuronal Blood Biomarkers Compared With Clinical Decision Rules in Assessing the Need for Computed Tomography in Patients With Mild Traumatic Brain Injury. *JAMA Netw. Open* **2022**, *5*, e221302. [CrossRef]
82. Graham, N.S.N.; Zimmerman, K.A.; Moro, F.; Heslegrave, A.; Maillard, S.A.; Bernini, A.; Miroz, J.-P.; Donat, C.K.; Lopez, M.Y.; Bourke, N.; et al. Axonal marker neurofilament light predicts long-term outcomes and progressive neurodegeneration after traumatic brain injury. *Sci. Transl. Med.* **2021**, *13*, eabg9922. [CrossRef]
83. Lind, N.M.; Moustgaard, A.; Jelsing, J.; Vajta, G.; Cumming, P.; Hansen, A.K. The use of pigs in neuroscience: Modeling brain disorders. *Neurosci. Biobehav. Rev.* **2007**, *31*, 728–751. [CrossRef]
84. Østergaard, K.; Holm, I.E.; Zimmer, J. Tyrosine hydroxylase and acetylcholinesterase in the domestic pig mesencephalon: An immunocytochemical and histochemical study. *J. Comp. Neurol.* **1992**, *322*, 149–166. [CrossRef]
85. Namjoshi, D.R.; Cheng, W.H.; McInnes, K.A.; Martens, K.M.; Carr, M.; Wilkinson, A.; Fan, J.; Robert, J.; Hayat, A.; Cripton, P.A.; et al. Merging pathology with biomechanics using CHIMERA (Closed-Head Impact Model of Engineered Rotational Acceleration): A novel, surgery-free model of traumatic brain injury. *Mol. Neurodegener.* **2014**, *9*, 55. [CrossRef] [PubMed]
86. Sauerbeck, A.D.; Fanizzi, C.; Kim, J.H.; Gangolli, M.; Bayly, P.V.; Wellington, C.L.; Brody, D.L.; Kummer, T.T. modCHIMERA: A novel murine closed-head model of moderate traumatic brain injury. *Sci. Rep.* **2018**, *8*, 7677. [CrossRef] [PubMed]
87. Ahmadian, A.; Mizzi, A.; Banasiak, M.; Downes, K.; Camporesi, E.M.; Thompson Sullebarger, J.; Vasan, R.; Mangar, D.; Van Loveren, H.R.; Agazzi, S. Cardiac manifestations of subarachnoid hemorrhage. *Heart Lung Vessel.* **2013**, *5*, 168–178. [PubMed]
88. Koza, Y.; Aydin, N.; Aydin, M.D.; Koza, E.A.; Bayram, E.; Atalay, C.; Altas, E.; Kursad, H.; Kabalar, M.E. Neurogenic Stress Cardiomyopathy Following Subarachnoid Hemorrhage Is Associated with Vagal Complex Degeneration: First Experimental Study. *World Neurosurg.* **2019**, *129*, e741–e748. [CrossRef]
89. Norberg, E.; Odenstedt-Herges, H.; Rydenhag, B.; Oras, J. Impact of Acute Cardiac Complications After Subarachnoid Hemorrhage on Long-Term Mortality and Cardiovascular Events. *Neurocrit. Care* **2018**, *29*, 404–412. [CrossRef]
90. Feldstein, E.; Dominguez, J.F.; Kaur, G.; Patel, S.D.; Dicpinigaitis, A.J.; Semaan, R.; Fuentes, L.E.; Ogulnick, J.; Ng, C.; Rawanduzy, C.; et al. Cardiac arrest in spontaneous subarachnoid hemorrhage and associated outcomes. *Neurosurg. Focus* **2022**, *52*, E6. [CrossRef]

91. Kerro, A.; Woods, T.; Chang, J.J. Neurogenic stunned myocardium in subarachnoid hemorrhage. *J. Crit. Care* **2017**, *38*, 27–34. [CrossRef]
92. Al-Mufti, F.; Morris, N.; Lahiri, S.; Roth, W.; Witsch, J.; Machado, I.; Agarwal, S.; Park, S.; Meyers, P.M.; Connolly, E.S.; et al. Use of Intra-aortic- Balloon Pump Counterpulsation in Patients with Symptomatic Vasospasm Following Subarachnoid Hemorrhage and Neurogenic Stress Cardiomyopathy. *J. Vasc. Interv. Neurol.* **2016**, *9*, 28–34.
93. Frontera, J.A. Clinical Trials in Cardiac Arrest and Subarachnoid Hemorrhage: Lessons from the Past and Ideas for the Future. *Stroke Res. Treat.* **2013**, *2013*, 263974. [CrossRef]
94. Zachariah, J.; Stanich, J.A.; Braksick, S.A.; Wijdicks, E.F.; Campbell, R.L.; Bell, M.R.; White, R. Indicators of Subarachnoid Hemorrhage as a Cause of Sudden Cardiac Arrest. *Clin. Pract. Cases Emerg. Med.* **2016**, *1*, 132–135. [CrossRef]
95. Shin, J.; Kim, K.; Lim, Y.S.; Lee, H.J.; Lee, S.J.; Jung, E.; Kim, J.; Yang, H.J.; Kim, J.J.; Hwang, S.Y. Incidence and clinical features of intracranial hemorrhage causing out-of-hospital cardiac arrest: A multicenter retrospective study. *Am. J. Emerg. Med.* **2016**, *34*, 2326–2330. [CrossRef]
96. Levin, H.S.; Meyers, C.A.; Grossman, R.G.; Sarwar, M. Ventricular Enlargement After Closed Head Injury. *Arch. Neurol.* **1981**, *38*, 623–629. [CrossRef]
97. Levine, B.; Fujiwara, E.; O'Connor, C.; Richard, N.; Kovacevic, N.; Mandic, M.; Restagno, A.; Easdon, C.; Robertson, I.H.; Graham, S.J.; et al. *In Vivo* Characterization of Traumatic Brain Injury Neuropathology with Structural and Functional Neuroimaging. *J. Neurotrauma* **2006**, *23*, 1396–1411. [CrossRef]
98. Meyers, C.A.; Levin, H.S.; Eisenberg, H.M.; Guinto, F.C. Early versus late lateral ventricular enlargement following closed head injury. *J. Neurol. Neurosurg. Psychiatry* **1983**, *46*, 1092–1097. [CrossRef]
99. Poca, M.A.; Sahuquillo, J.; Mataro, M.; Benejam, B.; Arikan, F.; Baguena, M. Ventricular Enlargement after Moderate or Severe Head Injury: A Frequent and Neglected Problem. *J. Neurotrauma* **2005**, *22*, 1303–1310. [CrossRef]
100. Oishi, K.; Lyketsos, C.G. Alzheimer's Disease and the Fornix. *Front. Aging Neurosci.* **2014**, *6*, 241. [CrossRef]
101. Mielke, M.M.; Okonkwo, O.C.; Oishi, K.; Mori, S.; Tighe, S.; Miller, M.I.; Ceritoglu, C.; Brown, T.; Albert, M.; Lyketsos, C.G. Fornix integrity and hippocampal volume predict memory decline and progression to Alzheimer's disease. *Alzheimer's Dement.* **2012**, *8*, 105–113. [CrossRef]
102. Douet, V.; Chang, L. Fornix as an Imaging Marker for Episodic Memory Deficits in Healthy Aging and in Various Neurological Disorders. *Front. Aging Neurosci.* **2015**, *6*, 343–362. [CrossRef]
103. Metzler-Baddeley, C.; Jones, D.K.; Belaroussi, B.; Aggleton, J.P.; O'Sullivan, M.J. Frontotemporal Connections in Episodic Memory and Aging: A Diffusion MRI Tractography Study. *J. Neurosci.* **2011**, *31*, 13236–13245. [CrossRef]
104. Fletcher, E.; Raman, M.; Huebner, P.; Liu, A.; Mungas, D.; Carmichael, O.; DeCarli, C. Loss of Fornix White Matter Volume as a Predictor of Cognitive Impairment in Cognitively Normal Elderly Individuals. *JAMA Neurol.* **2013**, *70*, 1389–1395. [CrossRef]
105. Chou, A.; Torres-Espín, A.; Huie, J.R.; Krukowski, K.; Lee, S.; Nolan, A.; Guglielmetti, C.; Hawkins, B.E.; Chaumeil, M.M.; Manley, G.T.; et al. Empowering Data Sharing and Analytics through the Open Data Commons for Traumatic Brain Injury Research. *Neurotrauma Rep.* **2022**, *3*, 139–157. [CrossRef] [PubMed]
106. Browne, K.D.; Chen, X.-H.; Meaney, D.; Smith, D.H. Mild Traumatic Brain Injury and Diffuse Axonal Injury in Swine. *J. Neurotrauma* **2011**, *28*, 1747–1755. [CrossRef] [PubMed]
107. O'Donnell, J.C.; Swanson, R.L.; Wofford, K.L.; Grovola, M.R.; Purvis, E.M.; Petrov, D.; Cullen, D.K. Emerging Approaches for Regenerative Rehabilitation Following Traumatic Brain Injury. In *Regenerative Rehabilitation: From Basic Science to the Clinic*; Greising, S.M., Call, J.A., Eds.; Physiology in Health and Disease; Springer International Publishing: Cham, Switzerland, 2022; pp. 409–459. ISBN 978-3-030-95884-8.

Disclaimer/Publisher's Note: The statements, opinions and data contained in all publications are solely those of the individual author(s) and contributor(s) and not of MDPI and/or the editor(s). MDPI and/or the editor(s) disclaim responsibility for any injury to people or property resulting from any ideas, methods, instructions or products referred to in the content.

Article

Exploring the Therapeutic Potential of Phosphorylated *Cis*-Tau Antibody in a Pig Model of Traumatic Brain Injury

Samuel S. Shin [1,*], Vanessa M. Mazandi [2], Andrea L. C. Schneider [1,3], Sarah Morton [2,4], Jonathan P. Starr [2,4], M. Katie Weeks [2,4], Nicholas J. Widmann [2,4], David H. Jang [4,5], Shih-Han Kao [2,4], Michael K. Ahlijanian [6] and Todd J. Kilbaugh [2,4]

[1] Division of Neurocritical Care, Department of Neurology, Perelman School of Medicine at the University of Pennsylvania, Philadelphia, PA 19104, USA; andrea.schneider@pennmedicine.upenn.edu

[2] Department of Anesthesiology and Critical Care Medicine, The Children's Hospital of Philadelphia, Perelman School of Medicine at the University of Pennsylvania, Philadelphia, PA 19104, USA; mazandiv@chop.edu (V.M.M.); mortons2@chop.edu (S.M.); starrjp@chop.edu (J.P.S.); weeksmk@chop.edu (M.K.W.); widmann@chop.edu (N.J.W.); kaos@chop.edu (S.-H.K.); kilbaugh@email.chop.edu (T.J.K.)

[3] Department of Biostatistics, Epidemiology and Informatics, University of Pennsylvania, Philadelphia, PA 19104, USA

[4] Resuscitation Science Center of Emphasis, The Children's Hospital of Philadelphia, Perelman School of Medicine at the University of Pennsylvania, Philadelphia, PA 19104, USA; david.jang@pennmedicine.upenn.edu

[5] Department of Emergency Medicine, Perelman School of Medicine at the University of Pennsylvania, Philadelphia, PA 19104, USA

[6] Pinteon Therapeutics, Inc., Newton, MA 02459, USA; mahlijanian@pinteon.com

* Correspondence: samuel.shin@pennmedicine.upenn.edu

Abstract: Traumatic brain injury (TBI) results in the generation of tau. As hyperphosphorylated tau (p-tau) is one of the major consequences of TBI, targeting p-tau in TBI may lead to the development of new therapy. Twenty-five pigs underwent a controlled cortical impact. One hour after TBI, pigs were administered either vehicle ($n = 13$) or PNT001 ($n = 12$), a monoclonal antibody for the *cis* conformer of tau phosphorylated at threonine 231. Plasma biomarkers of neural injury were assessed for 14 days. Diffusion tensor imaging was performed at day 1 and 14 after injury, and these were compared to historical control animals ($n = 4$). The fractional anisotropy data showed significant white matter injury for groups at 1 day after injury in the corona radiata. At 14 days, the vehicle-treated pigs, but not the PNT001-treated animals, exhibited significant white matter injury compared to sham pigs in the ipsilateral corona radiata. The PNT001-treated pigs had significantly lower levels of plasma glial fibrillary acidic protein (GFAP) at day 2 and day 4. These findings demonstrate a subtle reduction in the areas of white matter injury and biomarkers of neurological injury after treatment with PNT001 following TBI. These findings support additional studies for PNT001 as well as the potential use of this agent in clinical trials in the near future.

Keywords: traumatic brain injury; tau; phosphorylated tau; TBI; PNT001

1. Introduction

Tau proteins are highly expressed in neurons as well as astroglia and oligodendroglia [1] and play an important role in maintaining the stability of axonal microtubules [2]. In the setting of traumatic brain injury (TBI), axons undergo damage that includes the disruption of cytoskeletal structures. Specifically, tau proteins that are normally bound to microtubules dissociate and undergo aberrant post-translational modifications such as phosphorylation [3]. Pathologically phosphorylated tau is prone to aggregate and induce harmful effects such as mitochondrial injury, apoptosis, and neuronal death [3]. Once phosphorylated tau is formed in one region of the brain, it can spread to neighboring regions and has been previously demonstrated to transmit from the cortex to the hippocampus as well as to the contralateral side of the brain [4].

A major pathological consequence of TBI is chronic traumatic encephalopathy. Known to be a result of repetitive mild traumatic brain injury, hyperphosphrylated tau has been well described [5] to be broadly distributed across the brain regions of CTE subjects. Although Alzheimer's disease also shows prominent levels of hyperphosphorylated tau, there are subtle differences in the location of their accumulation. Specifically, in CTE, hyperphosphorylated tau has been prominently noted in axons and perivascular regions [6], whereas the tau pathology is diffusely distributed in these regions in AD [7]

A prior study showed that the *cis* conformer of tau phosphorylated at threonine 231 (T231 cis p-tau) is acutely produced by neurons following TBI [4]. In a mouse model of TBI, a monoclonal antibody targeting cis p-tau has been shown to prevent CTE-like pathological changes as well as cognitive impairment [8,9]. Given the damaging effect of cis p-tau to neurons in animals, this molecule has been considered to contribute to TBI pathophysiology. With this insight, we utilized a humanized monoclonal antibody specific to cis p-tau (PNT001, Pinteon Inc., Newton, MA, USA) [10] in order to explore its therapeutic potential in acute TBI using a clinically relevant pig model of TBI.

2. Materials and Methods

2.1. Animal Surgery

The experiments in this study were approved by the Institutional Animal Care and Use Committee of the University of Pennsylvania in accordance with the Guide for the Care and Use of Laboratory Animals. There were two groups in this study: the vehicle group ($n = 13$) and PNT001 group ($n = 12$), with pigs weighing approximately 30 kg (Figure 1). In order to minimize sex-specific effects, each group had approximately the same number of male and female pigs. In the placebo group, there were $n = 6$ males and $n = 7$ females, and in the PNT001 group, there were $n = 5$ males and $n = 7$ females. Surgical procedures were performed as previously described [11]. Briefly, intramuscular ketamine (20 mg/kg) and xylazine (2 mg/kg) were administered, followed by induction using 4% inhaled isoflurane. The pigs were then intubated and maintained on anesthesia with 1% isoflurane throughout the experiment. Bupivacaine analgesia was provided by injection into the subcutaneous tissue at the incision site, followed by prophylactic cefazolin intramuscular injection. Pigs then underwent central venous catheter placement in the cephalic veins terminating in the superior vena cava. Right-sided craniotomy overlying the rostral gyrus was performed, and the pigs underwent mild–moderate severity injury with a spring-loaded impactor velocity at 4 m/s. The pigs were then administered buprenorphine-SR for additional analgesia. Drug administration using 60 mg/kg of PNT001 or vehicle was performed by intravenous injection at 1 h after TBI. Sham pigs included in the data analysis for DTI underwent the same anesthesia exposures and all surgical procedures (skin incision) except craniotomy and controlled cortical impact. The vehicle solution was composed of 25 mM histidine/220 mM sucrose/0.02% (*w/v*) polysorbate 80 prepared in sterile water. To confirm the appropriate administration of PNT001, exposure was measured at several time points in serum and cerebrospinal fluid (CSF) in a subset of 8 treated animals (Figure S1). At the end of the experiment on day 14, the animals were anesthetized by intramuscular injection of ketamine/xylazine. They were then euthanized by intracardiac injection of pentobarbital at 150 mg/kg.

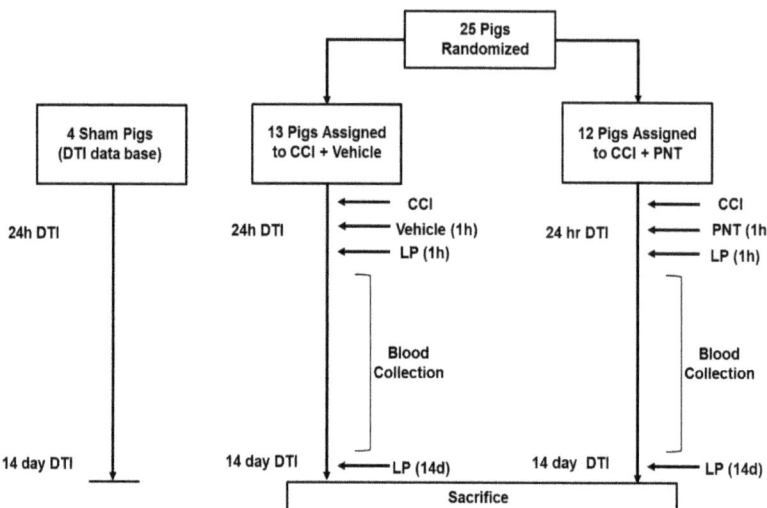

Figure 1. Schematic of the study. Data from DTI database of sham pigs at 24 h and 14 days were compared to the PNT001-treated CCI group and vehicle-treated CCI group. LP = lumbar puncture, PNT = PNT001.

2.2. Biomarker Study

Pigs in both groups underwent blood draws from the central line before the injury and then at the range of times as follows: 30 min, 2 h, 5 h, 1 d, 2 d, 4 d, 7 d, 10 d, and 14 d after TBI. For plasma isolation, the blood samples were centrifuged at $4400\times g$ for 5 min, and the supernatants were collected and then stored at $-80\ °C$ until analysis. Lumbar puncture was also performed at 1 h and 14 d after injury to collect CSF, which was also stored at $-80\ °C$ until analysis. In order to assess the plasma biomarker levels, a single-molecule array (Simoa) was used. Using a 4-Plex assay (Quanterix, Billerica, MA, USA) for glial fibrillary acidic protein (GFAP), neurofilament-light (NfL), ubiquitin C-terminal hydrolase L1 (UCHL-1), and tau, we processed the plasma and CSF samples. The samples were then analyzed by Quanterix Corp. (Billerica, MA, USA). Additionally, we performed exposure analysis using serum analysis. The blood samples were first allowed to clot for 15–30 min and were then centrifuged at $4400\times g$ for 5 min. These samples were stored at $-80\ °C$ until analysis. Both serum and CSF were tested for PNT001 levels using enzyme-linked immunoassay (ELISA) as previously described [10].

2.3. Diffusion Tensor Imaging

Magnetic resonance imaging with diffusion tensor imaging (DTI) sequencing was performed at two time points: 24 h and 14 days after injury. A 3T Tim Trio whole-body magnetic resonance scanner (Siemens, Munich, Germany) with a 12-channel phased array head coil was used for 64 noncolinear/noncoplanar direction scans with single-shot spin-echo, echo-planar imaging. The specifics of the DTI sequence were as follows: repetition time (TR) = 4200 mS, echo time (TE) = 103 msec, flip angle = 180 degrees, bandwidth = 1186 Hz/pixel, field of view (FOV) = 192 mm, slice thickness = 2 mm, number of slices = 24, voxel size = $2 \times 2 \times 2$ mm, and b-values = 0, 1000, and 2000 s/mm^2. Track-based spatial statistics were used for the analysis using the FSL software. Eddy current distortions as well as motion-induced distortions were corrected. From a composite white matter skeleton, the region of interest (ROI) for the corpus callosum, the ipsilateral (right) and contralateral (left) corona radiata, and the ipsilateral and contralateral cerebral peduncles were drawn. For each ROI, fractional anisotropy (FA) and mean diffusivity (MD) values were calculated. Heat maps of significant ($p < 0.05$) regions of decreased FA values between the placebo and PNT001 groups were displayed.

2.4. Statistical Analysis

Statistical analysis was performed for FA and MD values using multiple Mann-Whitney U tests. As there was no sham injury group for this study, historical data from our group with the same conditions (30 kg weight) that underwent the same surgical/anesthetic exposure but with only skin incision without TBI were used to normalize the FA and MD values. Plasma and CSF biomarker values were also assessed using multiple Mann-Whitney U tests. In the sensitivity analyses, we performed linear mixed-effects models with random intercepts for the time since first biomarker measurement after injury and an unstructured covariance matrix to estimate the association between the treatment group and the change in the natural-log-transformed biomarkers over 14 days post-injury. All biomarkers were ln transformed for statistical modeling, as the distributions were not normally distributed. To account for the non-linear association between plasma NfL and GFAP over time, a linear spline was used to model the time since the first biomarker measurement after injury was used with a knot at 168 h (7 days) for NfL and at 24 h (1 day) for GFAP. For all the tests in this study, $p < 0.05$ was used as a significant cut-off value. Stata SE Version 17 (College Station, TX, USA) and GraphPad Prism (San Diego, CA, USA) were used for the analysis.

3. Results

After the administration of the vehicle or PNT001(60 mg/kg, intravenous), we analyzed drug exposure in an initial subset of eight animals in both serum and CSF to validate the appropriate administration (Figure S1). This showed the appropriate drug levels in the serum and CSF of the pigs, which were in agreement with the previously demonstrated K_d of PNT [10]. Then, we analyzed the time course of white matter integrity using DTI at 1 day and 14 days. At 1 day following TBI, there were multiple areas of significant reduction in FA among both the PNT001- and vehicle-treated CCI pigs (Figure 2). Both the PNT001-treated and vehicle-treated injured animals exhibited reduced FA levels in the corona radiata. Whereas the PNT001-treated animals appeared to have smaller areas of FA reduction in the bilateral corona radiata, the vehicle-treated group appeared to have larger areas of FA reduction. Specifically, the contralateral (left) corona radiata and posterior portion of the corpus callosum showed an FA reduction.

For the quantification of the white matter integrity, we normalized the FA values to our historical sham animals' FA levels in each respective area, as shown in (Figure 3). In agreement with the heat map of the areas with FA reduction (Figure 2), there was a significant decrease in the FA values of the corona radiata for both the vehicle- and PNT001-treated groups compared to the sham pigs. The PNT001-treated pigs also displayed a reduction in the MD in the corpus callosum, while the vehicle-treated group did not. When DTI was performed again at the 14-day time point (Figure 4), there were significant reductions in the areas in which FA decreases were acutely observed. The quantification of the reductions in the FA in these regions at 14 days resulted in significant differences between the sham and vehicle groups but not the sham and PNT001 groups at the right corona radiata (Figure 5). Although the heat map showed small areas of FA reduction in the bilateral corona radiata, because the region of FA reduction was very small, the quantitative assessment showed no significant difference between the sham and vehicle or PNT001 groups in the left corona radiata. Similarly, the longitudinal study using linear mixed effects model showed no significant differences between the two groups for FA and MD (Table S1).

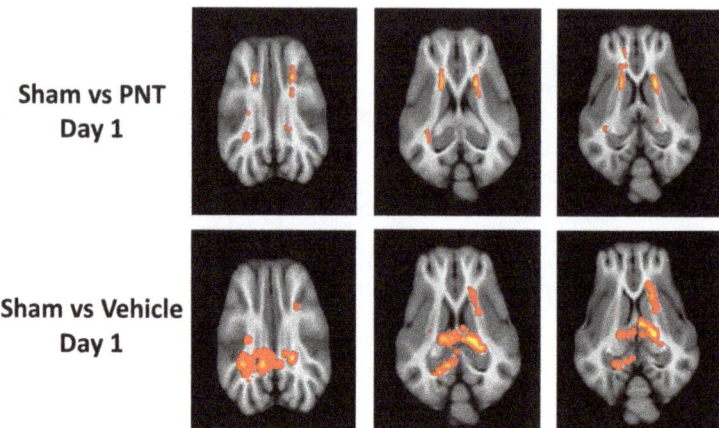

Figure 2. FA map showing difference between sham vs PNT-treated CCI group (**top**) and sham vs vehicle-treated CCI group (**bottom**) at 1 day following injury. The red–yellow regions show FA-reduced areas as compared to sham group.

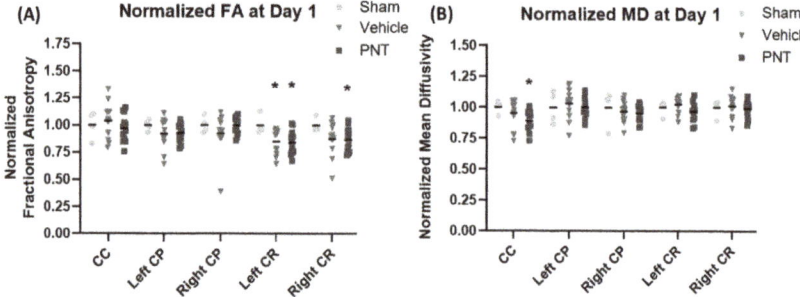

Figure 3. DTI parameter comparison between vehicle-treated CCI group and PNT-treated pigs normalized to sham group at 1 day. Region specific changes in FA (**A**) and MD (**B**) are shown here.

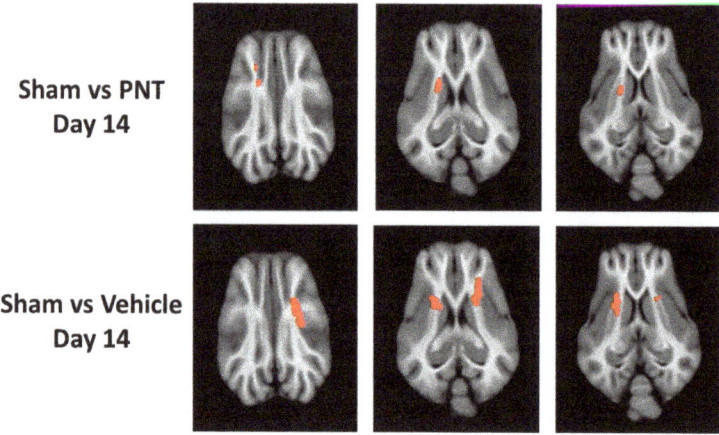

Figure 4. FA map showing difference between sham vs PNT001-treated CCI group (**top**) and sham vs vehicle-treated CCI group (**bottom**) at 14 days. The red regions show FA-reduced areas as compared to sham group. PNT = PNT001.

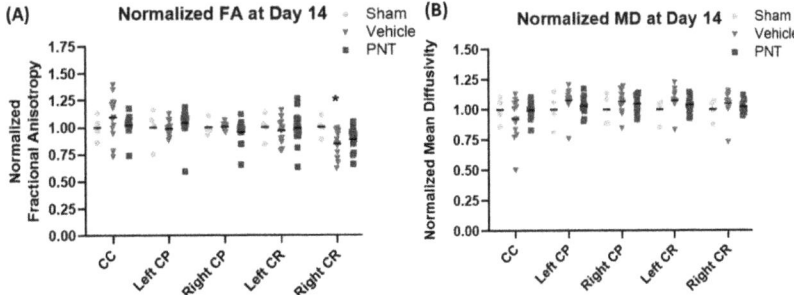

Figure 5. DTI parameter comparison between vehicle-treated and PNT001-treated CCI groups normalized to sham group at 14 days. PNT = PNT001. * $p < 0.05$. Region specific changes in FA (**A**) and MD (**B**) are shown here.

The biomarker studies on plasma showed an early peak elevation of GFAP at 1–2 days, while the elevation of NfL peaked at 7 days. There were no changes in UCHL1 throughout the study (Figure 6). Although there was no difference between the vehicle and PNT001 groups at the peak levels for NfL and GFAP (7 and 1 day post-injury, respectively), the GFAP concentrations at 2 days and 4 days post-injury showed notable changes in the PNT001 group compared to the vehicle group. The PNT001 group had a lower GFAP concentration at 2 days, although this was not statistically significant ($p = 0.051$). At 4 days, there was significant reduction in the GFAP concentration ($p < 0.01$) in the PNT001 group compared to the vehicle group. For NfL and UCH-L1, there were no differences throughout the 14 days. Similarly, the linear mixed effects model (Table S1) showed a significant difference between the vehicle and PNT001 groups for plasma GFAP levels over 14 days post-injury, but other biomarkers showed no significant differences. We also analyzed the CSF biomarker levels at 1 day and 14 days after injury (Figure 7). While the CSF NfL concentrations were unchanged at 1 h post-injury, a dramatic increase was observed (40-fold) 14 days post-injury. In contrast, the CSF concentrations of tau, UCHL1, and GFAP were increased as soon as 1 h post-injury. By 14 days post-injury, the GFAP levels had returned to the pre-injury levels, while the concentrations of tau and UCHL1 remained elevated. While numerical reductions in all the CSF biomarkers were observed following treatment with PNT001, the substantial variability of the biomarker concentrations precluded the achievement of any statistical significance for any analyte. The longitudinal analysis using a linear mixed effects model showed no difference between the vehicle and PNT001 groups for CSF biomarker levels.

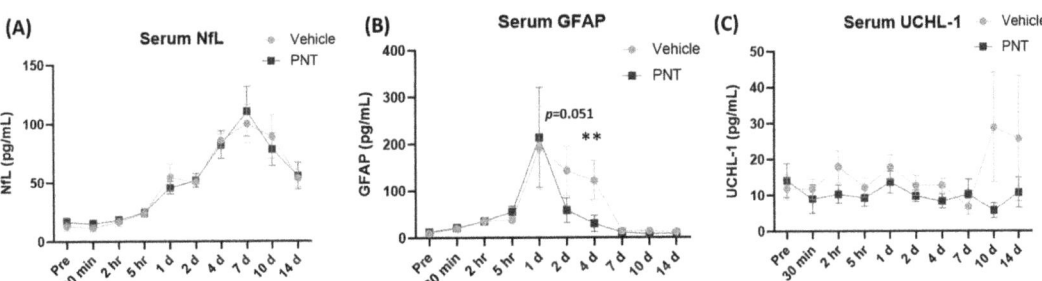

Figure 6. Plasma levels of biomarkers over 14 days. NfL (**A**), GFAP (**B**), and UCHL-1 (**C**) are displayed at 1–4-day intervals during the acute-to-subacute period following TBI. The biomarker assay lacked the sensitivity to detect pig plasma tau. ** $p < 0.01$.

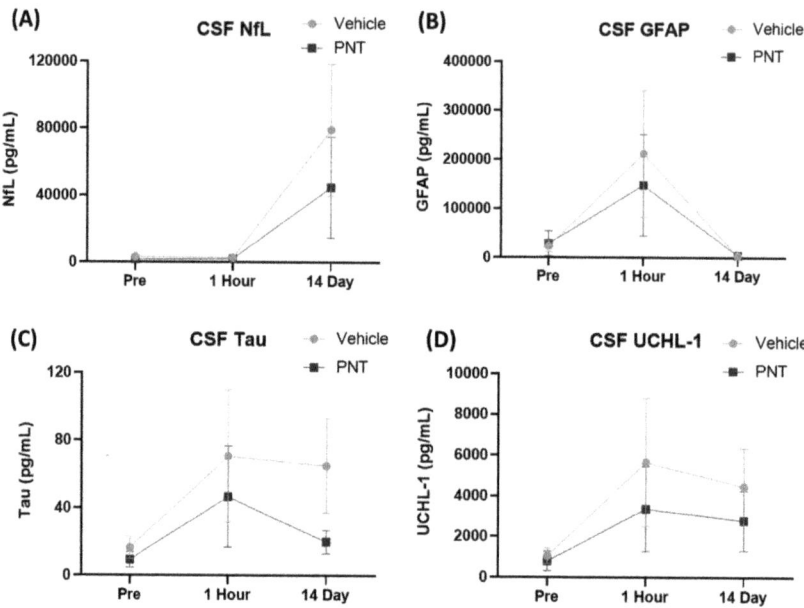

Figure 7. CSF levels of biomarkers over 14 days. NfL (**A**), GFAP (**B**), tau (**C**), and UCHL-1 (**D**) are displayed at two time points: 1 h and 14 days following TBI. PNT = PNT001.

4. Discussion

The results of this study showed a subtle reduction in the areas of white matter injury and biomarkers of injury in the PNT001-treated pigs compared to the vehicle-treated pigs after TBI. Given the well-documented link between TBI and Alzheimer's disease, [12,13] as well as CTE [14,15], the pathomolecular mechanisms that link these disease entities have been pursued by TBI clinicians and scientists over the years. The generation of cis p-tau after TBI [4] and its association with many neurodegenerative changes have been demonstrated [8]. Given this promising novel target for TBI, we utilized a pharmacological agent aimed at reducing the cis p-tau burden in the central nervous system following TBI using a pig model of injury.

In the DTI scans of pigs following TBI, specific areas of FA reduction were found in the bilateral corona radiata and splenium of the corpus callosum. These areas of injury were consistent with our previously reported DTI data on 30-day post-injury pigs [16], indicating the selective vulnerability of these regions to TBI. Changes in FA levels at the ipsilateral corona radiata were specifically correlated with rises in both NFL and GFAP also in this study, supporting the validity of these radiographical findings. Although the quantification of FA values at each location showed no significant differences between the vehicle-treated and PNT001-treated groups, the subtle differences in the size of the area with FA changes between the two groups was more clear in the heat maps. These data showed that while there was no major effect in reducing the white matter damage by PNT001, the subtle reduction, as shown by heat map, may indicate that optimal dose and administration times should be explored. Over the 14 days, the areas of FA reduction were significantly attenuated, as shown between Figures 2 and 4. This change may be partly due to the recovery process of white matter injury over time, but decreasing tissue edema may also account for this, as previously noted [16].

Among the vehicle-treated pigs, the plasma biomarker profile showed a time course similar to that of human data. As previously described [16], the acute rise in GFAP within the first day and the delayed rise in NfL over the course of 1 week replicated the pattern that was shown in clinical TBI patients [17–19]. The injury biomarker profiles showed decreases in serum GFAP levels when PNT001 was administered after TBI (Figure 6). Since GFAP is an intermediate filament specifically expressed by astrocytes and is considered to have an important role in astrocyte mobility and proliferation [20,21], a reduction in the GFAP level by PNT001 treatment may indicate a protection of glial injury. Although GFAP elevation in the tissue levels occurred following the injury, this time course was slower than with a peak brain tissue expression occurring at 3 days after injury [22]. The serum peak levels were earlier at 1 day post-injury, indicating that this may be due to the acute release of existing GFAP from the astroglia secondary to injury. The time course of cis p-tau generation after TBI has been detailed previously [4]: the elevation of cis p-tau began 12 h after a single moderate-to-severe TBI and peaked at 48 h. The exposure data (Figure S1) showed an appropriate elevation of PNT001 levels in the serum after the administration, starting at 2 h after injury. Given the expected delay of hours to days it would take for PNT001 to distribute to the interstitial space and brain parenchyma and interact with cis p-tau, the reduction in GFAP levels by the PNT001 treatment at 2 and 4 days in the current study seems consistent.

Meanwhile, the other biomarkers of injury (NFL and UCH-L1) did not show any significant difference between the two groups. This may mean that while glial injury may have been attenuated, other pathobiological components of TBI (delayed white matter injury and neuronal injury) were not directly mitigated by the PNT001 treatment. Additionally, the sensitivity of the Simoa assays were not optimized for porcine proteins and thus tau detection was not achieved in the serum samples, while higher levels of tau in the CSF resulted in significant detection (Figure 7).

The biomarker studies in the CSF showed no significant differences between the PNT001- and vehicle-treated animals, although the PNT001 group had minor trends with lower biomarker values. This limited detection at 1 day and 14 day unfortunately missed the appropriate monitoring windows for biomarkers such as NfL which peaks at approximately 7 days. In addition, as was demonstrated in the serum data that attenuation of GFAP elevation occurred at 2 and 4 days after the PNT001 treatment, the CSF data, which was not collected at these time points, missed this monitoring window. Given these few collection time points for CSF, the comparison of the plasma and CSF biomarker values to assess for their correlation was limited. As the biomarkers did not yet reach their peak elevation at 30 min to 1 h following injury, no significant correlation was acutely found between the CSF and plasma biomarker levels (Table S2). At 14 days, NfL did not show a significant correlation between the CSF and plasma levels ($p = 0.0670$).

While tau in its physiological state has a major function in the central nervous system, such as mediating neurite growth and stabilizing microtubules and thus the cytoskeleton of neurons, the pathological transformation of this protein can lead to toxic effects [23]. The hyperphosphorylation of monomeric tau detaches it from microtubules, leading to the binding of other detached monomers and the formation of oligomers [24]. Oligomers of tau can propagate through neighboring neurons over time and gradually over larger areas of the brain. Given the neurotoxicity of tau oligomers, their formation following TBI has been considered as a potential mechanism for the progression of injury aside from the primary damage of TBI. Larger oligomers can then form neurofibrillary tangles in various brain regions, which has also been demonstrated in TBI [25,26], Alzheimer's disease [27], and other neurodegenerative conditions.

Aside from the prevention of oligomers, the potential therapeutic effect of PNT001 may be through the mitigation of pathologies such as aquaporin-4 (AQP4) dysfunction. Since AQP4 is expressed in the astrocytic end-foot and is acutely upregulated following TBI [28], this has been considered as a mechanism of cerebral edema after TBI. Additionally, AQP4 is an important component of the glymphatic system, which functions for the clearance of pathological proteins such as hyperphosphorylated tau [29]. However, given that TBI damages and disrupts the normal function of this clearance system, the brain may become more vulnerable to the accumulation of pathological tau. In this setting, the neurotoxicity of pTau may become more pronounced. Thus, a monoclonal antibody against pTau may have a therapeutic effect by mitigating the effects of damage in the glymphatic system. In this initial study using PNT001 for acute administration following TBI, promising reductions in serum GFAP levels as well as potential small reductions in white matter injury were noted using DTI. Given these promising findings, future studies can be designed to explore a shorter window of administration and varied doses of PNT001 after TBI. Furthermore, low-severity TBI and different modalities of TBI such as rotational injury can be explored.

Supplementary Materials: The following supporting information can be downloaded at: https://www.mdpi.com/article/10.3390/biomedicines11071807/s1, Figure S1: Exposure data for serum (A) and CSF (B). Serum PNT001 levels are shown here, with expected decrease over time from 2 h to 14 days. Two time-point CSF levels of PNT001 were measured: prior to the injury and at 14 days. PNT = PNT001; Table S1: Difference in change in ln(biomarker) over 14 days post-injury between treatment groups; Table S2: Plasma and CSF biomarker correlation at acute time point (first row) and subacute time point (second row). Given the different time points in which samples were collected for plasma and CSF, we chose the closest matching times with plasma (30 min) and CSF (1 h) for the acute time point. The second row shows the correlation values at 14 days.

Author Contributions: Conceptualization, T.J.K. and M.K.A.; methodology, S.-H.K., M.K.W. and N.J.W.; software, M.K.W. and N.J.W.; validation, M.K.A.; formal analysis, S.S.S., S.-H.K., A.L.C.S. and M.K.A.; investigation, S.M. and J.P.S.; resources, T.J.K.; writing—original draft preparation, S.S.S.; writing—review and editing, S.S.S., V.M.M., A.L.C.S., D.H.J. and M.K.A.; visualization, T.J.K.; supervision, M.K.A. and T.J.K.; project administration, S.M.; funding acquisition, M.K.A. and T.J.K. All authors have read and agreed to the published version of the manuscript.

Funding: This study was funded by a Sponsored Research Agreement between Pinteon Therapeutics Inc., Newton, MA, USA and the Children's Hospital of Philadelphia (FP000032542).

Institutional Review Board Statement: Not applicable.

Informed Consent Statement: Not applicable.

Data Availability Statement: The data obtained in this study will be made available if requested by outside parties.

Acknowledgments: We would like to acknowledge Lucas Hobson and Yuxi Lin for their help in animal experiments and Norah Taraska for administrative support.

Conflicts of Interest: Michael Ahlijanian is an employee of Pinteon Therapeutics and received financial compensation.

References

1. Shin, R.W.; Iwaki, T.; Kitamoto, T.; Tateishi, J. Hydrated autoclave pretreatment enhances tau immunoreactivity in formalin-fixed normal and Alzheimer's disease brain tissues. *Lab. Investig.* **1991**, *64*, 693–702. [PubMed]
2. Cleveland, D.W.; Hwo, S.Y.; Kirschner, M.W. Purification of tau, a microtubule-associated protein that induces assembly of microtubules from purified tubulin. *J. Mol. Biol.* **1977**, *116*, 207–225. [CrossRef] [PubMed]
3. Tagge, C.A.; Fisher, A.M.; Minaeva, O.V.; Gaudreau-Balderrama, A.; Moncaster, J.A.; Zhang, X.L.; Wojnarowicz, M.W.; Casey, N.; Lu, H.; Kokiko-Cochran, O.N.; et al. Concussion, microvascular injury, and early tauopathy in young athletes after impact head injury and an impact concussion mouse model. *Brain* **2018**, *141*, 422–458. [CrossRef]
4. Kondo, A.; Shahpasand, K.; Mannix, R.; Qiu, J.; Moncaster, J.; Chen, C.H.; Yao, Y.; Lin, Y.M.; Driver, J.A.; Sun, Y.; et al. Antibody against early driver of neurodegeneration cis P-tau blocks brain injury and tauopathy. *Nature* **2015**, *523*, 431–436. [CrossRef]
5. Mez, J.; Daneshvar, D.H.; Kiernan, P.T.; Abdolmohammadi, B.; Alvarez, V.E.; Huber, B.R.; Alosco, M.L.; Solomon, T.M.; Nowinski, C.J.; McHale, L.; et al. Clinicopathological Evaluation of Chronic Traumatic Encephalopathy in Players of American Football. *JAMA* **2017**, *318*, 360–370. [CrossRef] [PubMed]
6. McKee, A.C.; Stern, R.A.; Nowinski, C.J.; Stein, T.D.; Alvarez, V.E.; Daneshvar, D.H.; Lee, H.S.; Wojtowicz, S.M.; Hall, G.; Baugh, C.M.; et al. The spectrum of disease in chronic traumatic encephalopathy. *Brain* **2013**, *136*, 43–64. [CrossRef] [PubMed]
7. Katsumoto, A.; Takeuchi, H.; Tanaka, F. Tau Pathology in Chronic Traumatic Encephalopathy and Alzheimer's Disease: Similarities and Differences. *Front. Neurol.* **2019**, *10*, 980. [CrossRef]
8. Albayram, O.; Kondo, A.; Mannix, R.; Smith, C.; Tsai, C.Y.; Li, C.; Herbert, M.K.; Qiu, J.; Monuteaux, M.; Driver, J.; et al. Cis P-tau is induced in clinical and preclinical brain injury and contributes to post-injury sequelae. *Nat. Commun.* **2017**, *8*, 1000. [CrossRef]

9. Qiu, C.; Albayram, O.; Kondo, A.; Wang, B.; Kim, N.; Arai, K.; Tsai, C.Y.; Bassal, M.A.; Herbert, M.K.; Washida, K.; et al. Cis P-tau underlies vascular contribution to cognitive impairment and dementia and can be effectively targeted by immunotherapy in mice. *Sci. Transl. Med.* **2021**, *13*, eaaz7615. [CrossRef]
10. Foster, K.; Manca, M.; McClure, K.; Koivula, P.; Trojanowski, J.Q.; Havas, D.; Chancellor, S.; Goldstein, L.; Brunden, K.R.; Kraus, A.; et al. Preclinical characterization and IND-enabling safety studies for PNT001, an antibody that recognizes cis-pT231 tau. *Alzheimers Dement* **2023**, *in press*. [CrossRef]
11. Margulies, S.S.; Kilbaugh, T.; Sullivan, S.; Smith, C.; Propert, K.; Byro, M.; Saliga, K.; Costine, B.A.; Duhaime, A.C. Establishing a Clinically Relevant Large Animal Model Platform for TBI Therapy Development: Using Cyclosporin A as a Case Study. *Brain Pathol.* **2015**, *25*, 289–303. [CrossRef]
12. Smith, D.H.; Johnson, V.E.; Stewart, W. Chronic neuropathologies of single and repetitive TBI: Substrates of dementia? *Nat. Rev. Neurol.* **2013**, *9*, 211–221. [CrossRef] [PubMed]
13. DeKosky, S.T.; Blennow, K.; Ikonomovic, M.D.; Gandy, S. Acute and chronic traumatic encephalopathies: Pathogenesis and biomarkers. *Nat. Rev. Neurol.* **2013**, *9*, 192–200. [CrossRef] [PubMed]
14. Omalu, B.I.; DeKosky, S.T.; Minster, R.L.; Kamboh, M.I.; Hamilton, R.L.; Wecht, C.H. Chronic traumatic encephalopathy in a National Football League player. *Neurosurgery* **2005**, *57*, 128–134. [CrossRef]
15. Goldstein, L.E.; Fisher, A.M.; Tagge, C.A.; Zhang, X.L.; Velisek, L.; Sullivan, J.A.; Upreti, C.; Kracht, J.M.; Ericsson, M.; Wojnarowicz, M.W.; et al. Chronic traumatic encephalopathy in blast-exposed military veterans and a blast neurotrauma mouse model. *Sci. Transl. Med.* **2012**, *4*, 134ra160. [CrossRef]
16. Shin, S.; Hefti, M.M.; Mazandi, V.M.; Issadore, D.A.; Meaney, D.; Christman Schneider, A.; Diaz-Arrastia, R.; Kilbaugh, T.J. Plasma Neurofilament Light and Glial Fibrillary Acidic Protein Levels over thirty days in a Porcine Model of Traumatic Brain Injury. *J. Neurotrauma* **2022**, *39*, 935–943. [CrossRef] [PubMed]
17. Graham, N.S.N.; Zimmerman, K.A.; Moro, F.; Heslegrave, A.; Maillard, S.A.; Bernini, A.; Miroz, J.P.; Donat, C.K.; Lopez, M.Y.; Bourke, N.; et al. Axonal marker neurofilament light predicts long-term outcomes and progressive neurodegeneration after traumatic brain injury. *Sci. Transl. Med.* **2021**, *13*, eabg9922. [CrossRef]
18. Shahim, P.; Zetterberg, H.; Tegner, Y.; Blennow, K. Serum neurofilament light as a biomarker for mild traumatic brain injury in contact sports. *Neurology* **2017**, *88*, 1788–1794. [CrossRef]
19. Shahim, P.; Tegner, Y.; Marklund, N.; Blennow, K.; Zetterberg, H. Neurofilament light and tau as blood biomarkers for sports-related concussion. *Neurology* **2018**, *90*, e1780–e1788. [CrossRef]
20. Ortiz-Rodriguez, A.; Arevalo, M.A. The Contribution of Astrocyte Autophagy to Systemic Metabolism. *Int. J. Mol. Sci.* **2020**, *21*, 2479. [CrossRef]
21. Middeldorp, J.; Hol, E.M. GFAP in health and disease. *Prog. Neurobiol.* **2011**, *93*, 421–443. [CrossRef]
22. Villapol, S.; Byrnes, K.R.; Symes, A.J. Temporal dynamics of cerebral blood flow, cortical damage, apoptosis, astrocyte-vasculature interaction and astrogliosis in the pericontusional region after traumatic brain injury. *Front. Neurol.* **2014**, *5*, 82. [CrossRef]
23. Castellani, R.J.; Perry, G. Tau Biology, Tauopathy, Traumatic Brain Injury, and Diagnostic Challenges. *J. Alzheimers Dis.* **2019**, *67*, 447–467. [CrossRef] [PubMed]
24. Shafiei, S.S.; Guerrero-Munoz, M.J.; Castillo-Carranza, D.L. Tau Oligomers: Cytotoxicity, Propagation, and Mitochondrial Damage. *Front. Aging Neurosci.* **2017**, *9*, 83. [CrossRef]
25. Yoshiyama, Y.; Uryu, K.; Higuchi, M.; Longhi, L.; Hoover, R.; Fujimoto, S.; McIntosh, T.; Lee, V.M.; Trojanowski, J.Q. Enhanced neurofibrillary tangle formation, cerebral atrophy, and cognitive deficits induced by repetitive mild brain injury in a transgenic tauopathy mouse model. *J. Neurotrauma* **2005**, *22*, 1134–1141. [CrossRef]
26. Bittar, A.; Bhatt, N.; Hasan, T.F.; Montalbano, M.; Puangmalai, N.; McAllen, S.; Ellsworth, A.; Carretero Murillo, M.; Taglialatela, G.; Lucke-Wold, B.; et al. Neurotoxic tau oligomers after single versus repetitive mild traumatic brain injury. *Brain Commun.* **2019**, *1*, fcz004. [CrossRef]
27. Ihara, Y. PHF and PHF-like fibrils--cause or consequence? *Neurobiol. Aging* **2001**, *22*, 123–126. [CrossRef]
28. Ren, Z.; Iliff, J.J.; Yang, L.; Yang, J.; Chen, X.; Chen, M.J.; Giese, R.N.; Wang, B.; Shi, X.; Nedergaard, M. 'Hit & Run' model of closed-skull traumatic brain injury (TBI) reveals complex patterns of post-traumatic AQP4 dysregulation. *J. Cereb. Blood Flow Metab.* **2013**, *33*, 834–845. [CrossRef] [PubMed]
29. Iliff, J.J.; Chen, M.J.; Plog, B.A.; Zeppenfeld, D.M.; Soltero, M.; Yang, L.; Singh, I.; Deane, R.; Nedergaard, M. Impairment of glymphatic pathway function promotes tau pathology after traumatic brain injury. *J. Neurosci.* **2014**, *34*, 16180–16193. [CrossRef] [PubMed]

Disclaimer/Publisher's Note: The statements, opinions and data contained in all publications are solely those of the individual author(s) and contributor(s) and not of MDPI and/or the editor(s). MDPI and/or the editor(s) disclaim responsibility for any injury to people or property resulting from any ideas, methods, instructions or products referred to in the content.

Article

The Central Fluid Percussion Brain Injury in a Gyrencephalic Pig Brain: Scalable Diffuse Injury and Tissue Viability for Glial Cell Immunolabeling following Long-Term Refrigerated Storage

Mark Pavlichenko [1] and Audrey D. Lafrenaye [1,2,*]

[1] Department of Anatomy and Neurobiology, Virginia Commonwealth University, Richmond, VA 23298-0709, USA
[2] Richmond Veterans Affairs Medical Center, Richmond, VA 23249-4915, USA
* Correspondence: audrey.lafrenaye@vcuhealth.org

Abstract: Traumatic brain injury (TBI) affects millions of people annually; however, our knowledge of the diffuse pathologies associated with TBI is limited. As diffuse pathologies, including axonal injury and neuroinflammatory changes, are difficult to visualize in the clinical population, animal models are used. In the current study, we used the central fluid percussion injury (CFPI) model in a micro pig to study the potential scalability of these diffuse pathologies in a gyrencephalic brain of a species with inflammatory systems very similar to humans. We found that both axonal injury and microglia activation within the thalamus and corpus callosum are positively correlated with the weight-normalized pressure pulse, while subtle changes in blood gas and mean arterial blood pressure are not. We also found that the majority of tissue generated up to 10 years previously is viable for immunofluorescent labeling after long-term refrigeration storage. This study indicates that a micro pig CFPI model could allow for specific investigations of various degrees of diffuse pathological burdens following TBI.

Keywords: traumatic brain injury; axonal injury; micro pig; diffuse pathology; microglia; aged tissue

Citation: Pavlichenko, M.; Lafrenaye, A.D. The Central Fluid Percussion Brain Injury in a Gyrencephalic Pig Brain: Scalable Diffuse Injury and Tissue Viability for Glial Cell Immunolabeling following Long-Term Refrigerated Storage. *Biomedicines* **2023**, *11*, 1682. https://doi.org/10.3390/biomedicines11061682

Academic Editors: John O'Donnell and Dmitriy Petrov

Received: 16 May 2023
Revised: 8 June 2023
Accepted: 9 June 2023
Published: 10 June 2023

Copyright: © 2023 by the authors. Licensee MDPI, Basel, Switzerland. This article is an open access article distributed under the terms and conditions of the Creative Commons Attribution (CC BY) license (https://creativecommons.org/licenses/by/4.0/).

1. Introduction

Traumatic brain injury (TBI) is a major health care concern that carries significant personal and societal burdens [1–4]. There are approximately 2 million reported cases of TBI annually in the United States, with global cases reaching 50 million each year, with the understanding that the real numbers are likely higher due to the vast majority of TBIs being mild and going unreported [4–7]. Although our knowledge of the complex pathologies associated with TBI has progressed, understanding of diffuse pathologies, which are associated with morbidity in the human population [8–10], is still limited.

Diffuse traumatic axonal injury (TAI) is one of the pathological hallmarks of TBI, and is commonly used in experimental mild TBI studies to convey the degree of injury [11–18]. Neuroinflammation has become another key pathological feature of brain injury in both human and animal studies [8,19–21]. Many recent studies have demonstrated the impact of inflammatory cascades in regulating behavioral morbidities, general pathology, and neuronal function in both the normal brain and in various disease states, including TBI [22–25]. Studies have demonstrated neuroinflammation in the human population following TBI in various brain regions, with consistent involvement of the thalamic domain [8,19–21]. Microglia, the innate immune cells of the brain, are critical mediators of these TBI-induced neuroinflammatory processes [26–33]. Astrocytes are key regulators of various processes, including maintenance of the blood–brain barrier, water movement, synaptic activity regulation, glucose storage, and neuroinflammatory processes [30,34–36].

Serum levels of glial fibrillary acidic protein (GFAP), a component of the astrocyte cytoskeleton, are also correlated with negative outcomes following TBI and is one of two currently FDA-approved biomarkers for TBI [37–39].

We currently do not have the technology to specifically investigate TBI-induced diffuse pathological progression at an individual cellular level in the living human brain; therefore, animal models are used to tease out potential targets for further study. Many therapeutics have shown promising results in rodent models of TBI, unfortunately, this efficacy has been limited when translated to humans [40–42]. The high reliance on lower-order species for preclinical TBI research has been implicated in this translational failure, in that some processes and mechanisms found in rodents do not occur in humans, and vice versa [41,43]. Therefore, there have recently been calls for more models using higher order mammals with inflammatory responses and brain cytoarchitecture more similar to humans, such as pigs, to better evaluate potential therapeutics prior to human translation [42,44–49].

With this goal in mind, we began characterization of a central fluid percussion injury (CFPI) model of mild diffuse TBI in the adult micro pig [50–53]. We found that CFPI resulted in diffusely dispersed injured axons and activated microglia in various brain regions without producing focal lesions or gross tissue damage [51]. We further found that this model recapitulated diffuse pathological and serum biomarker profiles of the human population [50–52]. As human injury occurs on a spectrum, the current investigation aimed to evaluate the potential for graded levels of injury-induced pathology in the micro pig model of CFPI. Additionally, due to the expense of generating large animal TBI models, we sought to determine the duration that our previously generated micro pig brain tissue would maintain viability for immunolabeling.

2. Materials and Methods

2.1. Animals

Data from animals used in this investigation have been published previously [51,52]. Tissue from one recently generated animal was used as freshly harvested control tissue for our immunolabeling studies. We used tissue from a total of twenty-two ~6-month-old adult male Yucatan micro pigs ($n = 3$ sham; $n = 18$ injured, $n = 1$ new tissue), weighing 15–25 kg at time of injury. Prior to injury induction, animals were housed in pairs on a 12 h light–dark cycle in pens with free access to food and water. Pens had epoxy flooring with a drain in the back of the pen and a mixture of solid cinderblock wall and fencing. Following injury, pigs were single housed in a separate pen within the same room. All experimental protocols were approved by the Virginia Commonwealth University and Richmond Veterans Affairs Medical Center Institutional Animal Care and Use Committees, AAALAC international-accredited and USDA-registered organizations. The Virginia Commonwealth University and Richmond Veterans Affairs Medical Center Institutional Animal Care and Use Committees adhere to all regulations outlined in the "Guide for the Care and Use of Laboratory Animals: 8th Edition" (National Research Council [54]).

2.2. Surgical Preparation and Injury Induction

As previously published [51], micro pigs were initially sedated with an intramuscular injection in the rear flank of 100 mg/mL Xylazine (2.2 mg/kg; AnaSed Injection, Shenondoah, IA, USA), 100 mg/mL Telazol (2.0 mg/kg; Tiletamine HCL and Zolazepam HCL; Pfizer, New York, NY, USA), and 0.01 mg/kg Glycopyrrolate. After initial sedatives took full effect (5–10 min post injection), sodium pentobarbital (60 mg/kg; Sigma-Aldrich, St. Louis, MO, USA) was administered intravenously through a superficial ear vein. Loss of the palpebral reflex, in which the canthus of the palpebral fissure of the eye was touched gently to elicit a reflexive eye twitch/closing, and jaw tone were used to verify the plane of anesthesia. Once 2–3 gentle touches to the canthus of the palpebral fissure did not elicit a response, and the jaw was easily manipulated, the micro pig was intubated with a size 6–7 endotracheal tube and ventilated with 1–2% isoflurane mixed in 100% oxygen throughout the experiment. Ophthalmic lubricant was applied to avoid damage or drying of the eye during surgery or recovery. Body temperature was monitored with a rectal thermometer and maintained at 37 °C by adjusting the level of the self-heating surgical table. Chux underpads and/or blankets were placed between the stainless-steel surface of the surgical table and the pig to reduce chances of burns. The head and inner thigh were shaved with electric clippers and sanitized with betadine and 70% ethanol wipes. Heparin-coated catheters were inserted within the right femoral artery and vein and sutured into place for continuous monitoring of mean arterial blood pressure (MABP; via the arterial line), assessment of blood gases via the arterial line, and replacement of fluids with Lactated Ringer's (Hospira, Lake Forest, IL, USA) via the venous line to maintain hydration. Following the canula placement, the animal was placed on its sternum and the animal's head was shaved, sterilized with betadine and 70% ethanol wipes, and draped. A midline incision was made from the supraorbital process to the nuchal crest and a 16 mm diameter circular craniotomy was trephined along the sagittal suture with a manual surgical trephine (Medicon 57-63-16), positioning the center of the craniotomy 14 mm anterior to lambda (on the nuchal crest) and leaving the Dura intact (Figure 1A). A custom stainless-steel threaded hub (Custom Design and Fabrication, Richmond, VA, USA) with an outer diameter of 17 mm and an inner diameter of 14 mm (Figure 1B) was screwed into the craniotomy site to a depth of ~4 mm. Dental acrylic (methyl-methacrylate; Hygenic Corp., Akron, OH) was applied around the hub to insure hub stability. The hub was filled with sterile saline and wet gauze was placed over the dental acrylic to serve as a heat sink for the exothermic reaction that occurs during hardening.

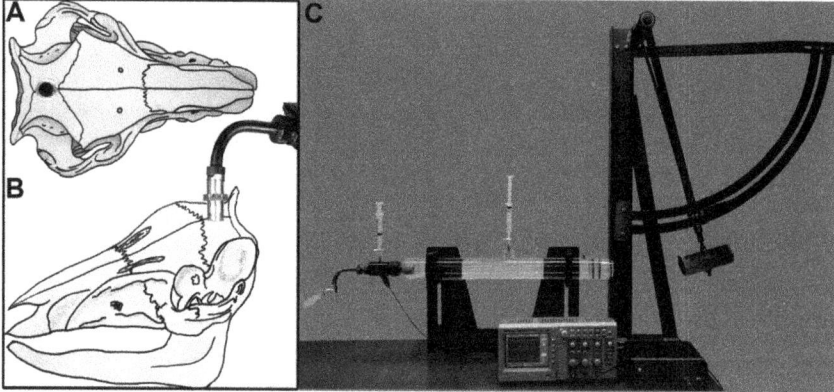

Figure 1. Adult male micro pigs were subjected to either sham injury or a mild central fluid percussion injury (CFPI). Representative drawings of (**A**) dorsal and (**B**) lateral view of the pig skull with the injury hub screwed into the skull along the sagittal suture. (**C**) Picture of the fluid percussion device used to induce CFPI in the micro pig. The injury hub was screwed into a 16 mm craniectomy placed 14 mm anterior to the nuchal crest. The hub was filled with sterile saline then connected to the fluid percussion device's L-shaped adaptor. Following connection to the device, the hammer was released, producing a fluid pressure pulse that transduced through the dura to the cerebrospinal fluid.

The induction of the central fluid percussion injury (cFPI) was carried out as described previously [50,51]. Following verification that the dental acrylic was fully hardened, anesthetized micro pigs were connected to the cFPI device that had been fitted with an L-shaped stainless steel adaptor onto which the injury hub locked (Figure 1B,C). This adaptor allowed for the redirection of the fluid pulse downward through the injury hub on the ventral surface of the pig's skull. Following connection and verification that bubbles hadn't been introduced to the closed fluid-filled system, micro pigs were injured with a fluid pressure pulse at a magnitude of 1.39–1.83 atmospheres and duration of 24–32 msec (Supplementary Table S1) measured by a transducer affixed to the injury device and displayed on an oscilloscope (Tektronix, Beaverton, OR, USA). Immediately after injury induction, animals were disconnected from the injury device, the set screws were unscrewed from the skull, and the dental acrylic, hub, and screws were removed as a unit. The hub was then removed from the surrounding dental acrylic for sterilization and reuse. While there was some sub-arachnoid bleeding following injury, the cFPI did not result in any breach of the Dura mater or induce subdural hematoma formation. Gel foam was placed over the craniotomy/injury site to alleviate small amounts of bone bleeding, and the scalp was sutured without replacing the bone to the craniotomy site.

2.3. Physiology

Systemic physiology was monitored and recorded throughout the duration of anesthesia, both prior to cFPI and throughout the post-injury monitoring period. Mean arterial blood pressure was measured through the canulated femoral artery, hemoglobin oxygen saturation was monitored via a pulse oximeter, and a rectal thermometer was used to measure body temperature. All physiological readings were measured and visualized by a Cardell® MAX-12HD (Sharn veterinary, Inc., Chicago, IL, USA) system. The partial pressures of oxygen and carbon dioxide in arterial blood, PaO_2, and $PaCO_2$, respectively, hematocrit (Hct), bicarbonate (HCO_3), hemoglobin (Hb), and pH values were assessed on blood draws from the femoral arterial cannula using a Stat Profile pHOx blood gas machine (NOVA Biomedical, Waltham, MA, USA). The summarized physiological results for these animals have previously been published [51]. However, this manuscript investigates individual animals' pre- and post-injury physiological readouts (Supplemental Table S1).

2.4. Tissue Processing

At terminal endpoints of 6 h post-cFPI, micro pigs were overdosed with 3 mL euthasol euthanasia-III solution (Henry Schein, Dublin, OH, USA) and immediately transcardially perfused with 6 L of 0.9% saline followed by 8 L of 4% paraformaldehyde/0.2% gluteraldehyde in Millonig's buffer (136 mM sodium phosphate monobasic/109 mM sodium hydroxide), as previously published [51]. After transcardial perfusion, the brains were removed from the skull and post-fixed in 4% paraformaldehyde/0.2% gluteraldehyde/Millonig's fixation buffer for an additional 48–72 h at 4 °C. Postfixed brains were blocked into 5 mm coronal segments throughout the rostral–caudal extent using a small pig brain slicer matrix (Cat. # PBMPBS050-1 Zivic Instruments, Pittsburgh, PA, USA). Segments containing the thalamus or corpus callosum were bisected at the midline, and the left side was further processed. All segmented brain tissue not used in active studies was cataloged, cryoprotected in 30% sucrose, and frozen in tissue tek Optimal Cutting Temperature (O.C.T.) compound (Tissue-Tek #4583; Sakura Finetek; Torrance, CA, USA) prior to freezing and storage at −80 °C for future use. The 5 mm coronal segments containing the left hemi-thalamus or corpus callosum were never frozen, rather, they were embedded in agarose and coronally sectioned in 0.1 M phosphate buffer with a vibratome (Leica, Banockburn, IL, USA) at a thickness of 40 μm. Sections were collected serially in 6-well plates (240 μm between sections in each well) and stored in Millonig's buffer at 4 °C. Thalamic tissue treated in this manner was stored in the refrigerator for multiple years prior to investigations of viability for immunobiological analysis. Buffer was changed and/or added every few years to avoid dehydration.

2.5. Immunohistochemistry

2.5.1. Previous Labeling of Thalamic and Corpus Callosal Tissue for Axonal Injury and Microglial Activity Index Assessments

All immunohistological labeling and analyses used to determine injury model scalability was carried out on freshly generated tissue during the studies published previously [51]. Mean thalamic data for injured and sham groups were previously published in Lafrenaye et al., 2015 [51]; however, corpus callosum data were not reported. For immunohistological labeling, a random well (1–6) was selected using a random number generator and six pieces of tissue containing the thalamic or twelve sections containing the corpus callosum were taken for immunolabeling. Tissue from all animals was processed concomitantly to reduce variability between animals.

To visualize injured axons, fluorescent immunohistochemistry against amyloid precursor protein (APP) was used to detect axonal transport issues indicative of axonal injury [13,14,55]. In this procedure, sections were blocked and permeabilized at room temperature in 5% normal goat serum (NGS), 2% bovine serum albumin (BSA), and 1.5% triton in phosphate-buffered saline for 2 h followed by overnight incubation with a rabbit antibody against the C-terminus of β-APP (1:700; Cat.# 51-2700, Life Technologies, Carlsbad, CA, USA) at 4 °C in 5% NGS/2% BSA. Tissue was washed with 1%NGS/1%BSA in PBS at least six times prior to secondary antibody incubation with Alexa Fluor 568-conjugated goat anti-rabbit IgG (1:700; Cat.# A-11011, Life Technologies, Carlsbad, CA, USA) in 1%NGS/1%BSA/PBS at room temperature for 1 h. Tissue was washed in 1%NGS/1%BSA in PBS at least six times. Tissue was mounted on slides using Vectashield hardset mounting medium with Dapi (Cat.#H-1500; Vector Laboratories, Burlingame, CA, USA).

For the visualization of microglia, chromatic immunohistochemistry against ionized calcium-binding adaptor molecule 1 (Iba-1) was carried out. Endogenous peroxidases were quenched with hydrogen peroxide prior to chromatic immunolabeling. Tissue was blocked and permeabilized in 1.5% triton/10% NGS/PBS for 2 h followed by incubation with a rabbit antibody against Iba-1 (1:1000; Cat.#019-19741 Wako; Richmond, VA, USA) in 10% NGS/PBS at 4 °C. Tissue was washed with 1%NGS/1%BSA in PBS at least six times prior to secondary antibody incubation with biotinylated goat anti-rabbit IgG (1:1000; Cat.# BA-1000, Vector Laboratories, Burlingame, CA, USA). To enhance the chromatic signal, tissue was incubated in avidin biotinylated enzyme complex using the Vectastain ABC kit (Vector Laboratories, Burlingame, CA, USA). The substrate used for the chromatic reaction to visualize microglia was 0.05% diaminobenzidene/0.01% H_2O_2/0.3% imidazole/PBS. Following cessation of the reaction, tissue was mounted, dehydrated in a series of alcohols, and cover-slipped with Permount mounting media (Fisher Scientific, Cat#SP15).

2.5.2. Labeling of Stored Thalamic Tissue for Immunohistochemistry Efficacy Assessments

Archived thalamic tissue from 19 different animals (generated between 25 July 2013 and 21 July 2015) that had been perfusion-fixed with 4% paraformaldehyde/0.2% glutaraldehyde followed by at least 3 days of post-fixation and stored at 4 °C were used for immunohistological labeling. Data from these animals have been published previously [51]. Fixed tissue used in this study was sectioned in 0.1 M dibasic sodium phosphate buffer within 3 months of being generated and stored long term at 4 °C in Millonigs buffer (136 mmol/L sodium phosphate monobasic/109 mmol/L sodium hydroxide). Glial fibrillary acidic protein (GFAP) was used to label astrocytic soma and processes, and Iba-1 was used to label microglial soma and processes. Tissue was sequentially co-labeled for both markers, with the majority of samples (n = 13) being initially labeled for GFAP followed by Iba-1, and the remaining samples (n = 6) being labeled for Iba-1 first followed by GFAP labeling. For this, tissue was washed in PBS 3 times for 5 min, blocked, and permeabilized for 1 h at room temperature in 5% NGS/2% BSA/1.5% Triton/PBS, and incubated overnight at 4 °C with either mouse anti-GFAP primary antibody (1:1000; Cat #MAB360; Millipore) or rabbit anti-Iba-1 primary antibody (1:1000; Waco). Tissue was then washed 6 times for 5 min in 1% NGS/1% BSA/PBS prior to 1 h incubation at room temperature with either goat anti-mouse Alexa 568 (1:500, Cat#;) for GFAP visualization, or goat anti-Rabbit Alexa 488 (1:500, Cat#;) for Iba-1 visualization. Tissue was washed again 4 times for 5 min in PBS. The single-labeled tissue was then blocked again for 1 h at room temperature in 5% NGS/2% BSA/PBS and immunolabeled with the other primary antibody, either rabbit anti-Iba-1 or mouse anti-GFAP, followed by the secondary antibody, either goat anti-rabbit Alexa 488 for Iba-1 visualization or goat anti-mouse Alexa 568 for GFAP visualization, as above. Tissue was mounted with Vectashield Hardset mounting medium with DAPI prior to imaging.

2.6. Image Analysis

The imaging of APP and microglia used for the correlation matrices was carried out prior to 2015 using a Nikon Eclipse 800 microscope (Nikon, Tokyo, Japan) equipped with an Olympus DP71 camera (Olympus, Center Valley, PA, USA). Image acquisition settings were held consistent for all animals and the regions of interest were restricted to the thalamus and corpus callosum using anatomical landmarks. Imaging was carried out by an investigator blinded to animal identity.

2.6.1. Quantitative Image Analysis of Diffuse Axonal Injury

For the APP assessments in the thalamus, a total of 60 images (10 images in each of the 6 sections assessed) were taken at $10\times$ magnification (0.72 mm^2 field) in a systematically random fashion starting at the dorsal lateral aspect of the thalamus. For the corpus callosum, a total of 24 images (2 images in each of 12 sections) were taken at $10\times$ magnification, which covered the majority of the region of interest. A nuclear DAPI label was used for field advancement and to verify focus as well as restriction within the region of interest. Thresholded masks of APP intensity were generated in ImageJ software (NIH, Bethesda, MD). The particle analysis tool in ImageJ was then used to assess the number of APP$^+$ axonal swellings in each image (particle analysis settings circularity = 0.2–1, size = 20-infinity). The number of APP$^+$ swellings per unit area was quantified for each image and averaged for each animal.

2.6.2. Quantitative Image Analysis of Microglial Activity Index

Assessment of microglia activation for the thalamus was published previously [51]. For this the entire thalamus or corpus callosum was assessed for each of the sections selected for each animal (6 sections for the thalamus and 12 sections for the corpus callosum). Identification of microglia activation was based on specific morphological criteria. Microglia with highly ramified fine process networks that were lightly labeled with Iba-1 were considered non-reactive, while microglia with thicker, shorter, or absent processes and darker Iba-1 labeling were identified as active/reactive [55–58]. The degree of microglia activation was assessed using an index from 0 to 5 in which 0 = no microglial activation observed, 1 = ramified microglia with thicker processes and darker Iba-1 labeling observed in ~5% of the region of interest, 2 = activated microglia observed in ~5–10% of the region of interest, 3 = activated microglia observed in ~10 < 25% of the region of interest, 4 = activated microglia observed in ~25 < 50% of the region of interest and 5 = activated microglia observed in >50% of the region of interest. Two blinded investigators analyzed all sections independently, and their scores were averaged for each animal.

2.6.3. Qualitative Analysis of Glial Labeling in Old Pig Tissue

Immunofluorescent labeling of microglia and astrocytes in pig tissue generated >7 years prior and stored at 4 °C was qualitatively assessed by investigators blind to the age of tissue. Images were taken with a Keyence BZ-X800 microscope (Keyence Corporation of America, Itasca, IL, USA) at $40\times$ magnification with all settings held constant for each sample. Experimenters randomly selected 6 areas of each section within the thalamic domain to evaluate the degree of labeling of Iba-1+ microglia and GFAP+ astrocytes. A high-degree label was associated with widely distributed and clearly visible cell soma and processes, a medium-degree label was associated with sporadically distributed and visible cell soma and/or processes, and a low/no-degree label was associated with no visible cell soma or processes. Age of tissue was assessed for each labeling category (high, medium, and low/no) to gain insight regarding the potential viability of tissue for immunohistochemistry over time.

2.7. Statistical Analysis

Two-tailed unpaired T-tests with equal variance not assumed were performed to assess corpus callosal axonal injury and microglial activation index between sham and cFPI animals. Individual animal correlation data were analyzed with a Spearman's Rho Correlation analysis due to its lack of reliance on the assumption of data normality. Kruskal–Wallis tests with a Dunn's post hoc were carried out for assessment of the age of tissue for each labeling category. Data are presented as mean ± standard error of the mean (SEM) or median ± quartiles, as indicated in the figure legends. Statistical significance was set at a p value < 0.05.

3. Results

3.1. Central Fluid Percussion Injury Induces Significant Diffuse Axonal Injury in the Thalamus and the Corpus Callosum That Correlate with Microglial Activation

In our previous study [51], we reported TAI in various brain regions, including the thalamus and corpus callosum. We focused our previous assessments for that study on the thalamus, finding a significant increase in the degree of axonal injury within the thalamic domain ($t_{18} = 4.46$, $p = 1.4 \times 10^{-5}$) as well as a significant increase in the microglial activation index ($t_4 = 0.51$, $p = 0.004$) [51]. For this study, we extended our evaluation of the degree of axonal injury and microglia activation to the corpus callosum, another brain region sustaining consistent TAI at 6 h following a central fluid percussion injury (CFPI) in our micro pig model [51]. We found that the degree of TAI, indicated as amyloid precursor protein (APP) accumulations at the proximal axonal swelling, within the corpus callosum was significantly higher at 6 h following CFPI ($n = 18$, 50.89 ± 19.26) compared with sham ($n = 3$, 0.64 ± 0.37; $t_{17} = 1.44$, $p = 0.018$; Figure 2A). Microglial activation, signified as an increase in the microglial activation index, was also significantly increased in the corpus callosum at 6 h following CFPI in our micro pig model (sham = 1.17 ± 0.13 CFPI = 3.96 ± 0.15; $t_9 = 2.51$, $p = 1.57 \times 10^{-7}$; Figure 2B), suggesting that the corpus callosum also sustains significant TAI and microglial activation within 6 h of CFPI.

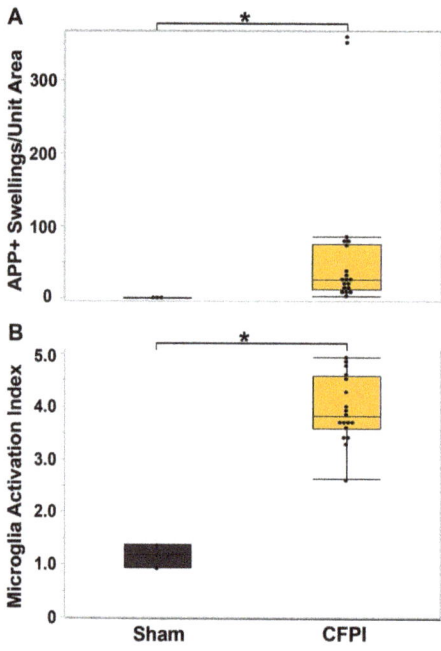

Figure 2. Central fluid percussion injury results in significant axonal injury and microglial activation

within the corpus callosum. Box plots of (**A**) APP+ Axonal swellings indicative of diffuse traumatic axonal injury and (**B**) microglial activation index at 6 h following sham (n = 3, grey boxes) or cFPI (n = 18, yellow boxes). Each black circle indicates an individual animal. Graphs depict the median ± quartile. * $p < 0.05$.

3.2. Central Fluid Percussion Injury Generates Scalable Diffuse Pathology That Does Not Significantly Impact Systemic Physiology

To investigate the potential of the micro pig cFPI model to scale the degree of pathology while maintaining a diffuse brain injury in which no cortical contusion was generated, we correlated the degree of TAI and microglial activation index within both the thalamus and the corpus callosum to the atmospheric pressure (ATM) induced in each individual animal generated in our 2015 study [51] (Figure 3). As the male micro pigs used in this study varied in terms of weight from 14.9 kg to 29.8 kg (Supplemental Table S1), we first normalized the ATM pressure to body weight. This normalized ATM pressure (ATM/kg) was significantly correlated with the raw ATM pressure transduced through the fluid percussion device (Rho = 0.96 $p < 0.0001$, Figure 3A), while allowing for specific weight-based refinement of the injury intensity.

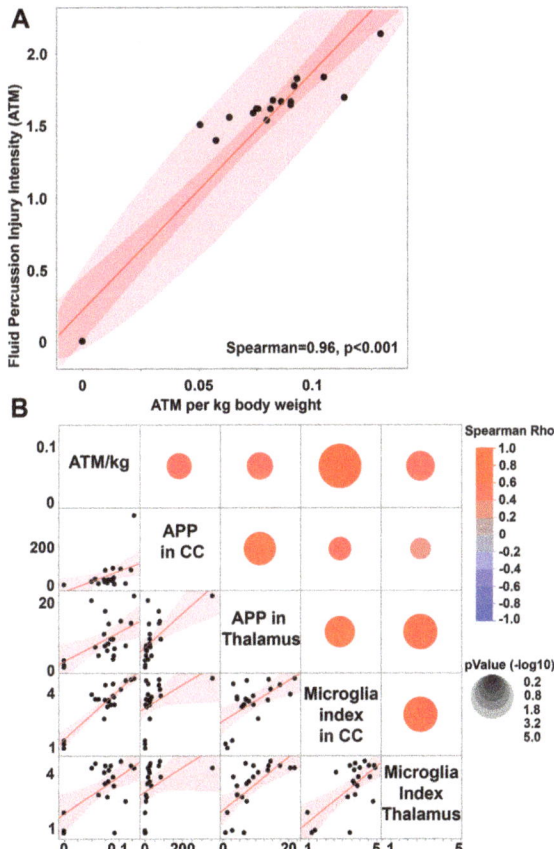

Figure 3. Weight-normalized central fluid percussion pressure pulse positively correlates with diffuse traumatic axonal injury and microglia activation in the thalamus and corpus callosum. (**A**) Scatter plot

of total fluid percussion ATM pressure and the ATM pressure normalized to kg weight. (**B**) Correlation matrix depicting the Spearman Rho correlation for the weight-normalized injury intensity (ATM/kg) with the degree of axonal injury in the corpus callosum (APP in CC) or thalamus (APP in Thalamus) as well as the degree of microglia activation in the corpus callosum (Microglia index in CC) or thalamus (Microglia index in Thalamus). Each pair of assessment values on the y and x axis is colored with red, indicating a positive correlation, or blue, indicating a negative correlation, with the correlation strength demonstrated by the circle's size, with larger circles representing lower p values. The lower part of the correlation matrix shows the correlation curves for each x and y comparison with each black dot representing data from an individual animal. Note that degree of axonal injury and microglial index values show positive correlations with ATM/kg, suggesting scalability between the pressure pulse inflicted and the degree of injury as well as the level of neuroinflammation in both regions.

The body-weight-normalized ATM pressure positively correlated with TAI within both the thalamus (Spearman Rho = 0.45 p = 0.04) and the corpus callosum (Spearman Rho = 0.60 p = 0.004; Figure 3B). The degree of axonal injury within the corpus callosum also significantly correlated with the degree of TAI found in the thalamus (Spearman Rho = 0.71, p = 0.0003; Figure 3B), indicating that both regions scale with the intensity of the fluid pressure pulse. As previously published, thalamic TAI significantly correlates with the microglia activation index within the thalamus (Spearman Rho = 0.76 p = <0.0001) [51], indicating an interplay between microglial activation and TAI in this region. Microglial activation within the corpus callosum also significantly and positively correlated with TAI within the region (Spearman Rho = 0.64 p = 0.0018), indicating that this association is not thalamus-specific.

The microglial activation index within the thalamus was found to be positively correlated with the microglial activation index within the corpus callosum (Spearman Rho = 0.54 p = 0.019; Figure 3B), suggesting that microglial activation might also scale with injury intensity in these two regions. While the microglial activation index within the thalamus was not correlated with the weight-normalized fluid pressure pulse (Spearman Rho = 0.29 p = 0.20), there was a significant positive correlation between ATM/kg and the microglial activation index within the corpus callosum (Spearman Rho = 0.68 p = 0.0008; Figure 3B).

Post-injury blood gases were also assessed to investigate the potential that the intensity of cFPI alters systemic physiology. As we previously published, neither the CFPI nor sham groups had blood gases that were outside of the normal range (Supplemental Table S1). None of the individual animal's blood measurements, including Oxygen saturation (O_2%), partial pressure of O_2 (PaO_2), partial pressure of CO_2 ($PaCO_2$), mean arterial blood pressure (MABP), blood pH, hematocrit (Hct), bicarbonate in the blood (HCO_3), or Hemoglobin (Hb), were correlated with weight-normalized pressure induction (Figure 4). This indicates that, while the diffuse pathology is correlated with the injury intensity within mild ranges, the overall systemic physiology reflected in the blood gases are not altered.

We did find significant positive correlations between diffuse pathology and subtle blood gas changes. The degree of TAI within the corpus callosum correlated with PaO_2 (Spearman Rho = 0.51 p = 0.018), HCO_3 (Spearman Rho = 0.47 p = 0.034), and blood pH (Spearman Rho = 0.46 p = 0.037). Thalamic TAI only correlated with blood pH (Spearman Rho = 0.44 p = 0.047), and thalamic microglia activation correlated with HCO_3 (Spearman Rho = 0.44 p = 0.045). However, the lack of correlation between injury intensity and blood gases suggests that blood gas readouts are not a good metric of injury intensity in this model.

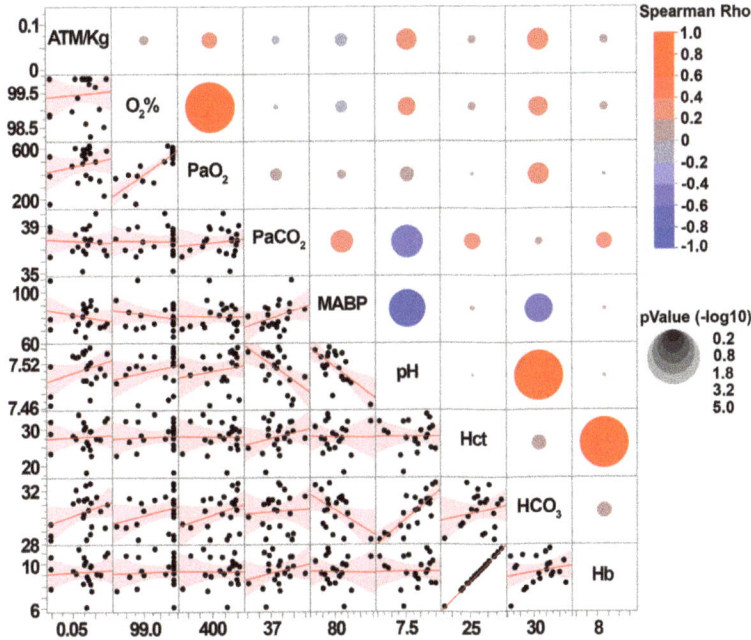

Figure 4. Blood gas readings do not correlate with cFPI injury intensity (ATM/kg) in micro pigs. Correlation matrices depicting the Spearman Rho correlation for the weight-normalized injury intensity (ATM/kg) with the various blood gas measurements and the mean arterial blood pressure (MABP) 6 h post-CFPI. Each pair of assessment values on the y and x axis is colored with red indicating a positive correlation, or blue indicating a negative correlation, with the correlation strength demonstrated by the circle's size. The lower part of the correlation matrix shows the correlation curves for each x and y comparison, with each black dot representing data from an individual animal.

3.3. Fixed Pig Tissue Maintains the Capacity for Immunohistochemical Labeling for Years after Refrigerated Storage

Fixed thalamic micropig tissue following long-term refrigerated storage was immunolabeled for astrocytes using GFAP and microglia, using Iba-1 to determine the tissue's viability for use in immunofluorescent microscopy. Tissue harvested within 2 months of labeling was used as a positive control (Figure 5A). The degree of labeling for aged tissue was distinguished by the visibility of cell soma and processes, as well as the density of labeling within the images that were assessed. Samples were categorized as having a high degree of labeling (densely labeled cell soma and processes), a medium degree of labeling (sporadically labeled cell soma or processes), or a low/no-degree of labeling (non-labeled cell soma and processes; Figure 5B–D). Most of the cases assessed had a high-degree of labeling for both GFAP (58%) and Iba-1 (47%; Figure 5E). Medium-degree labelling was more prevalent in Iba-1 labelling (32%) versus GFAP (21%; Figure 5E). Only 21% of tissue did not label for GFAP or Iba-1, indicating that this tissue maintains some degree of immunofluorescent labeling capacity.

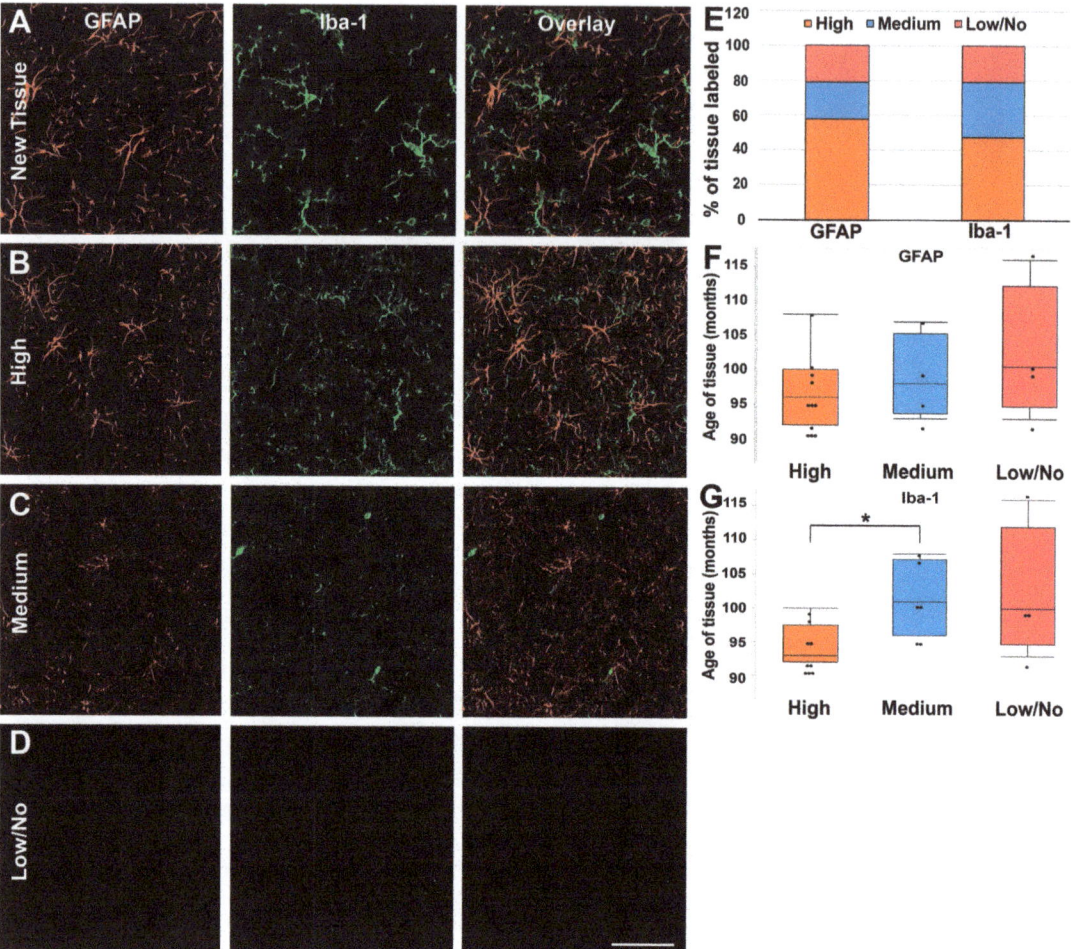

Figure 5. The majority of long-term stored labeled tissues exhibited a high degree of labeling of both GFAP and Iba-1. (**A–D**) Representative epifluorescent micrographs show astrocytes labeled with GFAP (first column; red) and microglia with Iba-1 (second column; green). The third column shows an overlay of GFAP and Iba-1. Each row shows representative images of (**A**) freshly harvested new tissue as a positive labeling control, and tissued stored for >7 years with (**B**) high-degree, (**C**) medium-degree, and (**D**) low/no-degree labeling of GFAP and IBA1. Scale bar: 50 µm. (**E**) Stacked bar graphs show the proportional number of tissues that were categorized into high-degree, medium-degree, and low/no-degree labeling for both GFAP and IBA1. (**F,G**) Box plots depicting the interquartile range and median age of tissue of each animal labeled with (**F**) GFAP or (**G**) IBA1 are categorized by degree of labeling. Note that the mean ages of GFAP-labeled tissue are less variable between labeling degrees, while the mean age of IBA1 tissue increases significantly from high-degree to medium-degree-labeled tissue. Graphs depict the median ± quartile. Black dots denote individual tissue samples. * $p < 0.05$.

As there was a difference in the efficacy of immunolabeling for GFAP and Iba-1, we further assessed the potential that the degree of immunolabeling was associated with the age of the tissue for each marker. The age of tissue did not significantly change when stratified by labeling degree for GFAP ($X^2(2) = 2.10$, $p = 0.35$; Figure 5F). In contrast, the age of the tissue was significantly different when stratified by the degree of labeling for Iba-1 ($X^2(2) = 7.55$, $p = 0.023$; Figure 5G). Specifically, the age of tissue with a high-degree of Iba-1 labeling was lower (94.8 ± 1.03 months) than tissue with a medium degree of Iba-1 labeling (101.5 ± 2.1 months; $p = 0.011$), and trended toward being lower than tissue with low/no Iba-1 labeling (102.3 ± 4.9; $p = 0.065$), suggesting that the age of the tissue impacts the labeling efficacy of Iba-1.

4. Discussion

Mild TBI is associated with diffuse pathological progressions, including, but not limited to, TAI and neuroinflammation, which are associated with negative outcomes and morbidities in the human population [59–62]. Although classified as mild clinically based on their Glasgow comma score, mild TBI injuries occur on a spectrum, which is still not fully understood, but involves changes in the degree of diffuse pathology. As diffuse axonal injury as a clinical diagnosis can only be confirmed using histological assessment in postmortem tissue [17,18,63], detailed investigations into the spectrum of mild injury requires animal modeling. The cytoarchitecture, inflammatory system, and metabolism of pigs are highly consistent with humans [44,47,64]; therefore, we utilized a pig model to investigate the spectrum of diffuse injury produced by a CFPI model of mild TBI.

In our initial study, we found significant axonal injury and microglia activation within the thalamic domain, which is consistent with findings in the human population [20,51,65–67]. We had noted TAI within the corpus callosum in that study as well; however, we had not previously quantitatively assessed the burden of TAI within the corpus callosum. In this study we found that our model of CFPI in the micro pig produced significant TAI and microglial activation within the corpus callosum (Figure 2). This is consistent with what is seen in the human population, with microscopic damage to the corpus callosum being one of the primary indicators of grade 1 diffuse axonal injury in the Adams diffuse axonal injury classification [68].

The burden of axonal injury within the corpus callosum and thalamus were significantly correlated with one another (Figure 3), demonstrating that in the pig CFPI model of mild TBI, both grey and white matter regions are impacted to a similar degree, allowing versatility for pathological assessments in this model. Axonal injury within both regions also demonstrated significant positive correlations to microglia activation following CFPI, indicating that microglia activation likely maps to areas of axonal injury. While we previously showed this correlation between TAI and microglia activation to be due to microglia process convergence onto injured axonal swellings in the thalamus in the pig following CFPI [50,51], further studies are needed to investigate the physical relationship between TAI and microglia activation in the corpus callosum.

Importantly, there was a significant correlation between the intensity of the fluid pulse and the degree of TAI within both the thalamus and corpus callosum, showing that this pig model of mild TBI allows assessment of subtle changes in both grey and white matter areas along the spectrum of mild TBI. This finding is consistent with clinical investigations in which the degree of diffuse injury has been associated with severity of injury and progressive morbidity following TBI in experimental and clinical studies [8,9,59].

While the degree of TAI in both the corpus callosum and thalamus significantly correlated with fluid pulse intensity, only corpus callosal microglia activation was significantly correlated with the pressure pulse (Figure 3). As the microglia activation indices within the thalamus and corpus callosum were found to be significantly correlated, the lack of significance of thalamic microglia activation with injury intensity is likely an artifact of the assessment strategy. The assessment within the corpus callosum involved twice as many sections for analysis for each animal since the corpus callosum occupies a much smaller area than the thalamus. The lower degree of sampling within the thalamus of each animal could have introduced a higher degree of inter-animal variability. This difference between the correlation of the corpus callosum and thalamic microglia activation index with injury intensity could also be due to the more homogenous fiber orientations found within the corpus callosum compared to the thalamus which has a larger overall area and more heterogeneous axonal orientations.

Interestingly, while the injury intensity was not significantly correlated with blood gas readouts and all blood gas readouts remained within normal ranges, diffuse pathology was found to be correlated with some blood gas readouts. Specifically, the degree of TAI within the corpus callosum significantly correlated with blood gas readouts, including paO2, pH, and HCO3, and the degree of thalamic TAI correlated with blood pH. However, this might not indicate a specific relationship between the brain pathology and subtle blood gas changes. Axonal injury and these subtle blood gas changes could both be symptoms of mild TBI that occur together but do not necessarily drive one another. Rather, the changes in blood gases could indicate subtle changes in systemic organs following a mild TBI. Studies have recently shown that brain injury can induce changes within systemic organs systems, including both the kidney and lung, both regions involved in regulating the HCO3, pH, and PaO2 of the blood. Kidney and lung pathology has been shown to occur in rodent models of mild TBI [69–73]. Retrospective clinical studies also found that either acute kidney injury [74] or acute lung injury [75] were associated with worse outcomes in TBI patients, indicating that systemic organ changes could be occurring in the human population as well. These are intriguing possibilities that require further investigation; however, due to the blood gas readouts in the current study all being within normal ranges, and the lack of correlation between any blood gas readout and the pressure pulse (Figure 4), it does not appear that CFPI substantially changes the blood gas readouts. Therefore, it is likely that whatever potential changes might be occurring systemically, they are not extreme enough to break homeostasis.

In terms of the ability to label tissue years following generation, most of the tissue stored in long-term (7.5–9.6 years) refrigeration expressed a high degree of immunolabeling of both astrocytic GFAP and microglial IBA1 markers. These data suggests that the efficacy of GFAP labeling is not altered as the age of tissue increases. Iba-1, on the other hand, demonstrates more medium and low/no degree of labeling in older tissues, suggesting that the specific antibody is less effective in older tissues. The Wako Iba-1 antibody has been used in various species to identify microglia with a high degree of efficacy [76–79]. The foundational studies using this Iba-1antibody all appeared to use transcardial perfusion with 4% paraformaldehyde followed by post-fixation [77–79]. Our prior published studies with the same tissue indicates that the Iba-1 antibody did not suffer decreased labeling in newly generated tissues as it did with tissues from long-term storage. This may suggest that the Wako IBA-1 antibody loses efficacy in older tissues.

5. Conclusions

Overall, the current study shows that a mild model of central fluid percussion injury can produce scalable diffuse injury within multiple brain regions without significantly altering blood gases. Additionally, we found that the majority of tissue fixed with 4% paraformaldehyde/0.2% glutaraldehyde and stored in Millonig's buffer under refrigeration conditions is usable for immunofluorescent analysis up to 8 years following generation.

Supplementary Materials: The following supporting information can be downloaded at: https://www.mdpi.com/article/10.3390/biomedicines11061682/s1, Table S1: Full data set.

Author Contributions: Conceptualization, A.D.L. methodology, A.D.L.; formal analysis, M.P. and A.D.L.; investigation, A.D.L. and M.P; writing—original draft preparation, M.P. and A.D.L. writing—review and editing, M.P. and A.D.L.; supervision, A.D.L.; funding acquisition, A.D.L. All authors have read and agreed to the published version of the manuscript.

Funding: This research was funded by the US Army grant W81XWH-10-1-0623 and NINDS grants R56NS128104, R21NS126611.

Institutional Review Board Statement: The study was conducted in accordance with the Declaration of Helsinki, and approved by the Institutional Review Board of Virginia Commonwealth University

(Protocol number: AM10169 Approval Date: 14 March 2011) and Richmond Veterans Affairs Medical Center (Protocol number: 1715187 Approval Date: 19 January 2023).

Informed Consent Statement: Not applicable.

Data Availability Statement: Full data for each animals can be found in Supplementary Table S1.

Acknowledgments: We would like to thank Martina Hernandez, Karen Gorse, and Amanda Logan-Wesley for technical assistance and scientific discussion.

Conflicts of Interest: The authors declare no conflict of interest.

References

1. James, S.L.; Theadom, A.; Ellenbogen, R.G.; Bannick, M.S.; Montjoy-Venning, W.; Lucchesi, L.R.; Abbasi, N.; Abdulkader, R.; Abraha, H.N.; Adsuar, J.C.; et al. Global, Regional, and National Burden of Traumatic Brain Injury and Spinal Cord Injury, 1990–2016: A Systematic Analysis for the Global Burden of Disease Study 2016. *Lancet Neurol.* **2019**, *18*, 56–87. [CrossRef] [PubMed]
2. Ortiz-Prado, E.; Mascialino, G.; Paz, C.; Rodriguez-Lorenzana, A.; Gómez-Barreno, L.; Simbaña-Rivera, K.; Diaz, A.M.; Coral-Almeida, M.; Espinosa, P.S. A Nationwide Study of Incidence and Mortality Due to Traumatic Brain Injury in Ecuador (2004-2016). *Neuroepidemiology* **2020**, *54*, 33–44. [CrossRef] [PubMed]
3. Brau, R.H.; Acevedo-Salas, Y.; Giovannetti, K. Epidemiological Trends of Traumatic Brain and Spinal Cord Injury in Puerto Rico from November 10th, 2006, through May 24th, 2011. *Puerto Rico Health Sci. J.* **2018**, *37*, 67–77.
4. Faul, M.; Coronado, V. Chapter 1—Epidemiology of Traumatic Brain Injury. In *Handbook of Clinical Neurology*; Grafman, J., Salazar, A.M., Eds.; Traumatic Brain Injury, Part I; Elsevier: Amsterdam, The Netherlands, 2015; Volume 127, pp. 3–13.
5. Maas, A.I.; Roozenbeek, B.; Manley, G.T. Clinical Trials in Traumatic Brain Injury: Past Experience and Current Developments. *Neurotherapeutics* **2010**, *7*, 115–126. [CrossRef] [PubMed]
6. Taylor, C.A.; Bell, J.M.; Breiding, M.J.; Xu, L. Traumatic Brain Injury–Related Emergency Department Visits, Hospitalizations, and Deaths—United States, 2007 and 2013. *MMWR Surveill. Summ.* **2017**, *66*, 1–16. [CrossRef]
7. Gardner, R.C.; Yaffe, K. Epidemiology of Mild Traumatic Brain Injury and Neurodegenerative Disease. *Mol. Cell. Neurosci.* **2015**, *66*, 75–80. [CrossRef]
8. Coughlin, J.M.; Wang, Y.; Munro, C.A.; Ma, S.; Yue, C.; Chen, S.; Airan, R.; Kim, P.K.; Adams, A.V.; Garcia, C.; et al. Neuroinflammation and Brain Atrophy in Former NFL Players: An in Vivo Multimodal Imaging Pilot Study. *Neurobiol. Dis.* **2015**, *74*, 58–65. [CrossRef]
9. Grossman, E.J.; Inglese, M. The Role of Thalamic Damage in Mild Traumatic Brain Injury. *J. Neurotrauma* **2016**, *33*, 163–167. [CrossRef]
10. Little, D.M.; Kraus, M.F.; Joseph, J.; Geary, E.K.; Susmaras, T.; Zhou, X.J.; Pliskin, N.; Gorelick, P.B. Thalamic Integrity Underlies Executive Dysfunction in Traumatic Brain Injury. *Neurology* **2010**, *74*, 558–564. [CrossRef]
11. Gennarelli, T.A.; Thibault, L.E.; Adams, J.H.; Graham, D.I.; Thompson, C.J.; Marcincin, R.P. Diffuse Axonal Injury and Traumatic Coma in the Primate. *Ann. Neurol.* **1982**, *12*, 564–574. [CrossRef]
12. Povlishock, J.T. Pathobiology of Traumatically Induced Axonal Injury in Animals and Man. *Ann. Emerg. Med.* **1993**, *22*, 980–986. [CrossRef] [PubMed]
13. Sherriff, F.E.; Bridges, L.R.; Sivaloganathan, S. Early Detection of Axonal Injury after Human Head Trauma Using Immunocytochemistry for Beta-Amyloid Precursor Protein. *Acta Neuropathol.* **1994**, *87*, 55–62. [CrossRef]
14. Gentleman, S.M.; Nash, M.J.; Sweeting, C.J.; Graham, D.I.; Roberts, G.W. β-Amyloid Precursor Protein (BAPP) as a Marker for Axonal Injury after Head Injury. *Neurosci. Lett.* **1993**, *160*, 139–144. [CrossRef] [PubMed]
15. Smith, D.H.; Meaney, D.F. Axonal Damage in Traumatic Brain Injury. *Neuroscientist* **2000**, *6*, 483–495. [CrossRef]
16. Scheid, R.; Walther, K.; Guthke, T.; Preul, C.; von Cramon, D.Y. Cognitive Sequelae of Diffuse Axonal Injury. *Arch. Neurol.* **2006**, *63*, 418–424. [CrossRef] [PubMed]
17. Chen, Q.; Chen, X.; Xu, L.; Zhang, R.; Li, Z.; Yue, X.; Qiao, D. Traumatic Axonal Injury: Neuropathological Features, Postmortem Diagnostic Methods, and Strategies. *Forensic Sci. Med. Pathol.* **2022**, *18*, 530–544. [CrossRef]
18. Mckee, A.C.; Daneshvar, D.H. The Neuropathology of Traumatic Brain Injury. In *Handbook of Clinical Neurology*; Elsevier: Amsterdam, The Netherlands, 2015; Volume 127, pp. 45–66.
19. Velázquez, A.; Ortega, M.; Rojas, S.; González-Oliván, F.J.; Rodríguez-Baeza, A. Widespread Microglial Activation in Patients Deceased from Traumatic Brain Injury. *Brain Inj.* **2015**, *29*, 1126–1133. [CrossRef]
20. Ramlackhansingh, A.F.; Brooks, D.J.; Greenwood, R.J.; Bose, S.K.; Turkheimer, F.E.; Kinnunen, K.M.; Gentleman, S.; Heckemann, R.a; Gunanayagam, K.; Gelosa, G.; et al. Inflammation after Trauma: Microglial Activation and Traumatic Brain Injury. *Ann. Neurol.* **2011**, *70*, 374–383. [CrossRef]
21. Zhou, Y.; Lui, Y.W.; Zuo, X.-N.; Milham, M.P.; Reaume, J.; Grossman, R.I.; Ge, Y. Characterization of Thalamocortical Association Using Amplitude and Connectivity of FMRI in Mild Traumatic Brain Injury. *J. Magn. Reson. Imaging* **2014**, *39*, 1558–1568. [CrossRef]

22. Morganti-Kossmann, M.C.; Satgunaseelan, L.; Bye, N.; Kossmann, T. Modulation of Immune Response by Head Injury. *Injury* **2007**, *38*, 1392–1400. [CrossRef]
23. Nizamutdinov, D.; Shapiro, L.A. Overview of Traumatic Brain Injury: An Immunological Context. *Brain Sci.* **2017**, *7*, 11. [CrossRef] [PubMed]
24. Kelley, B.J.; Lifshitz, J.; Povlishock, J.T. Neuroinflammatory Responses after Experimental Diffuse Traumatic Brain Injury. *J. Neuropathol. Exp. Neurol.* **2007**, *66*, 989–1001. [CrossRef] [PubMed]
25. Das, M.; Mohapatra, S.; Mohapatra, S.S. New Perspectives on Central and Peripheral Immune Responses to Acute Traumatic Brain Injury. *J. Neuroinflamm.* **2012**, *9*, 236. [CrossRef] [PubMed]
26. Mannix, R.C.; Whalen, M.J. Traumatic Brain Injury, Microglia, and Beta Amyloid. *Int. J. Alzheimer's Dis.* **2012**, *2012*, 608732. [CrossRef]
27. Neumann, H.; Kotter, M.R.; Franklin, R.J.M. Debris Clearance by Microglia: An Essential Link between Degeneration and Regeneration. *Brain J. Neurol.* **2009**, *132*, 288–295. [CrossRef]
28. Smith, C. Review: The Long-Term Consequences of Microglial Activation Following Acute Traumatic Brain Injury. *Neuropathol. Appl. Neurobiol.* **2013**, *39*, 35–44. [CrossRef]
29. Ransohoff, R.M.; Perry, V.H. Microglial Physiology: Unique Stimuli, Specialized Responses. *Annu. Rev. Immunol.* **2009**, *27*, 119–145. [CrossRef]
30. Karve, I.P.; Taylor, J.M.; Crack, P.J. The Contribution of Astrocytes and Microglia to Traumatic Brain Injury. *Br. J. Pharmacol.* **2016**, *173*, 692–702. [CrossRef]
31. Tay, T.L.; Hagemeyer, N.; Prinz, M. The Force Awakens: Insights into the Origin and Formation of Microglia. *Curr. Opin. Neurobiol.* **2016**, *39*, 30–37. [CrossRef]
32. Kigerl, K.a; Gensel, J.C.; Ankeny, D.P.; Alexander, J.K.; Donnelly, D.J.; Popovich, P.G. Identification of Two Distinct Macrophage Subsets with Divergent Effects Causing Either Neurotoxicity or Regeneration in the Injured Mouse Spinal Cord. *J. Neurosci. Off. J. Soc. Neurosci.* **2009**, *29*, 13435–13444. [CrossRef]
33. Loane, D.J.; Kumar, A. Microglia in the TBI Brain: The Good, the Bad, and the Dysregulated. *Exp. Neurol.* **2016**, *275*, 316–327. [CrossRef] [PubMed]
34. Chen, Y.; Swanson, R.A. Astrocytes and Brain Injury. *J. Cereb. Blood Flow Metab.* **2003**, *23*, 137–149. [CrossRef] [PubMed]
35. Lafrenaye, A.D.; Simard, J.M. Bursting at the Seams: Molecular Mechanisms Mediating Astrocyte Swelling. *Int. J. Mol. Sci.* **2019**, *20*, 330. [CrossRef] [PubMed]
36. Laird, M.D.; Vender, J.R.; Dhandapani, K.M. Opposing Roles for Reactive Astrocytes Following Traumatic Brain Injury. *NeuroSignals* **2008**, *16*, 154–164. [CrossRef] [PubMed]
37. Mondello, S.; Muller, U.; Jeromin, A.; Streeter, J.; Hayes, R.L.; Wang, K.K. Blood-Based Diagnostics of Traumatic Brain Injuries. *Expert Rev. Mol. Diagn.* **2011**, *11*, 65–78. [CrossRef]
38. Czeiter, E.; Mondello, S.; Kovacs, N.; Sandor, J.; Gabrielli, A.; Schmid, K.; Tortella, F.; Wang, K.K.W.; Hayes, R.L.; Barzo, P.; et al. Brain Injury Biomarkers May Improve the Predictive Power of the IMPACT Outcome Calculator. *J. Neurotrauma* **2012**, *29*, 1770–1778. [CrossRef]
39. Papa, L. Potential Blood-Based Biomarkers for Concussion. *Sport. Med. Arthrosc. Rev.* **2016**, *24*, 108–115. [CrossRef]
40. Schumacher, M.; Denier, C.; Oudinet, J.P.; Adams, D.; Guennoun, R. Progesterone Neuroprotection: The Background of Clinical Trial Failure. *J. Steroid Biochem. Mol. Biol.* **2016**, *160*, 53–66. [CrossRef]
41. Dai, J.X.; Ma, Y.B.; Le, N.Y.; Cao, J.; Wang, Y. Large Animal Models of Traumatic Brain Injury. *Int. J. Neurosci.* **2018**, *128*, 243–254. [CrossRef]
42. Kochanek, P.M.; Jackson, T.C.; Jha, R.M.; Clark, R.S.B.; Okonkwo, D.O.; Bayır, H.; Poloyac, S.M.; Wagner, A.K.; Empey, P.E.; Conley, Y.P.; et al. Paths to Successful Translation of New Therapies for Severe Traumatic Brain Injury in the Golden Age of Traumatic Brain Injury Research: A Pittsburgh Vision. *J. Neurotrauma* **2020**, *37*, 2353–2371. [CrossRef]
43. Seok, J.; Warren, H.S.; Cuenca, A.G.; Mindrinos, M.N.; Baker, H.V.; Xu, W.; Richards, D.R.; McDonald-Smith, G.P.; Gao, H.; Hennessy, L.; et al. Genomic Responses in Mouse Models Poorly Mimic Human Inflammatory Diseases. *Proc. Natl. Acad. Sci. USA* **2013**, *110*, 3507–3512. [CrossRef] [PubMed]
44. Sauleau, P.; Lapouble, E.; Val-Laillet, D.; Malbert, C.H. The Pig Model in Brain Imaging and Neurosurgery. *Animal* **2009**, *3*, 1138–1151. [CrossRef] [PubMed]
45. Wernersson, R.; Schierup, M.H.; Jørgensen, F.G.; Gorodkin, J.; Panitz, F.; Staerfeldt, H.-H.; Christensen, O.F.; Mailund, T.; Hornshøj, H.; Klein, A.; et al. Pigs in Sequence Space: A 0.66X Coverage Pig Genome Survey Based on Shotgun Sequencing. *BMC Genom.* **2005**, *6*, 70. [CrossRef] [PubMed]
46. Fairbairn, L.; Kapetanovic, R.; Sester, D.P.; Hume, D. a The Mononuclear Phagocyte System of the Pig as a Model for Understanding Human Innate Immunity and Disease. *J. Leukoc. Biol.* **2011**, *89*, 855–871. [CrossRef] [PubMed]
47. Lind, N.M.; Moustgaard, A.; Jelsing, J.; Vajta, G.; Cumming, P.; Hansen, A.K. The Use of Pigs in Neuroscience: Modeling Brain Disorders. *Neurosci. Biobehav. Rev.* **2007**, *31*, 728–751. [CrossRef] [PubMed]
48. Marklund, N.; Hillered, L. Animal Modelling of Traumatic Brain Injury in Preclinical Drug Development: Where Do We Go from Here? *Br. J. Pharmacol.* **2011**, *164*, 1207–1229. [CrossRef]
49. Statler, K.D.; Jenkins, L.W.; Dixon, C.E.; Clark, R.S.; Marion, D.W.; Kochanek, P.M. The Simple Model versus the Super Model: Translating Experimental Traumatic Brain Injury Research to the Bedside. *J. Neurotrauma* **2001**, *18*, 1195–1206. [CrossRef]

50. Gorse, K.M.; Lafrenaye, A.D. The Importance of Inter-Species Variation in Traumatic Brain Injury-Induced Alterations of Microglial-Axonal Interactions. *Front. Neurol.* **2018**, *9*. [CrossRef]
51. Lafrenaye, A.D.; Todani, M.; Walker, S.A.; Povlishock, J.T. Microglia Processes Associate with Diffusely Injured Axons Following Mild Traumatic Brain Injury in the Micro Pig. *J. Neuroinflamm.* **2015**, *12*, 186. [CrossRef]
52. Lafrenaye, A.D.; Mondello, S.; Wang, K.K.; Yang, Z.; Povlishock, J.T.; Gorse, K.; Walker, S.; Hayes, R.L.; Kochanek, P.M. Circulating GFAP and Iba-1 Levels Are Associated with Pathophysiological Sequelae in the Thalamus in a Pig Model of Mild TBI. *Sci. Rep.* **2020**, *10*, 13369. [CrossRef]
53. Lafrenaye, A.; Mondello, S.; Povlishock, J.; Gorse, K.; Walker, S.; Hayes, R.; Wang, K.; Kochanek, P.M. Operation Brain Trauma Therapy: An Exploratory Study of Levetiracetam Treatment Following Mild Traumatic Brain Injury in the Micro Pig. *Front. Neurol.* **2021**, *11*, 586958. [CrossRef] [PubMed]
54. National Research Council. Guide Laboratory Animals for the Care and Use. In *Committee for the Update of the Guide for the Care and Use of Laboratory Animals Institute for Laboratory Animal Research Division on Earth and Life Studies*, 8th ed.; National Academies Press: Washington, DC, USA, 2011; ISBN 978-0-309-15400-0.
55. Bramlett, H.M.; Kraydieh, S.; Green, E.J.; Dietrich, W.D. Temporal and Regional Patterns of Axonal Damage Following Traumatic Brain Injury: A Beta-Amyloid Precursor Protein Immunocytochemical Study in Rats. *J. Neuropathol. Exp. Neurol.* **1997**, *56*, 1132–1141. [CrossRef] [PubMed]
56. Taetzsch, T.; Levesque, S.; McGraw, C.; Brookins, S.; Luqa, R.; Bonini, M.G.; Mason, R.P.; Oh, U.; Block, M.L. Redox Regulation of NF-KB P50 and M1 Polarization in Microglia. *Glia* **2015**, *63*, 423–440. [CrossRef] [PubMed]
57. Byrnes, K.R.; Loane, D.J.; Stoica, B.a; Zhang, J.; Faden, A.I. Delayed MGluR5 Activation Limits Neuroinflammation and Neurodegeneration after Traumatic Brain Injury. *J. Neuroinflamm.* **2012**, *9*, 43. [CrossRef] [PubMed]
58. Haynes, S.E.; Hollopeter, G.; Yang, G.; Kurpius, D.; Dailey, M.E.; Gan, W.-B.; Julius, D. The P2Y12 Receptor Regulates Microglial Activation by Extracellular Nucleotides. *Nat. Neurosci.* **2006**, *9*, 1512–1519. [CrossRef]
59. van Eijck, M.M.; Schoonman, G.G.; van der Naalt, J.; de Vries, J.; Roks, G. Diffuse Axonal Injury after Traumatic Brain Injury Is a Prognostic Factor for Functional Outcome: A Systematic Review and Meta-Analysis. *Brain Inj.* **2018**, *32*, 395–402. [CrossRef]
60. Davceva, N.; Sivevski, A.; Basheska, N. Traumatic Axonal Injury, a Clinical-Pathological Correlation. *J. Forensic Leg. Med.* **2017**, *48*, 35–40. [CrossRef]
61. Vieira, R.d.C.A.; Paiva, W.S.; de Oliveira, D.V.; Teixeira, M.J.; de Andrade, A.F.; de Sousa, R.M.C. Diffuse Axonal Injury: Epidemiology, Outcome and Associated Risk Factors. *Front. Neurol.* **2016**, *7*, 178. [CrossRef]
62. Graham, N.S.N.; Jolly, A.; Zimmerman, K.; Bourke, N.J.; Scott, G.; Cole, J.H.; Schott, J.M.; Sharp, D.J. Diffuse Axonal Injury Predicts Neurodegeneration after Moderate–Severe Traumatic Brain Injury. *Brain* **2020**, *143*, 3685–3698. [CrossRef]
63. Mesfin, F.B.; Gupta, N.; Hays Shapshak, A.; Taylor, R.S. Diffuse Axonal Injury. In *StatPearls*; StatPearls Publishing: Treasure Island, FL, USA, 2023.
64. Hoffe, B.; Holahan, M.R. The Use of Pigs as a Translational Model for Studying Neurodegenerative Diseases. *Front. Physiol.* **2019**, *10*, 838. [CrossRef]
65. Singh, P.; Sinha, S.; Suri, V.; Agarwal, D.; Bisht, A.; Garg, K.; Gupta, D.; Kakkar, A.; Kale, S.; Lalwani, S.; et al. Histological Changes in Thalamus in Short Term Survivors Following Traumatic Brain Injury: An Autopsy Study. *Neurol. India* **2013**, *61*, 599. [CrossRef] [PubMed]
66. Kim, J.; Parker, D.; Whyte, J.; Hart, T.; Pluta, J.; Ingalhalikar, M.; Coslett, H.B.; Verma, R. Disrupted Structural Connectome Is Associated with Both Psychometric and Real-World Neuropsychological Impairment in Diffuse Traumatic Brain Injury. *J. Int. Neuropsychol. Soc.* **2014**, *20*, 887–896. [CrossRef] [PubMed]
67. Banks, S.D.; Coronado, R.A.; Clemons, L.R.; Abraham, C.M.; Pruthi, S.; Conrad, B.N.; Morgan, V.L.; Guillamondegui, O.D.; Archer, K.R. Thalamic Functional Connectivity in Mild Traumatic Brain Injury: Longitudinal Associations with Patient-Reported Outcomes and Neuropsychological Tests. *Arch. Phys. Med. Rehabil.* **2016**, *97*, 1254–1261. [CrossRef] [PubMed]
68. Adams, J.; Doyle, D.; Ford, I.; Gennarelli, T.; Graham, D.; McLellan, D. Diffuse Axonal Injury in Head Injury: Definition, Diagnosis and Grading. *Histopathology* **1989**, *15*, 49–59. [CrossRef] [PubMed]
69. Prus, R.; Pokotylo, P.; Logash, M.; Zvir, T. Morphological Particularities and Morphometry of Rats' Kidneys under the Effect of Experimental Mild Traumatic Brain Injury. *Folia Morphol.* **2021**, *80*, 310–316. [CrossRef] [PubMed]
70. Tekin Neijmann, Ş.; Kural, A.; Sever, N.; Dogan, H.; Sarıkaya, S. Evaluation of Renal Function in Rats with Moderate and Mild Brain Trauma. *Ulus. Travma Acil Cerrahi Derg.* **2022**, *28*, 1–7. [CrossRef] [PubMed]
71. Ruan, F.; Chen, J.; Yang, J.; Wang, G. Mild Traumatic Brain Injury Attenuates Pneumonia-Induced Lung Injury by Modulations of Alveolar Macrophage Bactericidal Activity and M1 Polarization. *Shock* **2022**, *58*, 400. [CrossRef]
72. Vermeij, J.-D.; Aslami, H.; Fluiter, K.; Roelofs, J.J.; van den Bergh, W.M.; Juffermans, N.P.; Schultz, M.J.; Van der Sluijs, K.; van de Beek, D.; van Westerloo, D.J. Traumatic Brain Injury in Rats Induces Lung Injury and Systemic Immune Suppression. *J. Neurotrauma* **2013**, *30*, 2073–2079. [CrossRef]
73. Lim, S.H.; Jung, H.; Youn, D.H.; Kim, T.Y.; Han, S.W.; Kim, B.J.; Lee, J.J.; Jeon, J.P. Mild Traumatic Brain Injury and Subsequent Acute Pulmonary Inflammatory Response. *J. Korean Neurosurg. Soc.* **2022**, *65*, 680–687. [CrossRef]
74. Li, N.; Zhao, W.-G.; Zhang, W.-F. Acute Kidney Injury in Patients with Severe Traumatic Brain Injury: Implementation of the Acute Kidney Injury Network Stage System. *Neurocrit. Care* **2011**, *14*, 377–381. [CrossRef]

75. Holland, M.C.; Mackersie, R.C.; Morabito, D.; Campbell, A.R.; Kivett, V.A.; Patel, R.; Erickson, V.R.; Pittet, J.-F. The Development of Acute Lung Injury Is Associated with Worse Neurologic Outcome in Patients with Severe Traumatic Brain Injury. *J. Trauma Acute Care Surg.* **2003**, *55*, 106. [CrossRef] [PubMed]
76. Ito, D.; Imai, Y.; Ohsawa, K.; Nakajima, K.; Fukuuchi, Y.; Kohsaka, S. Microglia-Specific Localisation of a Novel Calcium Binding Protein, Iba1. *Mol. Brain Res.* **1998**, *57*, 1–9. [CrossRef] [PubMed]
77. Rodriguez-Callejas, J.D.; Fuchs, E.; Perez-Cruz, C. Evidence of Tau Hyperphosphorylation and Dystrophic Microglia in the Common Marmoset. *Front. Aging Neurosci.* **2016**, *8*, 315. [CrossRef]
78. Ahn, J.-H.; Choi, J.-H.; Park, J.-H.; Yan, B.-C.; Kim, I.-H.; Lee, J.-C.; Lee, D.-H.; Kim, J.-S.; Shin, H.-C.; Won, M.-H. Comparison of Alpha-Synuclein Immunoreactivity in the Spinal Cord between the Adult and Aged Beagle Dog. *Lab. Anim. Res.* **2012**, *28*, 165–170. [CrossRef]
79. Gaigé, S.; Bonnet, M.S.; Tardivel, C.; Pinton, P.; Trouslard, J.; Jean, A.; Guzylack, L.; Troadec, J.-D.; Dallaporta, M. C-Fos Immunoreactivity in the Pig Brain Following Deoxynivalenol Intoxication: Focus on NUCB2/Nesfatin-1 Expressing Neurons. *Neurotoxicology* **2013**, *34*, 135–149. [CrossRef] [PubMed]

Disclaimer/Publisher's Note: The statements, opinions and data contained in all publications are solely those of the individual author(s) and contributor(s) and not of MDPI and/or the editor(s). MDPI and/or the editor(s) disclaim responsibility for any injury to people or property resulting from any ideas, methods, instructions or products referred to in the content.

Article

SmartPill™ Administration to Assess Gastrointestinal Function after Spinal Cord Injury in a Porcine Model—A Preliminary Study

Chase A. Knibbe [1,*], Rakib Uddin Ahmed [1], Felicia Wilkins [1], Mayur Sharma [1], Jay Ethridge [1], Monique Morgan [1], Destiny Gibson [1], Kimberly B. Cooper [1], Dena R. Howland [1,2], Manicka V. Vadhanam [3], Shirish S. Barve [3], Steven Davison [4], Leslie C. Sherwood [4], Jack Semler [5], Thomas Abell [3] and Maxwell Boakye [1]

1. Department of Neurological Surgery, Kentucky Spinal Cord Injury Research Center, University of Louisville, Louisville, KY 40202, USA; rakibuddin.ahmed@louisville.edu (R.U.A.); fwilkins@metrohealth.org (F.W.); sharm983@umn.edu (M.S.); jay@icord.org (J.E.); monique.morgan@emory.edu (M.M.); destiny.gibson@louisville.edu (D.G.); brookecooper1@hotmail.com (K.B.C.); dena.howland@louisville.edu (D.R.H.); maxwell.boakye@uoflhealth.org (M.B.)
2. Research Service, Robley Rex Veterans Affairs Medical Center, Louisville, KY 40206, USA
3. Division of Gastroenterology, Hepatology and Nutrition, Department of Internal Medicine, University of Louisville, Louisville, KY 40202, USA; manicka.vadhanam@louisville.edu (M.V.V.); shirish.barve@louisville.edu (S.S.B.); thomas.abell@louisville.edu (T.A.)
4. Comparative Medicine Research Unit, University of Louisville, Louisville, KY 40202, USA; steven.davison@louisville.edu (S.D.); leslie.sherwood@louisville.edu (L.C.S.)
5. Medtronic Inc., Minneapolis, MN 55432, USA; jackrsemler@outlook.com
* Correspondence: caknib01@louisville.edu; Tel.: +1-(859)-468-1109

Citation: Knibbe, C.A.; Ahmed, R.U.; Wilkins, F.; Sharma, M.; Ethridge, J.; Morgan, M.; Gibson, D.; Cooper, K.B.; Howland, D.R.; Vadhanam, M.V.; et al. SmartPill™ Administration to Assess Gastrointestinal Function after Spinal Cord Injury in a Porcine Model—A Preliminary Study. *Biomedicines* 2023, 11, 1660. https://doi.org/10.3390/biomedicines11061660

Academic Editors: John O'Donnell and Dmitriy Petrov

Received: 15 May 2023
Revised: 2 June 2023
Accepted: 4 June 2023
Published: 7 June 2023

Copyright: © 2023 by the authors. Licensee MDPI, Basel, Switzerland. This article is an open access article distributed under the terms and conditions of the Creative Commons Attribution (CC BY) license (https://creativecommons.org/licenses/by/4.0/).

Abstract: Gastrointestinal (GI) complications, including motility disorders, metabolic deficiencies, and changes in gut microbiota following spinal cord injury (SCI), are associated with poor outcomes. After SCI, the autonomic nervous system becomes unbalanced below the level of injury and can lead to severe GI dysfunction. The SmartPill™ is a non-invasive capsule that, when ingested, transmits pH, temperature, and pressure readings that can be used to assess effects in GI function post-injury. Our minipig model allows us to assess these post-injury changes to optimize interventions and ultimately improve GI function. The aim of this study was to compare pre-injury to post-injury transit times, pH, and pressures in sections of GI tract by utilizing the SmartPill™ in three pigs after SCI at 2 and 6 weeks. Tributyrin was administered to two pigs to assess the influences on their gut microenvironment. We observed prolonged GET (Gastric Emptying Time) and CTT (Colon Transit Time), decreases in contraction frequencies (Con freq) in the antrum of the stomach, colon, and decreases in duodenal pressures post-injury. We noted increases in Sum amp generated at 2 weeks post-injury in the colon, with corresponding decreases in Con freq. We found transient changes in pH in the colon and small intestine at 2 weeks post-injury, with minimal effect on stomach pH post-injury. Prolonged GETs and CTTs can influence the absorptive profile in the gut and contribute to pathology development. This is the first pilot study to administer the SmartPill™ in minipigs in the context of SCI. Further investigations will elucidate these trends and characterize post-SCI GI function.

Keywords: spinal cord injury; gastrointestinal motility; SmartPill™; wireless motility capsule; porcine model

1. Introduction

SCI is associated with significant gastrointestinal (GI), urologic, dermatologic, cardiac, pulmonary, and kinematic dysfunctions that contribute to poor outcomes and often require costly medical intervention [1–8]. Mid-thoracic SCIs are associated with neurologic and GI deficits below the level of injury and can result in chronic dysfunction that frequently becomes acutely pathologic [5,9–11]. The impaired motility and changes in microenvironment can contribute to the deterioration of GI function, with treatment largely limited to symptom management [5,12–21].

Scintigraphy studies that utilize radioisotopes, anorectal, and colon manometry are commonly used to assess motility; however, scintigraphy is rarely used pre-clinically due to its high cost [22,23]. Substances such as charcoal or dyes can be fed to subjects and exit times noted to assess GI transit times, but are quantitatively limited to time [24,25]. Other investigations of GI function include electrogastrography and carbohydrate breath tests [26,27]. The SmartPill™ (Medtronic Inc., Minneapolis, MN, USA) is a noninvasive method for measuring gastrointestinal motility using a wireless motility capsule (WMC) that is ingested [28–33]. The WMC is a relatively new modality FDA approved for clinical assessment of gastroparesis, constipation, and general motility [34]. Pre-clinically, the WMC is used to assess pathological GI conditions and pharmacological absorption parameters [31,34–38]. Few studies have used the SmartPill™ in the context of SCI with humans or animal models [39].

The porcine spinal cord is similar to that of humans, with comparable gray to white matter ratios, cortical structure, sacral enlargement, cord dimensions/structure/organization, and metabolic demands [40–45]. This porcine SCI model serves to assess critical GI parameters, a major comorbidity in SCI patients, compare to human functionality, and consider interventions to improve GI function. The objective of this study was to assess transit times, pH, and pressure of the GI tract in three pigs that underwent SCI through administration of the SmartPill™ pre-injury at 2 and 6 weeks post-injury. We hypothesize that there will be delays in gastric emptying time (GET), colonic transit time (CTT), decreases in pressures in the colon, and increases in gastric and colonic pH at post-injury time points.

2. Materials and Methods

This study was conducted in accordance with applicable institutional and national research guidelines and regulations for the care and use of animals in research [46]. All experimental procedures were approved by the Institutional Animal Care and Use Committee of the University of Louisville (UofL) (approval number-20845). The UofL's Animal Care and Use Program is fully accredited by the American Association for Accreditation of Laboratory Animal Care (AAALAC), International.

2.1. Animal and Surgical Procedures

Three 20–25 kg female Yucatan minipigs (Sinclair Bio-resources, Columbia, MO, USA) were pair-housed in floor pens and provided with a minimum of 10 ft2 per animal on 5–10 cm of Cellu-nest™ bedding (Shepherd Specialty Papers; Watertown, TN, USA) on top of 0.95 cm thick 1.22 × 1.22 m interlocking Rubber Gym Tiles (Rubber Flooring Inc., Mesa, AZ, USA). The animals were provided environmental enrichment consisting of toys, videos, and mirrors. The environmental conditions were maintained at 20.0–22.0 °C, 30–70% humidity, and a 12 h light/dark cycle. Pens were cleaned daily, and the animals were provided filtered tap water ad libitum.

Anesthesia was induced with Ketamine HCL (Zoofarm, Austintown, OH, USA, 5 mg/kg, i.m.), Dexmedetomidine (Dexmedesed®, KS, USA, 0.04 mg/kg, i.m.), and Glycopyrrolate (Piramal critical care®, PA, USA, 0.01 mg/kg, i.m.) mixed in the same syringe. Following the induction of anesthesia, Meloxicam (Covetrus®, UK, 0.4 mg/kg, i.v.) was administered for analgesia. Sustained release Bupivacaine SR (Nocita®, IN, USA, 2 mg/kg, s.q.) was divided into several locations adjacent to both sides of the planned incision site, providing local analgesia for up to 72 h. The pigs were endotracheally intubated and maintained with Propofol (PropoFlo®, NJ, USA, 8–20 mg/kg/h, i.v.) and Fentanyl (Fentanyl transdermal system, NJ, USA, 10–45 mcg/kg/h) for the entire procedure. Prior to surgery, an indwelling urinary catheter (8 French foley) was manually inserted and left in place post-operatively until the animal demonstrated continence. Standard intraoperative anesthetic monitoring recording heart and respiratory rate, blood pressure, end-tidal carbon dioxide, oxygen saturation, and urine output was conducted. The fluid status was monitored and maintained with Lactated Ringer's solution and 5% Dextrose i.v. to maintain normoglycemia and continued post-operatively until the animals were capable of drinking independently. Dextrose was discontinued once animals were eating well. The temperature was measured by a rectal probe and maintained at 37.3–39.4 °C by a heating pad (Bair Hugger Model 775, 3M, Saint Paul, MN, USA).

With the animals in ventral recumbency on the operating table, the location of T10 was confirmed with a dorsoventral radiograph. A dorsal midline incision was made between T8 and T13. The spinous processes, laminae, and pedicles of T8 and T13 were exposed using electrocautery dissection. A second radiograph was acquired to confirm the appropriate spinal and vertebral level. Laminectomy was performed at the 10th thoracic vertebrae level to expose the dura and spinal cord and widened to a diameter of approximately 1.2 cm to ensure unimpeded impact. Two 3.5 × 14 mm multi-axial cervical screws were inserted into the T11 and T13 pedicles. Titanium rods were secured to the articulating arm of the impactor and subsequently to the pedicular screws and secured with locking caps. This fixed the T11–13 vertebral segments and secured the impactor in place and leveled it; the custom impactor was provided by University of British Columbia researchers [41]. Rocuronium was administered to mitigate animal movement during electrocautery dissection. A bolus of Propofol (10 mL bolus, equating to 100 mcg from initial concentration of 10 mcg/mL) was given five minutes prior to injury and the breath was held for impact. A 50 g cylindrical weight was dropped via a triggering mechanism onto exposed dura from a randomized height. An additional mass of 100 g was added immediately for five minutes. Following compression, the weight drop apparatus was removed and the incision was closed. Post-operatively, transdermal Fentanyl patches (Fentanyl transdermal system, NJ, USA, 1.5–5 mcg/kg/h, t.d.) and Meloxicam (Covetrus®, UK, 0.2–0.4 mg/kg, p.o.) were continued post-operatively and administered for 5 and 7 days, respectively. Maropitant (Zoofarm, Austintown, OH, USA, 1 mg/kg, p.o.) was administered pre-operatively and continued once daily for the duration of Fentanyl administration. The animals were monitored 24/7 and individually housed in open top crates in intensive care until cleared by UofL CMRU veterinarian(s) for their return to normal housing.

2.2. Food Diet for Study

The animals were fed Purina LabDiet™ 5081 (Purina Inc., St. Louis, MO, USA) twice daily at 1% body weight per feeding for the duration of the experiment. Their food was withheld the morning of injury, resumed post-operatively initially with a/d wet dog food (Hill's Inc., Topeka, KS, USA), and transitioned back to Purina LabDiet™ 5081. The SmartPill™ was administered orally via a balling gun with the morning feeding. Twice daily feedings continued for the duration that the SmartPill™ dwelled in the animal.

2.3. Wireless Motility Capsule and Data Analysis

The WMC was administered randomly in the female Yucatan minipigs the week prior to initial injury (N = 2), 2 weeks post-injury (N = 3), and 6 weeks post-injury (N = 3). Female minipigs were used because of the ease of maintenance of the urinary bladder after the surgery. The WMC actively measured and transmitted pressure, pH, and temperature to a receiver secured to the animal. The pigs were monitored until capsule expulsion. The data were downloaded and the proprietary software MotiliGI®(version 2.5, Medtronic Inc., Minneapolis, MN, USA) was used to view the initial study. Proprietary Gastrointestinal Motility Software (GIMS®, version 3.0.0, Medtronic Inc., Minneapolis, MN, USA) refined the descriptive statistical analysis. The WMC was calibrated and the function confirmed using the MotiliGI®software before administration, and all of the WMCs were successfully retrieved.

The raw data generated by the MotiliGI® software are displayed in (Supplemental Figures S1–S8). A drop in pH to 1–3 indicated presence in the stomach. GET was identified by a permanent rise in pH above 4. WMC transition to the small bowel was indicated by a gradual increase in pH to approximately 7–9. A sharp decline to approximately 6–8 pH with concurrent increases in pressure indicated WMC passage through the ileocecal valve and entry into the colon. The measurements were refined and validated by investigators using a GIMS® data viewer for descriptive statistical analysis and stratified by anatomical section within that GI section and by time quartile.

2.4. Tributyrin Administration

Oral Tributyrin (Alfa Aesar Inc., Heysham, UK) at a dose of 1 g/kg/day BID was administered beginning the first day post-injury for 8 weeks as a liquid formulation mixed with flavoring emulsion (LorAnn Oils Inc., Lansing, MI, USA) in pigs 1 and 2. Previous pharmacological studies found Tributyrin to be protective against diarrhea and intestinal permeability by enhancing colonic tight junction gene expression in weaned piglets [47].

3. Results

3.1. Transit Times

Pre-injury transit times were observed for pigs 1 and 2, respectively: GETs of 4:06:30 and 7:21:10, small intestine transit times (SITTs) of 2:28:35 and 2:07:20, and CTTs of 19:55:55 and 17:27:30 (Figure 1). We observed times at 2 weeks post-injury in pigs 1, 2, and 3, respectively: GETs of 12:06:35, 99:07:37, 8:36:10, SITTs of 1:38:47, 2:14:00, and 2:39:25, and CTTs of 21:58:18, 26:25:03, and 18:08:05 (Figure 2). At 6 weeks post-injury, we recorded the following times for pigs 1, 2, and 3, respectively: GETs of 6:58:00, 8:13:51, and 8:41:40, SITTs of 2:58:05, 2:10:32, and 3:00:55, and CTTs of 36:29:15, 22:00:27, and 86:13:05 (Figure 3). These findings are clear in (Table 1), which shows delays in CTT, at both post-injury time points, and mild delays in GET in pigs 1 and 2, with indications that pig 3 was following the same trend.

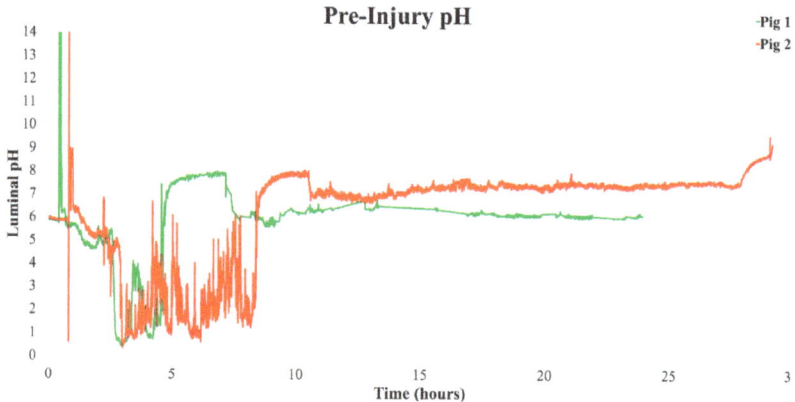

Figure 1. Whole gut pH recorded by the SmartPill™ at pre-injured condition. The green line depicts pig 1 and the red line pig 2. Graphs include raw data recorded by the SmartPill™. A drop in mean pH to between 1 and 3 indicates that the wireless motility capsule (WMC) has been ingested and resides in the stomach. A permanent rise in mean pH to above 4 to between approximately 7 and 9, with a concomitant increase in pressure, indicates passage through the pyloric sphincter GET and into the small intestine. A subsequent decrease in pH to approximately 6 to 8, with a congruent increase in pressure, indicates passage through the ileocecal valve SITT into the colon, indicating colon arrival time (CAT) and the start of CTT. A permanent and continuous rise above pH 7 indicates that the WMC exited the body, completing whole gut transit time (WGTT).

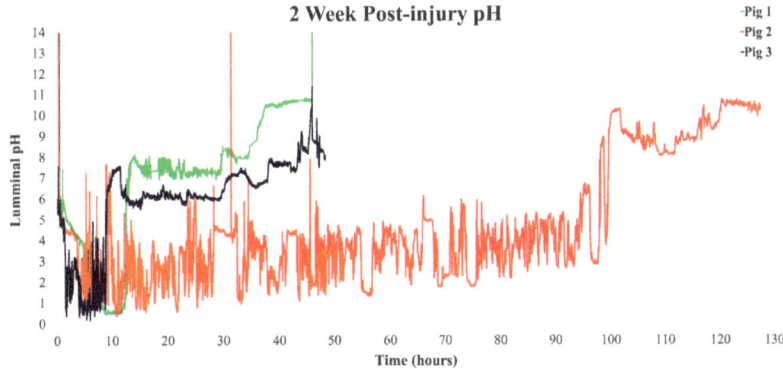

Figure 2. Whole gut pH recorded by the SmartPill™ at week 2 post-injured condition. The green line is pig 1, the red line is pig 2, and the dark blue is pig 3. Graphs include raw data recorded by the SmartPill™, with previously described graph analysis.

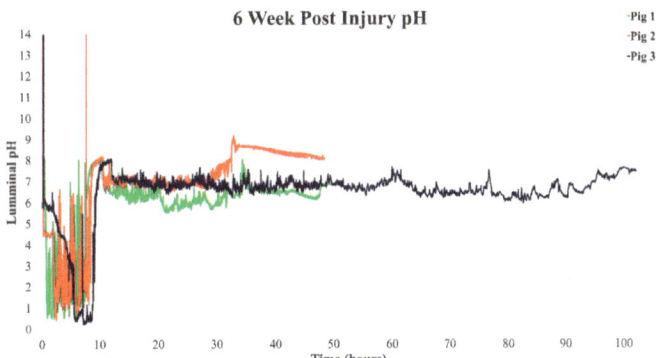

Figure 3. Six weeks post-injury whole gut pH recorded by the SmartPill™. The green line is pig 1, the red line is pig 2, and the dark blue is pig 3. Graphs include raw data recorded by the SmartPill™, with previously described graph analysis.

Table 1. Summary of transit times recorded in hours by the SmartPill® at each time point for each animal. [1] Colon Transit Time, [2] Gastric Emptying Time, [3] Small Intestine Transit Time, [4] Colon arrival time, [5] Whole Gut Transit Time.

Pig	Time Frame Relative to Injury	[1] GET	[2] SITT	[3] CTT	[4] CAT	[5] WGTT
1	Pre-injury	4:06:30	2:28:35	19:55:55	6:35:05	26:29:00
	2 weeks post	12:06:35	1:38:47	21:58:18	13:45:22	35:43:40
	6 weeks post	6:58:00	2:58:05	36:29:15	11:34:15	48:03:30
2	Pre-injury	7:21:10	2:07:20	17:27:30	9:28:30	26:56:00
	2 weeks post	99:09:57	2:09:50	26:26:53	101:21:47	127:48:40
	6 weeks post	8:10:35	2:10:30	22:12:55	10:21:05	32:43:00
3	2 weeks post	8:36:10	2:39:25	18:08:05	11:15:35	29:23:40
	6 weeks post	8:41:40	3:00:55	86:13:05	11:42:35	97:55:40

3.2. pH Changes

Changes in post-injury pH were noted in the small intestine and colon. We observed an overall decrease in minimum (min) and median (med) pH in quartile one of the small intestine (Table 2), but the remaining gastric (Supplemental Table S1) and small intestine sections displayed no appreciable trends between pre-injury and post-injury. Colonic pH interestingly increased in both min and med pH across all quartiles and anatomical regions at 2 weeks post-injury, but returned to normality by 6 weeks (Table 3).

Table 2. Small intestine min and med pH recordings stratified by time quartiles and specific anatomical colonic regions of interest for each animal at each time point after WMC administration.

Pig	Time	Duodenum		Ileum		Quartile 1		Quartile 2		Quartile 3		Quartile 4	
		Min	Med	Min	Med	Min	Med	Min	Med	Min	Med	Min	Med
1	Pre-injury	1.95	7.58	7.73	7.9	1.95	7.46	7.6	7.72	7.67	7.82	7.73	7.9
	2 weeks post	1.37	7.47	6.9	7.9	1.37	6.14	6.05	7.25	7.65	7.82	7.75	8.01
	6 weeks post	1.43	6.71	7.9	7.99	1.43	6.56	7.29	7.71	7.82	7.96	7.9	7.97
2	Pre-injury	2.12	7.84	7.74	7.88	2.1	7.08	7.42	7.69	7.74	7.86	7.74	7.88
	2 weeks post	7.86	9.99	10.12	10.24	7.86	9.42	10.01	10.12	10.12	10.18	10.22	10.26
	6 weeks post	2.22	7.74	7.84	8.07	2.22	7.42	7.73	7.86	7.84	7.99	7.95	8.09
3	2 weeks post	1.29	6.52	7.29	7.46	1.29	6.35	6.42	6.84	7.12	7.35	7.39	7.48
	6 weeks post	2.26	6.67	7.82	7.97	2.26	6.33	7.01	7.63	7.71	7.88	7.88	7.99

Table 3. Colonic min and med pH recordings, further stratified by quartile of colon and specific anatomical location for each animal at each time point after WMC was administered.

Pig	Time	Caecum		Sigmoid		Quartile 1		Quartile 2		Quartile 3		Quartile 4	
		Min	Med	Min	Med	Min	Med	Min	Med	Min	Med	Min	Med
1	Pre-injury	5.9	6.69	6.07	6.23	5.61	6.14	6.13	6.39	6.01	6.13	5.95	6.11
	2 weeks post	7.43	7.63	8.2	8.28	6.93	7.56	6.94	7.29	6.95	7.29	7.67	8.12
	6 weeks post	6.48	6.77	6.15	6.21	5.95	6.48	5.5	5.95	5.67	6.48	6.15	6.49
2	Pre-injury	6.76	7.08	7.35	7.5	6.67	6.99	7.11	7.39	7.14	7.43	7.31	7.5
	2 weeks post	9.05	9.22	10.03	10.24	8.27	8.89	8.03	8.56	8.65	9.61	9.92	10.41
	6 weeks post	6.59	6.97	7.22	8.01	6.59	7.01	6.63	6.92	6.56	6.96	6.63	7.41
3	2 weeks post	6.05	6.27	5.92	6.01	5.52	5.97	5.95	6.16	5.71	6.09	5.92	6.05
	6 weeks post	6.88	7.24	6.99	7.18	6.25	6.9	6.37	6.78	6.22	6.56	5.95	6.47

3.3. Pressure Changes

GIMS® software reported contraction frequencies (Con freq) and sum amplitudes (Sum amp), the latter sum of the peak pressure of each contraction. We observed an initial decrease in Con freq and Sum amp in the stomach at 2 weeks post-injury and, interestingly, a slight increase in the same parameters at 6 weeks post-injury. These observations illustrate the pathologic timeline and indicate potential for recovery (Table 4). The small intestine appears unimpeded by SCI, with the only notable trend in pressure being the decrease in duodenal Con freq and Sum amp that was obtained at both post-injury time points (Table 5). Decreases in colonic Con freq were observed for all quartiles at both post-injury time points, with a corresponding increase in Sum amp. In other words, there were less frequent but more forceful contractions (Table 6).

Table 4. Gastric pressure recordings stratified by time quartiles and anatomical regions of interest for each animal at each time point after WMC administration. [1] Contraction Frequency, [2] Summation of the Amplitudes.

Pig	Time	Antrum [1] Con Freq	Antrum [2] Sum Amp	Quartile 1 Con Freq	Quartile 1 Sum Amp	Quartile 2 Con Freq	Quartile 2 Sum Amp	Quartile 3 Con Freq	Quartile 3 Sum Amp	Quartile 4 Con Freq	Quartile 4 Sum Amp
1	Pre-injury	12.6	13,541.92	6.96	6224.81	5.17	3701.71	5.85	4125.89	12.25	13,542.63
1	2 weeks post	2.64	7204.76	3.14	9589.59	4.29	12,143.88	4.55	14,139.2	2.82	13,267.17
1	6 weeks post	7.71	12,609.13	3.17	5185.17	2.77	7703.32	4.05	16,758.34	7.37	20,221.82
2	Pre-injury	6.1	8624.08	2.84	4896.96	3.21	10,021.52	2.29	5614.76	4.95	13,141.26
2	2 weeks post	1.47	5920.24	2.42	79,795.96	2.26	87,694.51	1.89	64,920.47	1.82	53,130.43
2	6 weeks post	2.88	5928.46	1.46	2362.98	3.32	16,949.41	4.47	12,145.5	3.85	11,257.91
3	2 weeks post	2.75	11,208.56	5.1	10,271.68	3.27	7069.23	3.82	9694.13	3.39	23,596.8
3	6 weeks post	4.03	7358.97	3.32	7400.25	2.29	5747.71	3.67	9257.94	3.37	11,169.59

Table 5. Small intestine pressure recordings further stratified by time quartiles and specific regions for each animal at each time point after WMC administration.

Pig	Time	Duodenum Con Freq	Duodenum Sum Amp	Ileum Con Freq	Ileum Sum Amp	Quartile 1 Con Freq	Quartile 1 Sum Amp	Quartile 2 Con Freq	Quartile 2 Sum Amp	Quartile 3 Con Freq	Quartile 3 Sum Amp	Quartile 4 Con Freq	Quartile 4 Sum Amp
1	Pre-injury	5.49	4806.65	1.77	1073.03	7.59	3948.92	2.29	1286.37	2.57	823.44	1.49	714.42
1	2 weeks post	5.08	5123.13	5.35	4140.09	3.34	2105.76	7.03	2459.54	7.25	1614.31	3.66	1736.16
1	6 weeks post	1.44	1839.88	3.9	3740.14	0.81	941.3	4.54	3386.94	1.81	1568.18	4.87	3440.62
2	Pre-injury	2.2	2456.44	0.59	844.1	3.5	2132.12	0.96	380.9	0.84	688.16	0.42	241.96
2	2 weeks post	0.89	920.92	0.41	421.82	1.24	695.06	0.53	266.8	0.19	123.72	0.68	325.22
2	6 weeks post	1.1	1835.98	0.47	456	1.35	1566.22	0.73	293.76	0.49	202.56	0.44	275.52
3	2 weeks post	1.67	1821.23	0.57	659.86	1.51	1166.52	1.97	1140.22	0.92	548.96	0.51	438.47
3	6 weeks post	2.2	3139.08	1.15	1145.38	2.23	2655.14	1.27	881.84	0.96	659.2	1.04	814.2

Table 6. Colonic pressure recordings further stratified by time quartiles and anatomical region for each animal at each time point after WMC administration.

Pig	Time	Quartile 1 Con Freq	Quartile 1 Sum Amp	Quartile 2 Con Freq	Quartile 2 Sum Amp	Quartile 3 Con Freq	Quartile 3 Sum Amp	Quartile 4 Con Freq	Quartile 4 Sum Amp
1	Pre-injury	1.59	950.81	1.76	34.86	1.84	1721.61	1.44	1052.29
1	2 weeks post	2.26	1820.64	3.09	3776.22	2.59	1910.86	1.34	614.88
1	6 weeks post	0.08	1656.46	0.04	660.02	0.16	2522.4	0.14	2912.1
2	pre-injury	0.56	356.5	0.49	614.1	0.59	499.56	1.67	1090.66
2	2 weeks post	0.08	911.7	0.05	508.76	0.31	3543.36	0.4	5234.8
2	6 weeks post	0.06	765.6	0.08	842.4	0.14	1218.24	0.42	4306.1
3	2 weeks post	0.51	1178.29	1.16	1563.24	2.22	2458.04	1.6	2547.76
3	6 weeks post	0.17	5298.34	0.05	1546.96	0.03	1050.62	0.09	3143.62

4. Discussion

To the best of our knowledge, this is the first study that has used the WMC in Yucatan minipigs in the context of SCI. The focus of this study was validation of the SmartPill™ in our SCI porcine model to help guide future experiments and compare existing pre-clinical and clinical GI motility studies. It is important to note that spinal injury at this level was relatively moderate and resulted in complete loss of hindlimb function. CTTs are generally longer in non-injured pigs when compared to humans, but similarly variable [37,48].

T10–T11 level SCI in pigs 1 and 2 (Table 1) was associated with disturbances of GET and CTTs, as hypothesized. Pre-injury data for pig 3 was unobtainable. It is well established that SCI at a T4–5 level is associated with significant GI disturbance including pathologic gastroparesis [5,16]. The majority of pre-clinical GI motility SCI studies have been performed in rodents, specifically rats, and depict a prolongation of both the gastric and colonic times after injury [16,49,50]. While the mechanism behind gastric and colonic delays remains unclear, previous findings indicate that autonomic dysfunction via alterations in vagal nerve signaling integrity post-injury plays a significant role [5,15,16,51–54]. This is further confirmed by Schneider et al., who also showed a prolongation of GET that was greater than CTTs or SITTs [37].

We noted overall decreases in colonic contraction frequencies across all quartiles but, interestingly, an increase in Sum amp. This indicates that the colonic and gastric motor complexes are in an aberrant state, potentially contributing to delays in transit times and leading to a constipation also likely exacerbated by opioid use for analgesia after SCI [55]. Another trend was the increase in colonic pH in all quartiles at 2 weeks post-injury and a return to baseline at 6 weeks. This indicates a disturbance in the microenvironment in the acute time frame, with potential for recovery. In non-injured conditions, the colonic pH in both male Landrace and our pre-injury female Yucatan minipigs was comparable to that of humans and generally ranged from 5 to 8 [35]. GI dysbiosis has recently gained attention as a source of GI disturbances. The gut microbiome's normal function significantly impacts cognition, digestion and absorption, and overall wellness. In the context of SCI, the gut microbiome is altered and affects the clinical prognosis [19,56,57]. Thus, pharmacological influence of the biome is of interest. We administered oral Tributyrin, a triglyceride that reportedly encourages natural gut flora, to assess its impact on motility and the colonic environment. We would expect that pigs receiving Tributyrin would display improved colonic pH and CTTs. However, the effect of Tributyrin is unclear in our study, likely due to the small sample size, and further studies are required.

We observed a significant decrease inCon freq and Sum amp in the antrum at 2 weeks post-injury and a slight improvement in these parameters at 6 weeks post-injury. These findings indicate that there is significant deterioration or a "shock" period of autonomic function in the acute setting that partially abates by 6 weeks. This change over time suggests potential for improvement. Maximum pressures in both non-injured humans and large canines have been reported to range from 100 to 500 mmHg in the fed state [37,58]. Male Landrace pigs' gastric pressures were reported to range from 100 to 350 mmHg in the fed state [35]. Another study conducted by Raunch et al. in Pietrain farm pigs reported pressures between 4 and 20 mmHg [59]. It is unclear why these gastric pressures are substantially lower; it is likely a result of the experimental conditions. Beagle dog maximum gastric pressures have also been reported to range from approximately 200 to 800 mmHg [48].

Closer examination of the 2 weeks post-injury graph for pig 2 (Figure 2) shows the WMC dwelling in the stomach for an extended time and multiple unsuccessful attempts to expel the capsule over four days. This event could have stemmed from repetitious feedings, which rhythmically close the pylorus under normal physiologic conditions. The unique U shape of the pig stomach can impede passage of larger, more solid objects by the pyloric sphincter [35,60]. Having observed decreases in antrum pressures after injury, this could further exacerbate the passage of the WMC and solid foods into the duodenum and contribute to pathology development. Pig 3 also displayed a significant delay in the colon compared to other animals. This is likely an extreme effect of injury on colonic motility; however, it could also stem from the unique spiral shape of the porcine colon. On the other hand, the literature utilizing porcine GI models indicates that this feature does not appreciably affect motility [61]. We considered the effect of fentanyl administration on GI motility; however, the published pharmacokinetic and pharmacodynamic parameters in humans indicate that relatively large doses of fentanyl (100 g/h) administered transdermally and continuously over a week were present in negligible amounts after 6 days. Given that the animals had fentanyl removed 5 days after injury, we concluded that fentanyl would be cleared from the system by 2 weeks post-injury and was not responsible for the changes from week 2 to week 6 of administration [62].

The small intestine was largely unaffected when comparing pre- and post-injury pressures, except for decreased contraction frequencies and maximum pressures in the duodenum and quartile 1. The Yucatan minipigs in this study and a previous Beagle canine study displayed shorter SITTs of 1–3 h when compared to humans and a marginally shorter relative to male Landrace pigs in the fed condition [35,37]. It is suggested that the shorter intestinal transit times can be correlated to different dimensions of intestine. In dogs, the small intestine is shorter compared to humans and pigs, but this does not appear to affect transit [31,60]. These data show that small intestine motility is essentially unaffected after SCI, and this agreed with previous observations in humans, rodents, and other species studies to date [35,37,48,58,59]. We observed transient increases in small intestine pH in all quartiles and anatomical locations supposed by GIMS® software at 2 weeks post-injury that returned to pre-injury conditions by 6 weeks. The small intestine pH in male Landrace pigs was marginally higher and comparable to our limited pre-injury Yucatan data [35]. Two recently published studies involving a WMC in large canines and beagles revealed a baseline colon and small intestinal pHs that were comparable to those of humans [31,58]. Changes in small intestinal absorptive and microenvironment physiology likely play a role in pathogenesis after SCI. This change is important to consider clinically as these changes influence nutrimental status and drug absorption, features that can directly affect prognosis of SCI patients [5,12,13]. Minimum gastric pH was similar across all dog and pig species and relative to humans, ranging from 0.1 to 1, with the exception of the male Gottingen minipig study by Suenderhauf et al., which reported min gastric pHs ranging from 1.2 to 6 [30,31,59,60,63].

Williams et al. administered SmartPill™ to 20 patients in a chronic cervical and thoracic SCI setting with a mean injury duration of 15 ± 11 years [39]. Patients were fed after a 12-h fast and were asked to swallow a smart pill to measure the GET and the CTT. They reported a prolonged GET of 10.6 ± 7.2 h in patients with SCIs versus 3.5 ± 1.0 h in control subjects. Similarly, CTT was 52.3 ± 42.9 h versus 14.2 ± 7.6 h in controls. Additionally, they reported no significant change in post-injury min gastric pH after injury. This study concluded that the SmartPill™ was safe and can be used to demonstrate gastrointestinal motility delays in patients with both cervical and thoracic SCI [39]. Delays in GET and CTT and general trends in GI motility after SCI from previous studies agree with our observations in the Yucatan minipig model post-injury. Further comparisons are limited by the paucity of published data on pressure and pH parameters beyond MotiliGI®.

Although this study was a preliminary study to evaluate the Smartpill™ function in the minipig model of SCI, a major limitation of this study was the sample size. The study was conducted to evaluate and test the feasibility of the Smartpill™ in a pre-clinical model of SCI. Moreover, in this study, we did not consider the representability of the minipig gut with human. This could give us more insight into the gastrointestinal transit time information. Another limitation of the study is that the anatomical structure of the minipig spinal cord was not compared in this study, while some pieces of information are available only for domestic pigs. The nerves that supply to the gastrointestinal tract from the spinal cord were not very well known in Yucatan minipigs. Additionally, the administration of tributyrin can modulate the transit time and pH. Further study will elucidate the effect of tributyrin on the microbiome.

5. Conclusions

This is the first pre-clinical study that has implemented the SmartPill™ in the context of SCI using a large animal model. Our trends post-injury are consistent with the limited data that exist, both in pre-clinical and human studies. There are delayed GETs, CTTs, Con freq, and Sum amp in both the antrum and colon and a transient increase in colonic pH 2 weeks after injury. The potential impact of Tributyrin remains unclear, and further studies are necessary to elucidate its influence post-injury. We acknowledge the limited power of this pilot study, and additional studies are required to establish larger trends. Further investigations are warranted to establish these trends in a larger cohort, at which point GI microenvironmental changes can be effectively modulated to improve gut function and overall morbidity.

Supplementary Materials: The following supporting information can be downloaded at: https://www.mdpi.com/article/10.3390/biomedicines11061660/s1, Figure S1: SmartPill®pre-injury graph generated by MotiliGI®software for pig 1, Figure S2: SmartPill®2 week post-injury graph generated by MotiliGI®software for pig 1; Figure S3: SmartPill®6-week post-injury graph generated by MotiliGI®software for pig 1; Figure S4: SmartPill®pre-injury graph generated by MotiliGI®software for pig 2; Figure S5: SmartPill®2-week post-injury graph generated by MotiliGI®software for pig 2; Figure S6: SmartPill®6-week post-injury graph generated by MotiliGI®software for pig 2; Figure S7: SmartPill®2-week post-injury graph generated by MotiliGI®software for pig 3; Figure S8: SmartPill®6-week post-injury graph generated by MotiliGI®software for pig 3; Table S1: Stomach 1 min and 2 med pH recordings stratified by time quartiles of stomach and anatomical regions similarly to other GI sections for each animal at each timepoint when the 3 WMC was administered.

Author Contributions: Conceptualization, C.A.K., D.R.H., M.V.V., S.S.B., J.S., T.A. and M.B.; Methodology, C.A.K., J.E., M.V.V., S.S.B., J.S., T.A., M.M., D.G., L.C.S., K.B.C. and M.B.; Surgery, C.A.K., J.E. and S.D.; Software, C.A.K. and J.S.; Results, C.A.K., R.U.A., F.W., J.S., T.A. and M.B.; Writing—original draft preparation, C.A.K. and R.U.A.; Writing—review and editing, C.A.K., R.U.A., M.S. and M.B.; Supervision, J.S., T.A. and M.B.; Project administration, M.B.; Funding acquisition, M.B. All authors have read and agreed to the published version of the manuscript.

Funding: Research reported in this publication was supported by the Helmsley Foundation grant# 2016PG-MED005, Department of Defense grant# W81XWH-18-1-0117, Kentucky Spinal Cord Injury Center, The Veterans Affairs Rehabilitation, Research and Development grant# RCSB92495, the Craig F. Neilsen Foundation grant# 546123, and Ole A. Mabel Wise & Wilma Wise Nelson and Rebecca F. Hammond Endowment.

Institutional Review Board Statement: This study was conducted under the guidelines and approval by the Institutional Animal Care and Use Committee of the University of Louisville (UofL). IUCUC approval codes: 18232, 15019.

Data Availability Statement: All data of this study are available from author upon request.

Acknowledgments: We would like to thank Jackson Gallagher and the University of Louisville Veterinary staff for assisting with data collection and maintenance of animals, respectively.

Conflicts of Interest: J.S. is a representative of Medtronic, which provided the SmartPills™ and Gastric Intestinal Motility Software for visualization of the data and to assist with analysis. Other authors declared no conflict of interest.

References

1. Nash, M.S.; Groah, S.L.; Gater, D.R., Jr.; Dyson-Hudson, T.A.; Lieberman, J.A.; Myers, J.; Sabharwal, S.; Taylor, A.J. Identification and Management of Cardiometabolic Risk after Spinal Cord Injury: Clinical Practice Guideline for Health Care Providers. *Top. Spinal Cord Inj. Rehabil.* **2018**, *24*, 379–423. [CrossRef] [PubMed]
2. Tate, D.G.; Wheeler, T.; Lane, G.I.; Forchheimer, M.; Anderson, K.D.; Biering-Sorensen, F.; Cameron, A.P.; Santacruz, B.G.; Jakeman, L.B.; Kennelly, M.J.; et al. Recommendations for evaluation of neurogenic bladder and bowel dysfunction after spinal cord injury and/or disease. *J. Spinal Cord Med.* **2020**, *43*, 141–164. [CrossRef] [PubMed]
3. French, D.D.; Campbell, R.R.; Sabharwal, S.; Nelson, A.L.; Palacios, P.A.; Gavin-Dreschnack, D. Health care costs for patients with chronic spinal cord injury in the Veterans Health Administration. *J. Spinal Cord Med.* **2007**, *30*, 477–481. [CrossRef]
4. James, S.L.; Theadom, A.; Ellenbogen, R.G.; Bannick, M. Global, regional, and national burden of traumatic brain injury and spinal cord injury, 1990–2016: A systematic analysis for the Global Burden of Disease Study 2016. *Lancet Neurol.* **2019**, *18*, 56–87. [CrossRef] [PubMed]
5. Holmes, G.M.; Blanke, E.N. Gastrointestinal dysfunction after spinal cord injury. *Exp. Neurol.* **2019**, *320*, 113009. [CrossRef]
6. Rubin-Asher, D.; Zeilig, G.; Klieger, M.; Adunsky, A.; Weingarden, H. Dermatological findings following acute traumatic spinal cord injury. *Spinal Cord* **2005**, *43*, 175–178. [CrossRef] [PubMed]
7. Berlowitz, D.J.; Wadsworth, B.; Ross, J. Respiratory problems and management in people with spinal cord injury. *Breathe* **2016**, *12*, 328–340. [CrossRef]

8. Furlan, J.C.; Fehlings, M.G. Cardiovascular complications after acute spinal cord injury: Pathophysiology, diagnosis, and management. *Neurosurg. Focus.* **2008**, *25*, E13. [CrossRef]
9. Ebert, E. Gastrointestinal involvement in spinal cord injury: A clinical perspective. *J. Gastrointest. Liver Dis.* **2012**, *21*, 75–82.
10. Han, T.R.; Kim, J.H.; Kwon, B.S. Chronic gastrointestinal problems and bowel dysfunction in patients with spinal cord injury. *Spinal Cord* **1998**, *36*, 485–490. [CrossRef]
11. Correa, G.I.; Rotter, K.P. Clinical evaluation and management of neurogenic bowel after spinal cord injury. *Spinal Cord* **2000**, *38*, 301–308. [CrossRef] [PubMed]
12. Bernardi, M.; Fedullo, A.L.; Bernardi, E.; Munzi, D.; Peluso, I.; Myers, J.; Lista, F.R.; Sciarra, T. Diet in neurogenic bowel management: A viewpoint on spinal cord injury. *World J. Gastroenterol.* **2020**, *26*, 2479–2497. [CrossRef]
13. Bigford, G.; Nash, M.S. Nutritional Health Considerations for Persons with Spinal Cord Injury. *Top. Spinal Cord Inj. Rehabil.* **2017**, *23*, 188–206. [CrossRef]
14. den Braber-Ymker, M.; Lammens, M.; van Putten, M.J.; Nagtegaal, I.D. The enteric nervous system and the musculature of the colon are altered in patients with spina bifida and spinal cord injury. *Virchows Arch.* **2017**, *470*, 175–184. [CrossRef] [PubMed]
15. Fajardo, N.R.; Pasiliao, R.V.; Modeste-Duncan, R.; Creasey, G.; Bauman, W.A.; Korsten, M.A. Decreased colonic motility in persons with chronic spinal cord injury. *Am. J. Gastroenterol.* **2003**, *98*, 128–134. [CrossRef] [PubMed]
16. Gondim, F.A.; de Oliveira, G.R.; Thomas, F.P. Upper gastrointestinal motility changes following spinal cord injury. *Neurogastroenterol. Motil.* **2010**, *22*, 2–6. [CrossRef]
17. Johns, J.S.; Krogh, K.; Ethans, K.; Chi, J.; Quérée, M.; Eng, J.J.; Spinal Cord Injury Research Evidence Team. Pharmacological Management of Neurogenic Bowel Dysfunction after Spinal Cord Injury and Multiple Sclerosis: A Systematic Review and Clinical Implications. *J. Clin. Med.* **2021**, *10*, 882. [CrossRef]
18. Kabatas, S.; Yu, D.; He, X.D.; Thatte, H.S.; Benedict, D.; Hepgul, K.T.; Black, P.M.; Sabharwal, S.; Teng, Y.D. Neural and anatomical abnormalities of the gastrointestinal system resulting from contusion spinal cord injury. *Neuroscience* **2008**, *154*, 1627–1638. [CrossRef]
19. Kigerl, K.A.; Mostacada, K.; Popovich, P.G. Gut Microbiota Are Disease-Modifying Factors After Traumatic Spinal Cord Injury. *Neurotherapeutics* **2018**, *15*, 60–67. [CrossRef]
20. Middleton, J.W.; Lim, K.; Taylor, L.; Soden, R.; Rutkowski, S. Patterns of morbidity and rehospitalisation following spinal cord injury. *Spinal Cord* **2004**, *42*, 359–367. [CrossRef]
21. Szarka, L. Dysmotility of the Small Intestine and Colon. In *Yamada's Textbook of Gastroenterology*; John Wiley & Sons, Ltd.: Oxford, UK, 2015; pp. 1154–1195.
22. Corsetti, M.; Costa, M.; Bassotti, G.; Bharucha, A.E.; Borrelli, O.; Dinning, P.; Di Lorenzo, C.; Huizinga, J.D.; Jimenez, M.; Rao, S.; et al. First translational consensus on terminology and definitions of colonic motility in animals and humans studied by manometric and other techniques. *Nat. Rev. Gastroenterol. Hepatol.* **2019**, *16*, 559–579. [CrossRef] [PubMed]
23. Kornum, D.S.; Terkelsen, A.J.; Bertoli, D.; Klinge, M.W.; Høyer, K.L.; Kufaishi, H.H.A.; Borghammer, P.; Drewes, A.M.; Brock, C.; Krogh, K. Assessment of Gastrointestinal Autonomic Dysfunction: Present and Future Perspectives. *J. Clin. Med.* **2021**, *10*, 1392. [CrossRef] [PubMed]
24. Feldman, E.R.; Singh, B.; Mishkin, N.G.; Lachenauer, E.R.; Martin-Flores, M.; Daugherity, E.K. Effects of Cisapride, Buprenorphine, and Their Combination on Gastrointestinal Transit in New Zealand White Rabbits. *J. Am. Assoc. Lab. Anim. Sci.* **2021**, *60*, 221–228. [CrossRef] [PubMed]
25. Holmes, G.M.; Hubscher, C.H.; Krassioukov, A.; Jakeman, L.B.; Kleitman, N. Recommendations for evaluation of bladder and bowel function in pre-clinical spinal cord injury research. *J. Spinal Cord Med.* **2020**, *43*, 165–176. [CrossRef] [PubMed]
26. Arbizu, R.A.; Rodriguez, L.A. Electrogastrography, Breath Tests, Ultrasonography, Transit Tests, and SmartPill. In *Pediatric Neurogastroenterology*; Faure, C., Thapar, N., Lorenzo, C.D., Eds.; Humana Press: Totowa, NJ, USA, 2017; pp. 169–179.
27. Marucci, S.; Zarzour, J.; Callaway, J. Anatomic and Physiologic Tests of Esophageal and Gastric Function. In *The SAGES Manual of Foregut Surgery*; Grams, J., Perry, K.A., Tavakkoli, A., Eds.; Springer: Cham, Switzerland, 2019; pp. 65–89.
28. Carter, D.; Bardan, E. The Wireless Motility Capsule. In *Gastrointestinal Motility Disorders a Point of Care Clinical Guide*; Bardan, E., Shaker, R., Eds.; Springer Nature: Cham, Switzerland, 2018; pp. 373–378.
29. Farmer, A.D.; Scott, S.M.; Hobson, A.R. Gastrointestinal motility revisited: The wireless motility capsule. *United Eur. Gastroenterol. J.* **2013**, *1*, 413–421. [CrossRef]
30. Hasler, W.L. The use of SmartPill for gastric monitoring. *Expert. Rev. Gastroenterol. Hepatol.* **2014**, *8*, 587–600. [CrossRef]
31. Koziolek, M.; Grimm, M.; Bollmann, T.; Schäfer, K.J.; Blattner, S.M.; Lotz, R.; Boeck, G.; Weitschies, W. Characterization of the GI transit conditions in Beagle dogs with a telemetric motility capsule. *Eur. J. Pharm. Biopharm.* **2019**, *136*, 221–230. [CrossRef]
32. Saad, R.J. The Wireless Motility Capsule: A One-Stop Shop for the Evaluation of GI Motility Disorders. *Curr. Gastroenterol. Rep.* **2016**, *18*, 14. [CrossRef]
33. Steiger, C.; Abramson, A.; Nadeau, P.; Chandrakasan, A.; Langer, R.; Traverso, G. Ingestible electronics for diagnostics and therapy. *Nat. Mater.* **2018**, *4*, 83–98. [CrossRef]
34. Lee, Y.Y.; Erdogan, A.; Rao, S.S. How to assess regional and whole gut transit time with wireless motility capsule. *J. Neurogastroenterol. Motil.* **2014**, *20*, 265–270. [CrossRef]

35. Henze, L.J.; Koehl, N.J.; Bennett-Lenane, H.; Holm, R.; Grimm, M.; Schneider, F.; Weitschies, W.; Koziolek, M.; Griffin, B.T. Characterization of gastrointestinal transit and luminal conditions in pigs using a telemetric motility capsule. *Eur. J. Pharm. Sci.* **2021**, *156*, 105627. [CrossRef] [PubMed]
36. Maqbool, S.; Parkman, H.P.; Friedenberg, F.K. Wireless capsule motility: Comparison of the SmartPill GI monitoring system with scintigraphy for measuring whole gut transit. *Dig. Dis. Sci.* **2009**, *54*, 2167–2174. [CrossRef] [PubMed]
37. Schneider, F.; Grimm, M.; Koziolek, M.; Modeß, C.; Dokter, A.; Roustom, T.; Siegmund, W.; Weitschies, W. Resolving the physiological conditions in bioavailability and bioequivalence studies: Comparison of fasted and fed state. *Eur. J. Pharm. Biopharm.* **2016**, *108*, 214–219. [CrossRef] [PubMed]
38. Wang, Y.T.; Mohammed, S.D.; Farmer, A.D.; Wang, D.; Zarate, N.; Hobson, A.R.; Hellström, P.M.; Semler, J.R.; Kuo, B.; Rao, S.S.; et al. Regional gastrointestinal transit and pH studied in 215 healthy volunteers using the wireless motility capsule: Influence of age, gender, study country and testing protocol. *Aliment. Pharmacol. Ther.* **2015**, *42*, 761–772. [CrossRef]
39. Williams, R.E., 3rd; Bauman, W.A.; Spungen, A.M.; Vinnakota, R.R.; Farid, R.Z.; Galea, M.; Korsten, M.A. SmartPill technology provides safe and effective assessment of gastrointestinal function in persons with spinal cord injury. *Spinal Cord* **2012**, *50*, 81–84. [CrossRef]
40. Kim, K.T.; Streijger, F.; Manouchehri, N.; So, K.; Shortt, K.; Okon, E.B.; Tigchelaar, S.; Cripton, P.; Kwon, B.K. Review of the UBC Porcine Model of Traumatic Spinal Cord Injury. *J. Korean Neurosurg. Soc.* **2018**, *61*, 539–547. [CrossRef]
41. Lee, J.H.; Jones, C.F.; Okon, E.B.; Anderson, L.; Tigchelaar, S.; Kooner, P.; Godbey, T.; Chua, B.; Gray, G.; Hildebrandt, R.; et al. A novel porcine model of traumatic thoracic spinal cord injury. *J. Neurotrauma* **2013**, *30*, 142–159. [CrossRef]
42. Schomberg, D.T.; Tellez, A.; Meudt, J.J.; Brady, D.A.; Dillon, K.N.; Arowolo, F.K.; Wicks, J.; Rousselle, S.D.; Shanmuganayagam, D. Miniature Swine for Preclinical Modeling of Complexities of Human Disease for Translational Scientific Discovery and Accelerated Development of Therapies and Medical Devices. *Toxicol. Pathol.* **2016**, *44*, 299–314. [CrossRef]
43. Toossi, A.; Bergin, B.; Marefatallah, M.; Parhizi, B.; Tyreman, N.; Everaert, D.G.; Rezaei, S.; Seres, P.; Gatenby, J.C.; Perlmutter, S.I.; et al. Comparative neuroanatomy of the lumbosacral spinal cord of the rat, cat, pig, monkey, and human. *Sci. Rep.* **2021**, *11*, 1955. [CrossRef]
44. Ziegler, A.; Gonzalez, L.; Blikslager, A. Large Animal Models: The Key to Translational Discovery in Digestive Disease Research. *Cell. Mol. Gastroenterol. Hepatol.* **2016**, *2*, 716–724. [CrossRef]
45. Ahmed, R.U.; Knibbe, C.A.; Wilkins, F.; Sherwood, L.C.; Howland, D.R.; Boakye, M. Porcine spinal cord injury model for translational research across multiple functional systems. *Exp. Neurol.* **2023**, *359*, 114267. [CrossRef]
46. National Research Council. *Guidance for the Description of Animal Research in Scientific Publications*; National Academies Press: Washington, DC, USA, 2011.
47. Feng, W.; Wu, Y.; Chen, G.; Fu, S.; Li, B.; Huang, B.; Wang, D.; Wang, W.; Liu, J. Sodium Butyrate Attenuates Diarrhea in Weaned Piglets and Promotes Tight Junction Protein Expression in Colon in a GPR109A-Dependent Manner. *Cell. Physiol. Biochem.* **2018**, *47*, 1617–1629. [CrossRef] [PubMed]
48. Koziolek, M.; Schneider, F.; Grimm, M.; Modeβ, C.; Seekamp, A.; Roustom, T.; Siegmund, W.; Weitschies, W. Intragastric pH and pressure profiles after intake of the high-caloric, high-fat meal as used for food effect studies. *J. Control. Release* **2015**, *220*, 71–78. [CrossRef] [PubMed]
49. Gondim, F.A.; Alencar, H.M.; Rodrigues, C.L.; da Graça, J.R.; dos Santos, A.A.; Rola, F.H. Complete cervical or thoracic spinal cord transections delay gastric emptying and gastrointestinal transit of liquid in awake rats. *Spinal Cord* **1999**, *37*, 793–799. [CrossRef] [PubMed]
50. Gondim, F.A.; Rodrigues, C.L.; da Graça, J.R.; Camurça, F.D.; de Alencar, H.M.; dos Santos, A.A.; Rola, F.H. Neural mechanisms involved in the delay of gastric emptying and gastrointestinal transit of liquid after thoracic spinal cord transection in awake rats. *Auton. Neurosci.* **2001**, *87*, 52–58. [CrossRef]
51. Faaborg, P.M.; Christensen, P.; Finnerup, N.; Laurberg, S.; Krogh, K. The pattern of colorectal dysfunction changes with time since spinal cord injury. *Spinal Cord* **2008**, *46*, 234–238. [CrossRef]
52. Frias, B.; Phillips, A.A.; Squair, J.W.; Lee, A.H.X.; Laher, I.; Krassioukov, A.V. Reduced colonic smooth muscle cholinergic responsiveness is associated with impaired bowel motility after chronic experimental high-level spinal cord injury. *Auton. Neurosci.* **2019**, *216*, 33–38. [CrossRef]
53. Gondim, F.A.; Lopes, A.C., Jr.; Cruz, P.R.; Medeiros, B.A.; Queiroz, D.A.; Santos, A.A.; Rola, F.H. On the complex autonomic changes involved in the inhibition of gastrointestinal motility after spinal cord injury (SCI). *Dig. Dis. Sci.* **2006**, *51*, 1136. [CrossRef]
54. Hou, S.; Rabchevsky, A.G. Autonomic consequences of spinal cord injury. *Compr. Physiol.* **2014**, *4*, 1419–1453. [CrossRef]
55. Round, A.M.; Joo, M.C.; Barakso, C.M.; Fallah, N.; Noonan, V.K.; Krassioukov, A.V. Neurogenic Bowel in Acute Rehabilitation Following Spinal Cord Injury: Impact of Laxatives and Opioids. *J. Clin. Med.* **2021**, *10*, 1673. [CrossRef]
56. Daniel, H.; Gholami, A.M.; Berry, D.; Desmarchelier, C.; Hahne, H.; Loh, G.; Mondot, S.; Lepage, P.; Rothballer, M.; Walker, A.; et al. High-fat diet alters gut microbiota physiology in mice. *ISME J.* **2014**, *8*, 295–308. [CrossRef] [PubMed]
57. Du, J.; Zayed, A.A.; Kigerl, K.A.; Zane, K.; Sullivan, M.B.; Popovich, P.G. Spinal Cord Injury Changes the Structure and Functional Potential of Gut Bacterial and Viral Communities. *mSystems* **2021**, *6*, e01356-20. [CrossRef]
58. Balsa, I.M.; Culp, W.T.N.; Drobatz, K.J.; Johnson, E.G.; Mayhew, P.D.; Marks, S.L. Effect of Laparoscopic-assisted Gastropexy on Gastrointestinal Transit Time in Dogs. *J. Vet. Intern. Med.* **2017**, *31*, 1680–1685. [CrossRef] [PubMed]

59. Rauch, S.; Johannes, A.; Zollhöfer, B.; Muellenbach, R.M. Evaluating intra-abdominal pressures in a porcine model of acute lung injury by using a wireless motility capsule. *Med. Sci. Monit.* **2012**, *18*, BR163–BR166. [CrossRef] [PubMed]
60. Suenderhauf, C.; Tuffin, G.; Lorentsen, H.; Grimm, H.P.; Flament, C.; Parrott, N. Pharmacokinetics of paracetamol in Göttingen minipigs: In vivo studies and modeling to elucidate physiological determinants of absorption. *Pharm. Res.* **2014**, *31*, 2696–2707. [CrossRef]
61. Laber, K.E.; Whary, M.T.; Bingel, S.A.; Goodrich, J.A.; Smith, A.C.; Swindle, M.M. Biology and Diseases of Swine. *Lab. Anim. Med.* **2002**, *52*, 615–673.
62. Nelson, L.; Schwaner, R. Transdermal fentanyl: Pharmacology and toxicology. *J. Med. Toxicol.* **2009**, *5*, 230–241. [CrossRef]
63. Coleman, K.A.; Boscan, P.; Ferguson, L.; Twedt, D.; Monnet, E. Evaluation of gastric motility in nine dogs before and after prophylactic laparoscopic gastropexy: A pilot study. *Aust. Vet. J.* **2019**, *97*, 225–230. [CrossRef]

Disclaimer/Publisher's Note: The statements, opinions and data contained in all publications are solely those of the individual author(s) and contributor(s) and not of MDPI and/or the editor(s). MDPI and/or the editor(s) disclaim responsibility for any injury to people or property resulting from any ideas, methods, instructions or products referred to in the content.

Article

A Touchscreen Device for Behavioral Testing in Pigs

Will Ao [1], Megan Grace [1], Candace L. Floyd [2] and Cole Vonder Haar [1,*]

[1] Injury and Recovery Laboratory, Department of Neuroscience, Ohio State University, 460 W 12th Ave, Columbus, OH 43210, USA
[2] Department of Physical Medicine and Rehabilitation, University of Utah, Salt Lake City, UT 84132, USA
* Correspondence: cole.vonderhaar@osumc.edu

Abstract: Pigs are becoming more common research models due to their utility in studying neurological conditions such as traumatic brain injury, Alzheimer's disease, and Huntington's Disease. However, behavioral tasks often require a large apparatus and are not automated, which may disinterest researchers in using important functional measures. To address this, we developed a touchscreen that pigs could be trained on for behavioral testing. A rack-mounted touchscreen monitor was placed in an enclosed, weighted audio rack. A pellet dispenser was operated by a radio frequency transceiver to deliver fruit-flavored sugar pellets from across the testing room. Programs were custom written in Python and executed on a microcomputer. A behavioral shaping program was designed to train pigs to interact with the screen and setup responses for future tasks. Pigs rapidly learned to interact with the screen. To demonstrate efficacy in more complex behavior, two pigs were trained on a delay discounting tasks and two pigs on a color discrimination task. The device held up to repeated testing of large pigs and could be adjusted to the height of minipigs. The device can be easily recreated and constructed at a relatively low cost. Research topics ranging from brain injury to pharmacology to vision could benefit from behavioral tasks designed to specifically interrogate relevant function. More work will be needed to develop tests which are of specific relevance to these disciplines.

Keywords: cognition; behavior; operant; brain injury; Alzheimer's disease; transgenic

1. Introduction

Historically, the rodent has been the model of choice for neuroscience research. There are several reasons for this, including the economical nature, both in cost and space compared to larger species. As such, an abundance of methods and tools have been developed for use in rodents. Behavioral assessment is particularly well-developed with a number of sensorimotor tests and more complex tasks which measure human-relevant functions such as decision-making, working memory, and self-control. However, there are also concerns with rodent models. The physiology is dissimilar to humans, particularly the brain structure and size. To effectively develop a translational pipeline for neurotherapeutics and to understand neuropathology, larger species with greater brain similarities to humans are needed.

One such existing model is the pig, already well-established for the study of circulatory, nervous, and respiratory function due to their physiological similarities to the human [1]. Pigs are especially attractive in neuroscience because of a high degree of similarity to human brains in the sulci and gyri, with gyrification far exceeding the rodent model and even exceeding a common non-human primate model, the rhesus macaque [2]. Given the greater anatomical homology, CNS diseases, injuries, and challenges in pigs are much more likely to cause pathology of greater similarity to humans relative to rodent models. For example, the pig cortex is more compressible than rodents [3]. Pig brains are also considerably larger than their rodent counterparts, weighing in at 95.3 g relative to less than 2.5 g in the case of the rat. This size again exceeds the rhesus macaque (90 g) [2].

Recently, several fields have increased the use of pig models. Pigs have become prominent in the field of traumatic brain injury (TBI), where they serve multiple purposes such as analyzing pathophysiology, understanding surgical management of injury, and even studies on recovery of function [4]. In addition, the recent development of transgenic Göttingen minipigs have created a strong model for studying the pathology of Alzheimer's disease (AD; amyloid precursor protein and presenilin-1 mutations) and Huntington's disease (HD; huntingtin mutation) [5,6], while a transgenic of the Minnesota minipig has been created for cancer research (floxed line for cre-dependent tumor expression) [7]. The area of early developmental challenge has adopted pigs for both prenatal (e.g., hypoxia, nutrient restriction) and postnatal (e.g., hypoxia/ischemia) challenges [8,9]. Notably, each of these conditions strongly impact behavioral function, however porcine cognitive assays are relatively limited. Simple discriminations, reversals, and working memory tasks have been used in the case of TBI [10,11]. Motivation or ability to uncover hidden treats has been used in the HD transgenic model [12] and a recognition memory task (novel object test) in the AD transgenic model [13]. However, even these tasks do not always distinguish the condition from control and small effect sizes of injury or disease may limit detection of treatment effects. In surveying this literature, we have identified three primary barriers to expanding cognitive testing in pigs: (1) specialized equipment requirements, (2) behavioral expertise of the experimenters, and (3) study time constraints. While item number 3 will be inherently study dependent, the first two challenges can be addressed to some degree with technology.

Historically, both rodents and primates had purpose-designed behavioral equipment. Perhaps most notable is the operant chamber, which is a modular, computer-controlled chamber in which many different cognitive tests can be assessed and has a small space footprint. In contrast, current functional assays for pig behavior often require large, room-sized apparatus and manual setup and scoring. Even traditional T-mazes or multi-arm mazes become a challenge due to the space constraints and are inherently low throughput. Manual tasks also introduce the possibility of unconscious experimenter bias. While uncommon, several researchers have adapted automated devices (including a primate operant chamber) for testing cognitive function in pigs. These devices have been used to assess complex cognitive behaviors such as behavioral inhibition [14], working memory [15], and behavioral flexibility [16]. Others have set up similarly sophisticated behavioral assays, including gambling-like behavior [17], choice impulsivity [18], and working memory [19] but required full manual administration and/or room-sized mazes. The heterogeneity in testing apparatus and manual nature of many are likely a leading reason for sparse adoption of pig cognitive outcomes. As such, the development of a small footprint, low-cost platform with the capability to perform high throughput operant measures is needed.

To develop such a device, we can make use of touchscreen technology, open-source software, and readily available equipment. This may provide the benefit of standardizing methods across species, including humans. A recent argument has been made that this will help close the gaps in fields that have struggled with translation, such as pharmacology [20], although task similarity is likely a more important component than test medium (e.g., touchscreen). In the current paper, we describe the development of a device to make behavioral research for the porcine model more accessible to a wide variety of researchers. We utilized relatively low-cost materials and open-source software to create a robust tool for behavioral analyses. This can be constructed in the average laboratory environment by ordering the commercial pieces and assembling or can be made even lower cost with some modifications noted in the methods.

2. Materials and Methods

2.1. Touchscreen Design Overview

The overall device design and finished product is shown in Figure 1. Two versions were constructed; the first iteration was susceptible to damage from the pigs and is only briefly described in the results for transparency (but can be viewed in Video S1). Subsequent methods will refer to the more durable, second iteration. A touchscreen monitor was attached to a microcomputer (Raspberry Pi) which ran custom behavioral programs. In this final iteration, a radio frequency (RF) transceiver was attached to the input/output pins of the Raspberry Pi. This allowed pellet delivery to be located across the room to reduce rooting behavior toward the screen. The RF transceiver communicated to a second Raspberry Pi with a receiver. The second Raspberry Pi input/output pins were hooked up to a printed circuit board (PCB). The PCB contained 8 inputs and 8 outputs capable of handling standard 28 V operant equipment. In the iteration described here, only a pellet dispenser and Sonalert tone generator were attached as outputs.

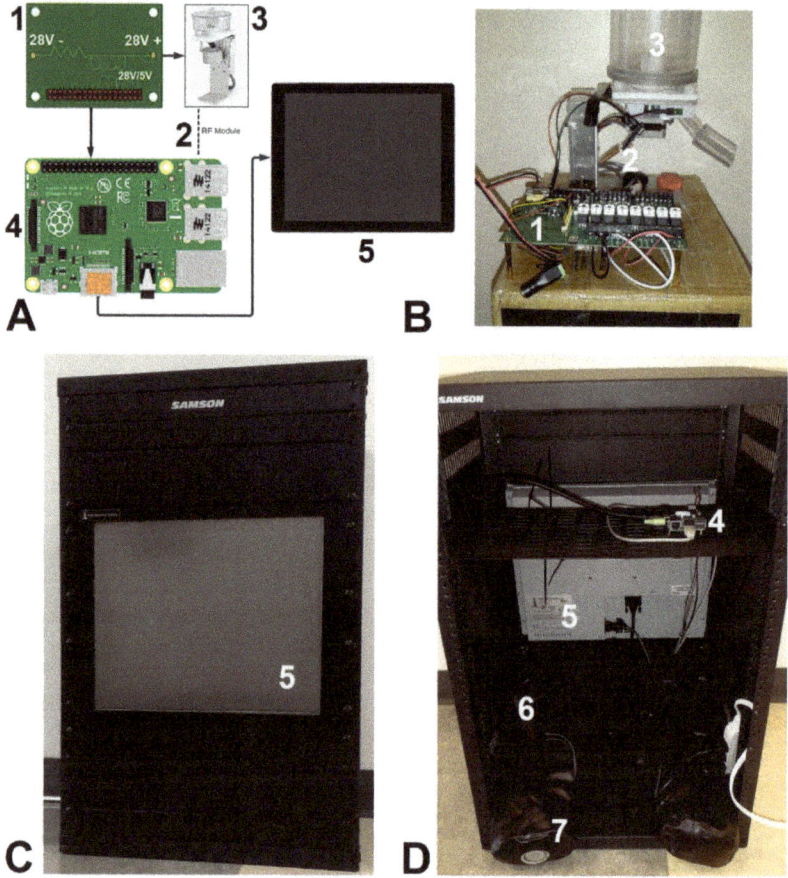

Figure 1. Schematic layout and actual device. (**A**) The conceptual schematic organization of the touchscreen device with pieces numbered in black, corresponding to white numbers on the actual device (panels (**B**–**D**)). A Raspberry Pi controls the touchscreen and records responses. Output is then taken from the Raspberry Pi I/O pins and put through an external printed circuit board to step the voltage up to the 28 V needed for peripheral components. Physical inputs were available but not used in the current studies. (**B**) An image of the remote-controlled dispenser, attached to the PCB. A raspberry Pi is attached underneath the PCB and not visible. Not in this picture is the sonalert tone generator which was also hooked up to the PCB. (**C**) Front of testing device showing screen. (**D**) Rear of testing device showing back of screen, Pi, and other components. Numbered items indicate: (1) PCB, (2) RF transceiver, (3) pellet dispenser, (4) Raspberry Pi, (5) touchscreen, (6) audio speakers, (7) weighted sandbags.

2.2. Physical Components and Construction

All components are detailed in Table 1 along with the cost to acquire at time of purchase. The below description is for the final iteration of the device.

Table 1. List of physical components and pricing as of purchase. Sums in each category are given in bold. * The board and components were purchased as a set of 5 to meet minimum order thresholds; price given is per board. ** A smaller rack would have been suitable even for the large Yorkshire pigs. Future builds will use the SRK-12 (12U height). *** The rack shelf was not explicitly necessary, but gave easy access to the Raspberry Pi.

Item	Parts	Price	Vendor
Computer		**$148**	
	Raspberry Pi 4B (2×)	$110	SparkFun
	Power supply (2×)	$16	SparkFun
	Misc. jumpers, etc.	$10	SparkFun
	RF Transceivers	$12	Amazon
PCB		**$39**	
	Board *	$19	PCBWay
	Misc. components *	$20	Digikey
Peripherals		**$799**	
	Pellet Dispenser	$729	Med-Associates
	Sonalert	$70	Med-Associates
	Desktop speakers	$0	in lab
Monitor		**$855**	
	19″ Rack-mounted Resistive	$835	Hope Industrial
	Screen protector	$20	Hope Industrial
Frame		**$238**	
	Samson SRK-16 **	$165	B&H Audio-Video
	1U blank panel (4×)	$18	B&H Audio-Video
	2U blank panels (2×)	$12	B&H Audio-Video
	rack shelf ***	$15	B&H Audio-Video
	Sandbags (2×)	$28	Amazon
Total		**$2079**	

2.2.1. Frame

Standard 19″ wide racking for servers or audio equipment was used to house the screen. It was enclosed on two sides and 31″ high (16U rack height), 18″ deep. Rack "blanks" were used to fill the front not occupied by the screen and the rear was left open but put against a wall. A rack-mounted shelf was put in the rear to hold the Raspberry Pi. The modular nature of the frame allowed the screen to be moved up or down to deal with smaller or larger pigs. A pair of sandbags (9 kg each) were purchased and filled to weigh down the device so that pigs could not move it.

2.2.2. Screen

A rack-mounted resistive touchscreen (Hope Industrial Systems, HIS-RL19-CTDH) designed for industrial use was mounted and adjusted for the height of the pigs. A plastic screen protector was placed over the screen to reduce scratches and for easier cleaning.

2.2.3. Circuit Board Interface

A PCB was custom-designed to take 28 V input/output to 3.3 V for interfacing to the Raspberry Pi. For outputs an optical switch was used to isolate the 3.3 V from 28 V. For inputs, resistors were used to step down the voltage. A pass-through so that the board could power the Pi was used.

2.2.4. RF Transmitter/Receiver

A simple 433 MHz radio frequency transceiver/receiver combination were purchased, and an antenna soldered to each. Software was modified to turn received signals into outputs (pellet, tone) on the receiving Raspberry Pi and to transmit them on the sending Raspberry Pi. The pellet dispenser was attached to a box which sat on top of a sink, raising it above the reach of the pigs.

2.2.5. Raspberry Pi

Two Raspberry Pi model 4B were used, along with power supplies and jumper cables to connect. These microcomputers allowed for a full range of function to be controlled, including input/output switches to the PCB, radio signals, and the software which recorded responses.

2.2.6. Peripheral Components

A pair of desktop computer speakers were placed in the bottom of the frame to provide auditory feedback. A pellet dispenser (ENV-203-1000; Med-Associates, St. Albans, VT, USA) was connected to the receiving raspberry pi via the PCB, located across the room. A Sonalert tone generator (ENV-223AM; 2900 Hz, 100 dB; Med-Associates) was also attached alongside the pellet dispenser.

2.2.7. Alternative Materials

The largest costs in construction are the monitor, pellet dispenser, and frame. Cheaper framing materials may be available from other vendors, although we recommend a reputable brand that will stand up to pigs interacting with it. Other rack-mounted screen options are also available, although price points vary widely. We recommend choosing one that has a reasonable impact rating and does not protrude from the racking as pigs may chew at the corners. For the dispenser, it is possible to 3D print and purchase small motors to operate it as described in papers for rats [21,22]. Alternate food delivery systems would likely also be suitable.

2.3. Software Components

2.3.1. Peripheral Device Control

A program to send and receive RF signals in Linux was modified from RFChat [23]. A program was written and ran on boot of the receiving Raspberry Pi. On receipt of a given cue (e.g., "1"), the pellet dispenser would cycle. On receipt of another cue (e.g., "2"), the tone would turn on for 1 s. Thus, as long as the board attached to the receiving Raspberry Pi was powered, it would control the peripheral devices. Behavioral programs on the sending Raspberry Pi used these commands on relevant events (i.e., reinforcement).

2.3.2. Graphical User Interface (GUI)

The Python Tkinter package was used to develop GUIs. For human user input, a pop-over box with options to change variables (e.g., subject number, training stage, etc.) in the underlying behavioral program populated at the start. A program for a touchscreen numpad was written to allow numbers to be input without a keyboard (a wireless keyboard may also be used). For pig responses, buttons housed within a full screen window were used. Buttons were designated to respond on initial touch (default is release of click or removal of touch).

2.3.3. Data Recording

Every response was recorded as a new line in a comma-separated value file with a separate file for each subject. Each behavioral program recorded information relating to the individual trial as well. Summary data were reported on the screen at the end for daily monitoring of progress.

2.3.4. Behavioral Testing Programs

Custom programs were written in Python according to the descriptions below. A common shaping program was used to train initial response to the screen and to smaller boxes within the screen.

2.3.5. Data Transfer To/From Raspberry Pi

An FTP server (vsftpd package) was setup on the raspberry pi with a folder to which a remote computer could read and write files. An FTP transfer utility (FileZilla) was used to move behavioral programs onto the Pi and pull data files for analysis.

2.4. Evaluation of Device Durability

The core goal of the current studies was to determine if a touchscreen device would withstand repeated testing with pigs. In our first iteration of the device, the screen was broken after 28 sessions with two pigs. In the second iteration, which is what is described above, the screen withstood testing throughout all 35 sessions and remains intact.

2.5. Subjects

Male (castrated) and female Yorkshire pigs were used in the described experiments. Two male pigs were 9 weeks of age at start of training, weighed 24–26 kg and were tested in experiment 1. Two female pigs were 12 weeks of age at start of training, weighed 45–52 kg and were tested in experiment 2. Pigs were obtained from the Ohio State University farms and were acclimated to the vivarium for one week prior to testing. Pigs were housed in individual pens adjacent to one another. Males and females were housed separately. All procedures were approved by the Ohio State University Institutional Animal Care and Use Committee.

2.6. Behavioral Training
2.6.1. Reinforcers

Mini-marshmallows were used as reinforcers during pre-training and 1 g fruit-flavored sucrose pellets (F05478, F05711; Bio-Serv, Flemington, NJ, USA) were delivered from the pellet dispenser while pigs interacted with the touchscreen.

2.6.2. Pre-Training of Pigs

The goal of this step was to familiarize pigs with the experimenter, get them used to leash walking, and traveling to the testing room using basic behavior shaping techniques with a clicker. In their home room, pigs were trained to associate a clicker with mini marshmallow delivery and approach the experimenter to receive the reinforcer. The experimenter then familiarized them with the leash by draping it across them, then wrapping it around them, while providing reinforcers. Once comfortable, a large dog harness was placed over the shoulders and clipped behind the legs. Pigs were then trained to walk on the lead in the home room while receiving reinforcers. Once comfortable, the outer hallway was blocked off (either physical blockade, or second researcher with a board) and pigs were taken back and forth down the hallway. Once pigs were responding well to the leash, they were led to the behavior testing room. This process could be accelerated with multiple sessions per day if needed.

2.6.3. Response Shaping

To shape responses to the touchscreen, a multistage procedure was followed. Pigs moved up in stages automatically within the program or were started at later stages if they had completed the prerequisite the session before. Audio speakers inside the device provided auditory feedback when the button was pressed (2900 Hz tone) and when a trial began (7500 Hz tone).

Stage 0 was a Pavlovian autoshaping procedure in which the entire screen illuminated (yellow color) and then a pellet was delivered 10 s following illumination. However, there was also a fixed ratio (FR)-1 schedule in effect such that at any given time, a press to anywhere on the screen would be reinforced. If a pig did not contact the screen, ketchup was wiped on the screen to motivate approach. After 20 presses, pigs moved to stage 1. Stage 1 made the response conditional-presses were only reinforced when the screen was illuminated (FR-1 schedule). After 15 presses, pigs moved to stage 2. Stage 2 reduced the size of the response box from the entire screen to a large, illuminated (yellow) rectangle occupying 1/3 of the screen. The rectangle was positioned randomly at one of three heights to shape responses to track the change in position. Responses to the rectangle were reinforced (FR-1 schedule). After 15 presses, pigs moved to stage 3. Stage 3 reduced the size of the response box further to a square (yellow, 40% of screen width) which was positioned randomly in one of five positions (just offset from each corner, and center). After 40 presses, pigs were considered ready for testing. A few optional manipulations may be considered at this point. If a higher response requirement will be needed for subsequent tasks, stage 3 may be increased to FR-3 or FR-5. If smaller response boxes will be needed, a stage 4 where the box size shrinks gradually over successive correct trials may help shape precision. Close attention should be paid throughout to the behavioral topology, or the way in which responses are made and how that may affect subsequent tasks. See results for qualitative descriptions of pig behavioral topology in interacting with the device.

2.7. Behavioral Assessment of Pigs

After shaping nose pokes to illuminated buttons, pigs were tested in one of two experiments. Experiment 1 used two male pigs, while Experiment 2 used two female pigs. Because a primary goal was to test the device, many minor adjustments were made throughout the experiments to optimize the pig's responses. Thus, these may serve better as proof of concept that pigs can be trained on a task rather than strong baseline data for either task.

2.7.1. Experiment 1—Delay Discounting Task

The goal of this behavior is to assess choice impulsivity [24]. This experiment was performed on the first iteration of the touchscreen device, which ultimately broke. After learning to respond to boxes on the screen, pigs were presented with a magnitude discrimination of two buttons. One delivered 4 pellets ("Large" button), and the other 1 pellet ("Small" button). 6 forced-choice and 6 free-choice trials were given. The first step was to train a magnitude discrimination such that pigs showed preference for the Large button. After that, delays to reinforcement on the Large button were then introduced progressively across the session every 12 trials (0, 5, 10, 20 s). Because pigs rapidly became delay averse, several behavioral manipulations were made to improve stability of choice (described in results). These extended modifications served to provide a long period of assessment for the device.

2.7.2. Experiment 2—Visual Discrimination

The goal of this behavior is to assess the ability to discriminate based on color. This experiment was performed on the second iteration of the touchscreen which is described fully in the methods. After learning to respond to buttons on the screen, pigs were presented with a yellow box on the center of the screen, and after a response, a yellow and blue box on the left and right side of the screen (pseudorandomly presented). Like in experiment 1, multiple behavioral manipulations were performed to improve performance and are described in the results.

3. Results

3.1. Device Durability

The first iteration of the touchscreen device was ultimately not strong enough and was not described above. An aluminum base and frame mounted a conventional capacitive (home/office grade) touchscreen monitor and was enclosed in a plexiglass covering.

Because pigs could get under the plexiglass box, they tended to root at it, and the frame was not heavy enough to prevent this. Ultimately, they broke the screen despite repeated attempts to shape behavior away from such rough interactions. Much of this damage came from lifting the frame and letting it fall, thus it may still be possible to use a capacitive touchscreen in the second iteration, however researchers should anticipate frequent replacements as a common screen protector will not be sufficient long term.

The second iteration, which is fully described in the methods, was much more robust. At the conclusion of the experiment, there was no obvious damage to the device, although pigs did begin to root at it more and additional weight in the bottom may have been beneficial to minimize this. A slightly smaller version of the frame (12U-rack) may have been suitable as well to reduce movement when pushed on and would still accommodate large Yorkshire pigs.

3.2. Behavioral Topology of Pigs on the Touchscreen

3.2.1. Capacitive vs. Resistive Touchscreen

Version 1 of this device used a capacitive touchscreen (similar to modern smartphones) while version 2 (described in methods) used a resistive touchscreen (similar to ordering kiosks). Pigs learned initial touch responses more rapidly to the capacitive touchscreen, but it is unlikely to hold up to long-term testing. An alternative might be for researchers to swap a capacitive screen in for the initial stages of training, then replace with a resistive screen.

3.2.2. Presses on Touchscreen

Overall, pigs were reasonably accurate at pressing buttons. However, it should be noted that the tendency is to press and then swipe upward at a slight diagonal. This resulted in many initial problems as the software was designed to record/act on button release rather than press. Even after fixing the buttons, pigs would often have inaccurate responses slightly above the button. For one (male) pig, this resulted in more pronounced swiping behavior. A future option may be to present a slightly larger but invisible response box which extends above the visible button. Care should be taken in considering the layout of buttons in a task.

3.2.3. Rooting

Pigs engaged in rooting behavior as noted above. This was largely mitigated in the second version by a frame which they could not get their nose under. However, there were still rooting behaviors present. This seems to be most mitigated by lack of device movement. If an object moves, the pig is more likely to root at it. Smaller/heavier objects are likely to be best for this. For the first experiment, which was performed with a weaker device, rooting behavior was mitigated by instigating a differential reinforcement of other (DRO) behavior schedule during the intertrial intervals. This involved periodically giving marshmallows in other locations of the room to reinforce exploratory behaviors instead of rooting at the screen. In the second version of the device, the pellet dispenser was attached to a remote to reduce rooting toward the screen.

3.2.4. Responsiveness to Tones

Tones were added to each response to help pigs discriminate when a response was made. Because there is no tactile feedback on the touchscreen, this helped to distinguish that a press had been recorded. Pigs robustly responded to tone presentation when paired with a reinforcer. A different pitch tone was also used to indicate the start of a trial and was reasonably successful.

3.2.5. Frustration and Sensitivity to Reinforcer Magnitude

Pigs were very sensitive to when reinforcement was withheld. This was evident on the few occasions the pellet dispenser failed to deliver: an experimenter would have to give a marshmallow in most cases before pigs would leave the pellet area and re-engage with the touchscreen. Experiment 1 also demonstrated immediate aversion to delayed reinforcement, and similar frustrations were seen on the transition from FR-1 to FR-3 in Experiment 2. Conversely, larger numbers of pellets motivated more rapid re-engagement with the task. Gradually changing response requirements or reinforcer density is recommended.

3.2.6. Competing Exploration

All pigs explored extensively during intertrial intervals and often during trials themselves. Items on the walls (e.g., sink, hose) competed for the attention of the pigs. This was more drastic in Experiment 2 which took place in a larger room. The addition of a tone to indicate a trial start helped pigs to orient to the device. Trial durations were increased to allow more time to respond. Intertrial intervals were kept short (<20 s). A reasonably small, plain room with minimal distractions is recommend.

3.3. Device Components & Evaluation

3.3.1. First Touchscreen Device

The first iteration of our touchscreen broke as described in Device Durability, above. From this, we identified that the standard capacitive (office/home grade) touchscreen was not strong enough. This was largely for two reasons, (1) food was delivered from just under the screen (hopper behind screen and dropped out just in front of screen) and (2) pigs could get under it to lift with their snout. These problems were both changed for the second iteration which is what is fully described in the methods. Other core components, including pellet dispenser, step up/down PCB, and raspberry pi for operation were satisfactory. Supplemental Video S1 shows pigs learning to push the buttons on this version.

3.3.2. Second Touchscreen Device

Performance of all components of the second iteration were satisfactory and are fully described in the methods. Some initial problems were found with the infrared remote control of the dispenser. Stray infrared signals will be picked up, so care should be taken to monitor which commands are received from other electronics in the area and select input numbers which will not occur from other electronics. It may be advisable to adjust the program such that it periodically turns off all outputs in the absence of a received "on" signal or to use encoding to secure against stray interference. We located the pellet dispenser (and tone generator, and Pi/PCB) on a box placed on top of a sink because it was a convenient method to elevate it above the pigs' reach. However, a wall-mounted shelf, or 3D-printed hanger would also be sufficient.

3.3.3. Extension to Rodents or Other Animals

We plugged the printed PCB and Pi into a Med-Associates rat operant testing chamber and were able to control 28 V inputs (two levers, one nose poke) and outputs (pellet dispenser, houselight, levers' extension, food hopper light) using the standard 7" Raspberry Pi touchscreen. Thus, the same equipment should be capable of running experiments for other animals. However, care should be taken when going across species. For instance, rats and mice do not respond well to a capacitive touchscreen in our experience. This may be why many current commercial solutions use infrared touchscreens in currently available equipment.

3.4. Pre-Training and Response Shaping

Pigs were habituated to the researchers through feeding of treats for two days. Pigs were then acclimated to leash walking to the testing room over the course of 2–5 days as described in the methods. The younger, male pigs (experiment 1) took longer, while the female pigs (experiment 2) were able to be put on the leash on the first day.

Pigs were then trained to respond to the touchscreen and gradually shaped to respond to a small, yellow box which moved around the screen. For the male pigs in Experiment 1, seven total sessions were required for them to accurately track the small box on the screen. For the female pigs in Experiment 2, the initial touch response took longer, presumably due to the resistive screen requiring a firm press compared to the capacitive screen in the first device. This was solved by modifying the program to allow any touch to the screen to be immediately reinforced and then by placing ketchup on the screen for one pig. After this, 4–6 sessions were required to shape the response to small boxes. Three sessions additional training was then performed to determine how small of a box the pigs could accurately respond to. Pigs were able to respond to a square occupying as little as 22% of the screen height. Supplemental Video S2 shows pigs progressing through the stages of response shaping.

Based on these results, we would recommend researchers habituating pigs in vivarium to experimenter handling, leash, and walking over the course of 5 sessions. The response shaping procedure could likewise be accomplished in 5–7 sessions. Though we tested daily, this timeline could be sped up with multiple sessions per day.

3.5. Experiment 1—Delay Discounting Task

After response shaping, pigs began the magnitude discrimination. Pigs rapidly acquired preference for the Large button across 3–5 sessions. Delays of 0, 5, 10, and 20 s (incremented every 12 trials) were then introduced to the Large button. This caused on increase in initiation latency (Figure 2A), but no large change in omissions (Figure 2B), which held mostly constant across subsequent testing. However, pigs became extremely averse to the delay as indicated by a drop in the Area Under the Curve (AUC) across five sessions (Figure 2C). The AUC represents the proportion of total choices of the large lever (minimum: 0; maximum: 1). By the final session, almost no choice of the Large lever was made, even at 0 s delay. Because pigs were so averse to these longer delays, the Buttons were reversed and delays decreased to 0, 2, 6, and 10 s. However, pigs again rapidly lost preference (5 sessions) for the Large lever, even at the lowest delays. To attempt to fix this, we reversed the Buttons again and implemented another magnitude discrimination. Within 3 sessions pigs strongly preferred the large lever with no delay. We then implemented another set of delays at 0, 1, 2, and 4 s. One pig tolerated this well and the other displayed some discounting but still preference for the large lever. At this point in the experiment, the first iteration of our device broke, and we were unable to further evaluate whether these delays could be titrated out further in a more gradual fashion.

3.6. Experiment 2—Color Discrimination

After the response shaping, pigs began the color discrimination. Choices of the yellow button were reinforced (blue was the comparison color). During the first phase free choices were made throughout the session. During the second phase, a correction trial was implemented such that if an incorrect choice was made, the same trial was immediately re-presented but the incorrect option could not be chosen (box appeared but as inactive). Once pigs were accurate at this, the third phase presented a more difficult conditional discrimination where a single-colored button (green or blue) was presented in the center and then only choices of that color were reinforced. As performance degraded in the third phase, an FR-3 requirement was put on the center (comparison) button and also on the choice buttons to try and increase salience. Pigs initiated trials within approximately 8 s, but latencies to start increased as performance went down in the conditional discrimination (Figure 3A). Similarly, number of trials completed was similar for each pig across all conditions until performance went down (Figure 3B). Omissions were initially high for one pig but remained low throughout the rest of testing (Figure 3C). Accuracy on the task was very low until the correction trial was implemented and then rapidly increased over 1–2 sessions. Performance dropped sharply during the conditional discrimination, suggesting additional training would be needed. The FR-3 requirement was not sufficient to rescue accuracy (Figure 3D). Pigs also had a tendency toward side biases (Figure 3E) and a bias toward the green color once the conditional discrimination was implemented (Figure 3F). Supplemental Video S3 shows pigs performing the discrimination task.

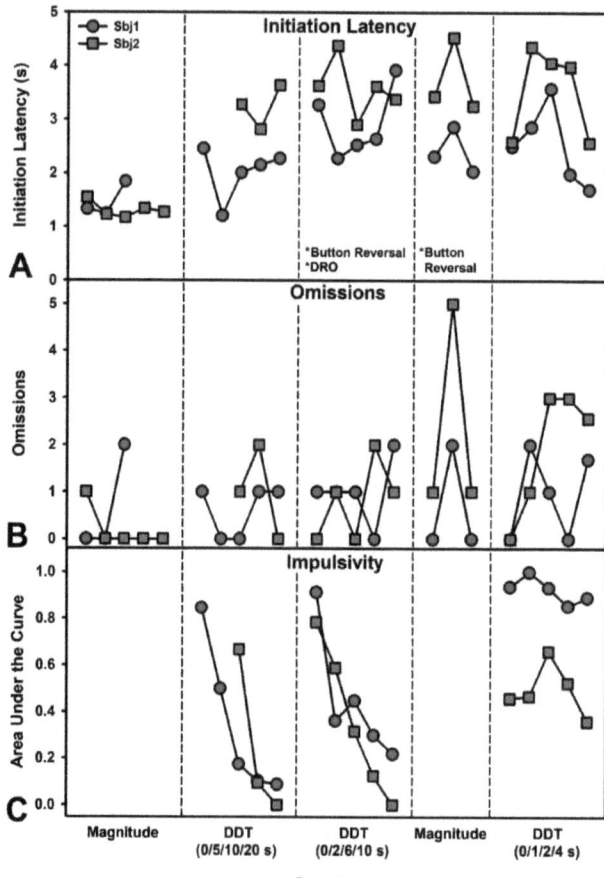

Figure 2. Performance on the Delay Discounting Task (DDT) for two male pigs (Experiment 1). Partitions from left to right indicate performance during a magnitude discrimination (only large versus small buttons, no delays), performance on the DDT with delays of 0, 5, 10, or 20 s (incrementing every 12 trials within the session), performance on the DDT with delays of 0, 2, 6, or 10 s after reversing the levers and giving marshmallows on a differential reinforcement of other (DRO) behavior schedule, performance on a second magnitude discrimination after reversing levers, and finally performance on the DDT with delays of 0, 1, 2 and 4 s. (**A**) Mean latency per session to initiate trials after a tone played indicating button availability. Latencies increased once delays were introduced. (**B**) Total omissions per session. Omissions remained relatively low throughout testing. (**C**) Area under the discounting curve which indicates the proportion of large/delayed/self-controlled choices made. Pigs made progressively more impulsive choices during the first set of delays but performed better at the low delays trained at the end.

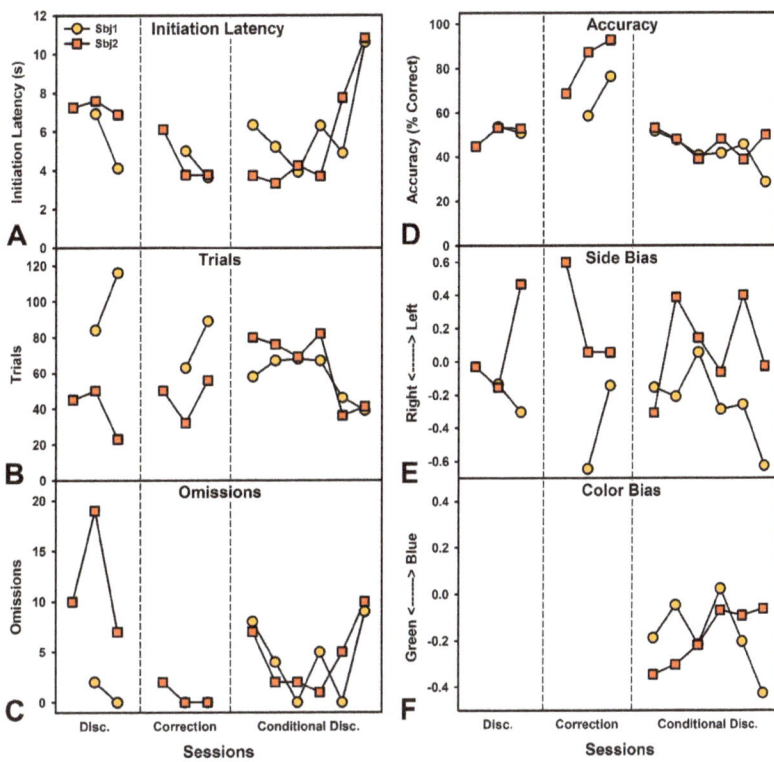

Figure 3. Performance on a Color Discrimination task for two female pigs (Experiment 2). Partitions from left to right indicate performance during an initial discrimination (yellow choices reinforced), the same discrimination but with correction trials which forces choice of the correct option after an incorrect choice, and finally a conditional discrimination where a blue or green box was shown in the middle and then two choice boxes presented and only choices of the presented color were reinforced. (**A**) Mean latency per session to initiate trials after a tone played indicating button availability. Latencies increased as performance worsened under the conditional discrimination. (**B**) Total trials completed per session. Trials remained relatively high throughout. (**C**) Total omissions per session. Omissions for one pig were large during initial performance, and then somewhat variable under the conditional discrimination. (**D**) Percent correct choices. Pigs performed poorly until the correction trials were implemented and then rapidly improved. Conditional discrimination performance was very poor. (**E**) Preference for left versus right (−1 to 1 scale with 0 indicating no preference). Pigs exhibited moderate side biases throughout testing. (**F**) Preference for green versus blue color (−1 to 1 scale with 0 indicating no preference). Pigs showed a strong preference toward green, which may be more similar to the previously reinforced yellow.

4. Discussion

In the current protocol, we provide a detailed description for constructing an operant touchscreen for pig behavioral testing and assessed two behavioral tasks, highlighting the utility of such devices. Previously, pig behavior had largely been limited to relatively simplistic measures or required room-sized equipment. The current touchscreen device is only limited by the time constraints of the researcher to train. Although pigs are not a common laboratory species for all disciplines, there are several key research areas that could benefit from such a device.

The fields with the strongest integration of pigs are those of pharmacology, cardiology, cancer, and vision. Pigs have long been used for safety testing of drugs and represented a critical translational step from rodent safety and efficacy studies [25]. For drugs affecting peripheral physiology, the pig has made a strong model due to the many physiological similarities. However, for psychoactive drugs, researchers have historically relied on monkeys to obtain both efficacy data and safety data. With a touchscreen device, pigs could become a viable alternative for testing psychotherapeutics. In vision research, pigs were classically used from the 1960s through the 1980s [26], and in recent years there has been a resurgence of interest in pigs to study retinal degeneration [27]. A recent study established an obstacle course behavioral test to assay gross visual impairment [28] but a touchscreen assay could provide much more detail on the nature and progression of deficits by systematically manipulating the salience of stimuli.

In the field of TBI, pigs have become much more common in recent years [29]. Their large and gyrencephalic brains make them ideal specimens for examining pathophysiology associated with rotational damage and unique cortical damage which cannot be observed in rodents. However, functional assessments have been much more limited. Now, functional assessments with great relevance to brain injury can be used. Behavioral flexibility, attentional impairments, and impulsivity are all symptoms of brain injury in patients and could not be readily assessed without a device such as this. In the current paper, we report an example of delay discounting, which could be used to measure impulsivity in a pig model of brain injury. This would provide crucial data about a relatively common psychiatric outcome of TBI [30]. More robust and extended behavioral batteries will need to be developed to meet the needs of the TBI field. In particular, both assays that are rapid (for acute studies) and those that can hold up to extended, repeat testing (for chronic studies) will be important.

Researchers studying early life stresses and challenges may also benefit from the inclusion of cognitive testing in pigs. During gestation, pigs are amenable to various types of translationally relevant hypoxic, diet, and surgical interventions to affect offspring while also offering researchers the ability to sample blood from both sow and fetus [8]. Shortly after birth, hypoxia may be induced to mimic perinatal asphyxia to study another common developmental challenge [9]. A large portion of such research has focused on acute therapeutics (e.g., hypothermia) and immediate outcomes (e.g., pathology) [31]. However, gaining an understanding of the potential efficacy of such treatments on long-term cognitive function could have great benefits. A range of conditions are associated with preterm birth (a potential indicator of pregnancy challenges), including speech and cognitive delays and potentially even ADHD risk [32].

Still other fields are being shaped by the recent advent of transgenic minipigs. The current device could be immediately adapted to the minipig by dropping the screen down lower on the frame and would require no other changes. It could feasibly be used in labs which work in both mini- and full-size species. Perhaps most immediately relevant to studying behavior are the transgenic Göttingen minipigs for HD [5] and for AD [6]. Memory tasks or others could be easily programmed for assessment. However, these same pigs likely have even broader applications. Systems for rapid gene editing (e.g., CRISPR) are now being used to reduce the cost of generating a unique transgenic minipig for a given question of interest [33,34]. Thus, a device such as this touchscreen, with the flexibility to design multiple behavioral assessments provides great utility for these research questions. A small but encouraging literature exists describing these types of assessments in pigs. Researchers have evaluated impulsivity, memory, and decision-making [14–19] using non-touch operant devices.

While there are numerous fields that could benefit from adopting behavioral testing using an apparatus such as the one described here, there are also several limitations to consider. First, while we have described a method for constructing a device with relative ease, individual researchers will still need to program relevant behavioral tests for their questions of interest. Perhaps more challenging is the need for behavioral expertise. For a physiology researcher with no background, it may be difficult to adapt to the needs of behavioral study (e.g., training time). However, colleagues in psychology and neuroscience departments may be readily available to provide such expertise. Despite stark differences in rat or human behaviors, many such researchers regularly program tasks such as the ones described in this paper. Our own lab is otherwise focused on rat behavior, but were able to design and program these tests for pigs.

Researchers must also be ready to recognize and adapt to the limitations of pig behavior. As an example, in the current study, we quickly determined that the method by which a pig pushed a button on a screen was not congruent with the default expectations for the software due to a swiping motion instead of click-and-release (see Results Section 3.2 on Behavioral Topology). Problems such as this can be mitigated by adjusting software commands (e.g., processing a response on depression of a button rather than release or making a taller response box to record presses). Similarly, we had challenges with rooting behaviors, and indeed our first device was broken by the pigs. Design refinements to reduce movement and access to areas under the screen as well as move the dispenser solved some of these problems. However, rooting is a problem that was noted as early as 1961 in a classic article titled "The Misbehavior of Organisms". This work noted (with pigs as one example) that there was often drift toward instinctual responses (e.g., rooting) after extended training with food reinforcement.

The current paper represents a start to automate and extend operant testing in pigs. There is still more work to be done to optimize behavioral training regimens which will reflect the needs of various fields. For example, rapid acquisition tasks for short time frames (e.g., <10 days) versus more extended in-depth repeated measurements of stable trait behaviors. Researchers may want to extend on this and integrate other peripheral elements or input devices. For instance, a physical button, a foot lever, or a lever which could be manipulated with the mouth can all be integrated using the device described here. It could even be taken into rats or other species if a suitable response device can be found (likely an infrared touchscreen instead of resistive or capacitive). Multiple reinforcer types could be used with a behavioral economics approach to tease out subtle aspects of preference and motivation. Or two screens could be used alongside one another (controlled by one Pi or coordinating multiple) to provide stronger spatial separation of choice boxes. The shaping program described in this paper will be available on the corresponding author's GitHub. Additional updates to this project and programs will be made available as they are developed.

Supplementary Materials: The following are available online at https://www.mdpi.com/article/10.3390/biomedicines10102612/s1. Three videos are available which show the first device (Video S1), acquisition of pressing behavior (Video S2), and the color discrimination (Video S3). A schematic for the PCB in provided as S4. A program and other online resources are provided on the corresponding author's GitHub repository, https://github.com/VonderHaarLab/.

Author Contributions: Conceptualization, C.V.H.; methodology, C.V.H. and C.L.F.; software, C.V.H., W.A. and M.G.; formal analysis, C.V.H.; investigation, W.A. and M.G.; writing—original draft preparation, W.A.; writing—review and editing, C.V.H. and C.L.F. All authors have read and agreed to the published version of the manuscript.

Funding: Funding for this project was provided by West Virginia University and Ohio State University.

Institutional Review Board Statement: The animal study protocol was approved by the Institutional Animal Care and Use Committee of Ohio State University (protocol 2021A00000018, 16 November 2021).

Informed Consent Statement: Not applicable.

Data Availability Statement: Resources to create and reproduce this device are available on the corresponding author's GitHub repository. Raw data will be made available upon request.

Acknowledgments: Special thanks to Gregory Lusk and Doug Mathess who helped with initial design aspects. Thanks to Injury and Recovery Lab members who assisted with behavioral testing.

Conflicts of Interest: The authors have no financial or competing interest in the outcome of this research.

References

1. Lunney, J.K.; Van Goor, A.; Walker, K.E.; Hailstock, T.; Franklin, J.; Dai, C. Importance of the pig as a human biomedical model. *Sci. Transl. Med.* **2021**, *13*, eabd5758. [CrossRef] [PubMed]
2. Pillay, P.; Manger, P.R. Order-specific quantitative patterns of cortical gyrification. *Eur. J. Neurosci.* **2007**, *25*, 2705–2712. [CrossRef] [PubMed]
3. MacManus, D.B.; Pierrat, B.; Murphy, J.G.; Gilchrist, M.D. Region and species dependent mechanical properties of adolescent and young adult brain tissue. *Sci. Rep.* **2017**, *7*, 13729. [CrossRef] [PubMed]
4. Kinder, H.A.; Baker, E.W.; West, F.D. The pig as a preclinical traumatic brain injury model: Current models, functional outcome measures, and translational detection strategies. *Neural Regen. Res.* **2019**, *14*, 413. [PubMed]
5. Baxa, M.; Hruska-Plochan, M.; Juhas, S.; Vodicka, P.; Pavlok, A.; Juhasova, J.; Motlik, J. A transgenic minipig model of Huntington's disease. *J. Huntingt. Dis.* **2013**, *2*, 47–68. [CrossRef] [PubMed]
6. Jakobsen, J.E.; Johansen, M.G.; Schmidt, M.; Liu, Y.; Li, R.; Callesen, H.; Jørgensen, A.L. Expression of the Alzheimer's disease mutations AβPP695sw and PSEN1M146I in double-transgenic Göttingen minipigs. *J. Alzheimer's Dis.* **2016**, *53*, 1617–1630. [CrossRef] [PubMed]
7. Schook, L.B.; Collares, T.V.; Hu, W.; Liang, Y.; Rodrigues, F.M.; Rund, L.A.; Schachtschneider, K.M.; Seixas, F.K.; Singh, K.; Wells, K.D.; et al. A genetic porcine model of cancer. *PLoS ONE* **2015**, *10*, e0128864. [CrossRef]
8. Gonzalez-Bulnes, A.; Astiz, S.; Parraguez, V.H.; Garcia-Contreras, C.; Vazquez-Gomez, M. Empowering translational research in fetal growth restriction: Sheep and swine animal models. *Curr. Pharm. Biotechnol.* **2016**, *17*, 848–855. [CrossRef]
9. Mallard, C.; Vexler, Z.S. Modeling ischemia in the immature brain: How translational are animal models? *Stroke* **2015**, *46*, 3006–3011. [CrossRef] [PubMed]
10. Williams, A.M.; Dennahy, I.S.; Bhatti, U.F.; Halaweish, I.; Xiong, Y.; Chang, P.; Nikolian, V.C.; Chtraklin, K.; Brown, J.; Zhang, Y. Mesenchymal stem cell-derived exosomes provide neuroprotection and improve long-term neurologic outcomes in a swine model of traumatic brain injury and hemorrhagic shock. *J. Neurotrauma* **2019**, *36*, 54–60. [CrossRef] [PubMed]
11. Sullivan, S.; Friess, S.H.; Ralston, J.; Smith, C.; Propert, K.J.; Rapp, P.E.; Margulies, S.S. Improved behavior, motor, and cognition assessments in neonatal piglets. *J. Neurotrauma* **2013**, *30*, 1770–1779. [CrossRef]
12. Baxa, M.; Levinska, B.; Skrivankova, M.; Pokorny, M.; Juhasova, J.; Klima, J.; Klempir, J.; Motli, K.J.; Juhas, S.; Ellederova, Z. Longitudinal study revealing motor, cognitive and behavioral decline in a transgenic minipig model of Huntington's disease. *Dis. Models Mech.* **2019**, *13*, dmm041293. [CrossRef]
13. Søndergaard, L.V.; Ladewig, J.; Dagnæs-Hansen, F.; Herskin, M.S.; Holm, I.E. Object recognition as a measure of memory in 1–2 years old transgenic minipigs carrying the APPsw mutation for Alzheimer's disease. *Transgenic Res.* **2012**, *21*, 1341–1348. [CrossRef] [PubMed]
14. Moustgaard, A.; Arnfred, S.M.; Lind, N.M.; Hemmingsen, R.; Hansen, A.K. Acquisition of visually guided conditional associative tasks in Göttingen minipigs. *Behav. Process.* **2005**, *68*, 97–102. [CrossRef] [PubMed]
15. Ferguson, S.A.; Gopee, N.V.; Paule, M.G.; Howard, P.C. Female mini-pig performance of temporal response differentiation, incremental repeated acquisition, and progressive ratio operant tasks. *Behav. Process.* **2009**, *80*, 28–34. [CrossRef] [PubMed]
16. Moustgaard, A.; Arnfred, S.M.; Lind, N.M.; Hansen, A.K.; Hemmingsen, R. Discriminations, reversals, and extra-dimensional shifts in the Göttingen minipig. *Behav. Process.* **2004**, *67*, 27–37. [CrossRef] [PubMed]
17. Zonderland, J.J.; Cornelissen, L.; Wolthuis-Fillerup, M.; Spoolder, H.A. Visual acuity of pigs at different light intensities. *Appl. Anim. Behav. Sci.* **2008**, *111*, 28–37. [CrossRef]
18. Melotti, L.; Thomsen, L.R.; Toscano, M.J.; Mendl, M.; Held, S. Delay discounting task in pigs reveals response strategies related to dopamine metabolite. *Physiol. Behav.* **2013**, *120*, 182–192. [CrossRef] [PubMed]
19. Nielsen, T.R.; Kornum, B.R.; Moustgaard, A.; Gade, A.; Lind, N.M.; Knudsen, G.M. A novel spatial delayed non-match to sample (DNMS) task in the Göttingen minipig. *Behav. Brain Res.* **2009**, *196*, 93–98. [CrossRef] [PubMed]
20. Palmer, D.; Dumont, J.R.; Dexter, T.D.; Prado, M.A.M.; Finger, E.; Bussey, T.J.; Saksida, L.M. Touchscreen cognitive testing: Cross-species translation and co-clinical trials in neurodegenerative and neuropsychiatric disease. *Neurobiol. Learn. Mem.* **2021**, *182*, 107443. [CrossRef]
21. Buscher, N.; Ojeda, A.; Francoeur, M.; Hulyalkar, S.; Claros, C.; Tang, T.; Terry, A.; Gupta, A.; Fakhraei, L.; Ramanathan, D.S. Open-source raspberry Pi-based operant box for translational behavioral testing in rodents. *J. Neurosci. Methods* **2020**, *342*, 108761. [CrossRef]
22. Escobar, R.; Gutiérrez, B.; Benavides, R. 3D-printed operant chambers for rats: Design, assembly, and innovations. *Behav. Process.* **2022**, *199*, 104647. [CrossRef] [PubMed]
23. Evans, P.; Özgür, S.; LaQua, M. Rfchat. Available online: https://github.com/mrpjevans/rfchat (accessed on 13 July 2022).
24. Vanderveldt, A.; Oliveira, L.; Green, L. Delay discounting: Pigeon, rat, human—Does it matter? *J. Exp. Psychol. Anim. Learn. Cogn.* **2016**, *42*, 141. [CrossRef] [PubMed]

25. Colleton, C.; Brewster, D.; Chester, A.; Clarke, D.O.; Heining, P.; Olaharski, A.; Graziano, M. The use of minipigs for preclinical safety assessment by the pharmaceutical industry: Results of an IQ DruSafe minipig survey. *Toxicol. Pathol.* **2016**, *44*, 458–466. [CrossRef] [PubMed]
26. Middleton, S. Porcine ophthalmology. *Vet. Clin. North Am. Food Anim. Pract.* **2010**, *26*, 557–572. [CrossRef]
27. Choi, K.E.; Anh, V.T.Q.; Kim, J.T.; Yun, C.; Cha, S.; Ahn, J.; Goo, Y.S.; Kim, S.W. An experimental pig model with outer retinal degeneration induced by temporary intravitreal loading of N-methyl-N-nitrosourea during vitrectomy. *Sci. Rep.* **2021**, *11*, 258. [CrossRef] [PubMed]
28. Barone, F.; Nannoni, E.; Elmi, A.; Lambertini, C.; Scorpio, D.G.; Ventrella, D.; Vitali, M.; Maya-Vetencourt, J.F.; Martelli, G.; Benfenati, F.; et al. Behavioral assessment of vision in pigs. *J. Am. Assoc. Lab. Anim. Sci.* **2018**, *57*, 350–356. [CrossRef]
29. Vink, R. Large animal models of traumatic brain injury. *J. Neurosci. Res.* **2018**, *96*, 527–535. [CrossRef] [PubMed]
30. Rochat, L.; Beni, C.; Billieux, J.; Azouvi, P.; Annoni, J.-M.; Van der Linden, M. Assessment of impulsivity after moderate to severe traumatic brain injury. *Neuropsychol. Rehabil.* **2010**, *20*, 778–797. [CrossRef]
31. Lee, J.K.; Santos, P.T.; Chen, M.W.; O'Brien, C.E.; Kulikowicz, E.; Adams, S.; Hardart, H.; Koehler, R.C.; Martin, L.J. Combining hypothermia and oleuropein subacutely protects subcortical white matter in a swine model of neonatal hypoxic-ischemic encephalopathy. *J. Neuropathol. Exp. Neurol.* **2021**, *80*, 182–198. [CrossRef] [PubMed]
32. Srinivas Jois, R. Neurodevelopmental outcome of late-preterm infants: A pragmatic review. *Aust. J. Gen. Pract.* **2018**, *47*, 776–781. [CrossRef]
33. Maxeiner, J.; Sharma, R.; Amrhein, C.; Gervais, F.; Duda, M.; Ward, J.; Mikkelsen, L.F.; Forster, R.; Malewicz, M.; Krishnan, J. Genomics Integrated Systems Transgenesis (GENISYST) for gain-of-function disease modelling in Göttingen Minipigs. *J. Pharmacol. Toxicol. Methods* **2021**, *108*, 106956. [CrossRef]
34. Berthelsen, M.F.; Riedel, M.; Cai, H.; Skaarup, S.H.; Alstrup, A.K.O.; Dagnæs-Hansen, F.; Luo, Y.; Jensen, U.B.; Hager, H.; Liu, Y.; et al. The CRISPR/Cas9 minipig-A transgenic minipig to produce specific mutations in designated tissues. *Cancers* **2021**, *13*, 3024. [CrossRef]

Review

The Pig as a Translational Animal Model for Biobehavioral and Neurotrauma Research

Alesa H. Netzley [1] and Galit Pelled [2,3,4,*]

1. Department of Biomedical Engineering, Michigan State University, East Lansing, MI 48824, USA; hughson3@msu.edu
2. Neuroscience Program, Michigan State University, East Lansing, MI 48824, USA
3. Department of Mechanical Engineering, Michigan State University, East Lansing, MI 48824, USA
4. Department of Radiology, Michigan State University, East Lansing, MI 48824, USA
* Correspondence: pelledga@msu.edu

Abstract: In recent decades, the pig has attracted considerable attention as an important intermediary model animal in translational biobehavioral research due to major similarities between pig and human neuroanatomy, physiology, and behavior. As a result, there is growing interest in using pigs to model many human neurological conditions and injuries. Pigs are highly intelligent and are capable of performing a wide range of behaviors, which can provide valuable insight into the effects of various neurological disease states. One area in which the pig has emerged as a particularly relevant model species is in the realm of neurotrauma research. Indeed, the number of investigators developing injury models and assessing treatment options in pigs is ever-expanding. In this review, we examine the use of pigs for cognitive and behavioral research as well as some commonly used physiological assessment methods. We also discuss the current usage of pigs as a model for the study of traumatic brain injury. We conclude that the pig is a valuable animal species for studying cognition and the physiological effect of disease, and it has the potential to contribute to the development of new treatments and therapies for human neurological and psychiatric disorders.

Keywords: neurotrauma; traumatic brain injury; translational neuroscience; large animal; behavior; swine; pig; cognition; physiology

1. Introduction

Translational neuroscience within biomedical research is a rapidly growing field, aiming to bridge the gap between basic science and clinical applications by using the findings from preclinical studies to develop more effective interventions and treatments for human neurological disorders and injuries [1,2]. The use of animal models is an essential step in the research process as these models provide a means to study the underlying mechanisms behind various disease states and the holistic effects of interventions in a living system [3]. The development of animal models that accurately replicate aspects of human neurological conditions is key to the success of translational neuroscience research. In this regard, pigs (sus scrofa) have emerged as a highly valuable model species due to notable similarities between pig and human neuroanatomy, physiology, and behavior [4–7].

Over the past several decades, the pig has gained significant attention as a model animal for translational and biobehavioral research, particularly in the fields of neurotrauma and cognitive neuroscience. The relative cognitive complexity, social behavior, and overall body composition of the pig make it an exceptional model organism for studying various disease states [8,9]. Pigs have been used extensively in studies of pharmacology and toxicology as a major translational model due to their vast biochemical similarities to humans [10]. Another important aspect of biomedical engineering is the development of new equipment and testing the diagnostic and prognostic potential of new methods and

therapies. The pig is instrumental in this regard, as its large body size allows the use of equipment and modalities with results that can be immediately applicable to humans [11].

This review will examine the different ways in which pigs are used to study cognition and behavior. We will highlight many commonly used behavioral tasks that pigs are capable of performing as well as several contemporary physiological assessments. We will also highlight the use of pigs for the study of traumatic brain injury as many pig models of brain injury have been developed, as well as how these models contribute to our understanding of brain function and dysfunction in humans. Finally, we will discuss the future directions of pig-based translational biobehavioral research, including the development of advanced techniques for assessing pig behavior, and the potential of pig models for studying novel treatments for neural injuries and neuropsychiatric disorders.

2. Review

2.1. Preclinical Research

Preclinical research is an essential step in advancing our understanding of neurological disorders and developing effective treatments. Animal models play a crucial role in preclinical neuroscience, providing valuable insights into the underlying mechanisms of brain function and dysfunction in a living system. There are several advantages to using animal models, including the ability to control experimental variables, genetic manipulations, perform invasive procedures, and analyze biological samples in ways that cannot be done in human subjects [12]. Over the past century, various animal species have been used to model diseases and conditions in neuroscience research [13]. Of these various animal model species, the most common and well-characterized are rodents, specifically mice and rats [14].

Rodent models have greatly advanced our understanding of fundamental neurological processes. Notably, rodent research has contributed to the study of neuronal function, synaptic plasticity, neurotransmission, and connectivity [15,16]. Rats and mice have been used extensively to model a wide range of neurological and psychiatric disorders, including neural injury, degenerative diseases, and psychiatric conditions such as depression, anxiety, and drug addiction [17]. By replicating these disorder-related conditions and phenotypes, researchers can investigate the underlying mechanisms behind the disorders, study disease progression, and develop and test the efficacy of potential therapeutic interventions. Preclinical rodent research continues to play a critical role in the development and evaluation of novel therapeutic interventions. Rodent models have been used to test pharmaceutical interventions, gene therapies, and various other treatments, with many showing great promise [18]. Rodents and humans share many physiological and genetic similarities, thus rodent studies allow for rigorous preclinical testing of interventional strategies for neurological disorders before moving on to human trials.

2.2. Translational Considerations

The use of rodent models in preclinical research has contributed significantly to the development of a wide variety of treatments for various neurological disorders. Although many potential treatments have shown promise in rodents, the vast majority of the therapeutics tested have failed in clinical trials [19,20]. Many researchers speculate that one major contributing factor toward the low rate of translational success is due to the anatomical and physiological differences between humans and rodents [6]. The anatomical differences between rodent and human brains are considerable. Rats and mice have small lissencephalic brains with a low white to gray matter ratio [6]. In this regard, rodents may be a less-than-ideal model organism for the study of diseases that have a major effect on white matter regions, such as Multiple Sclerosis [21]. Additionally, the lack of cerebral convolutions can be problematic for studies of brain injury as stress generated by mechanical insult tends to be distributed more evenly across surface areas in smooth brains [22]. In contrast, the presence of gyri and sulci in the human brain focuses mechanical stress toward the base of the sulci, deeper into the center of the brain and in close proximity to white matter regions [22].

While the overall organization of brain structures is fairly similar between primates and rodents, investigators are learning that the functions of these brain regions may not be as alike as previously thought [23]. A recent report by Hodge et al. (2019) has demonstrated considerable differences in the expression of genes among similar cell types within mouse and human brains [24]. In this study, researchers used single-cell transcriptomics to characterize various cell types in the cortex of mice and the human middle temporal gyrus. They then compared the expression of genes in homologous cell types and found that among similar cell types, there were major divergencies in the expression of neurotransmitter receptors, ion channels, extracellular matrix elements, and cell-adhesion molecules between species [24]. This divergent expression of key signaling elements could help to provide some justification as to why the translation rate of pharmacological treatments for neurological disorders has been so low.

2.3. Similarities of the Pig Brain

The pig brain shares many structural and functional similarities with the human brain, including comparable cortical organization, anatomical structure (as shown in Figure 1), gray and white matter ratio, and regional distribution of neurotransmitter systems [25,26]. They have a similar brain size and structure to humans, with analogous cortical folding, white matter tracts, and subcortical structures [27,28]. Additionally, pigs exhibit similar patterns of brain development, allowing researchers to investigate the underlying neurodevelopmental processes associated with developmental disorders [29].

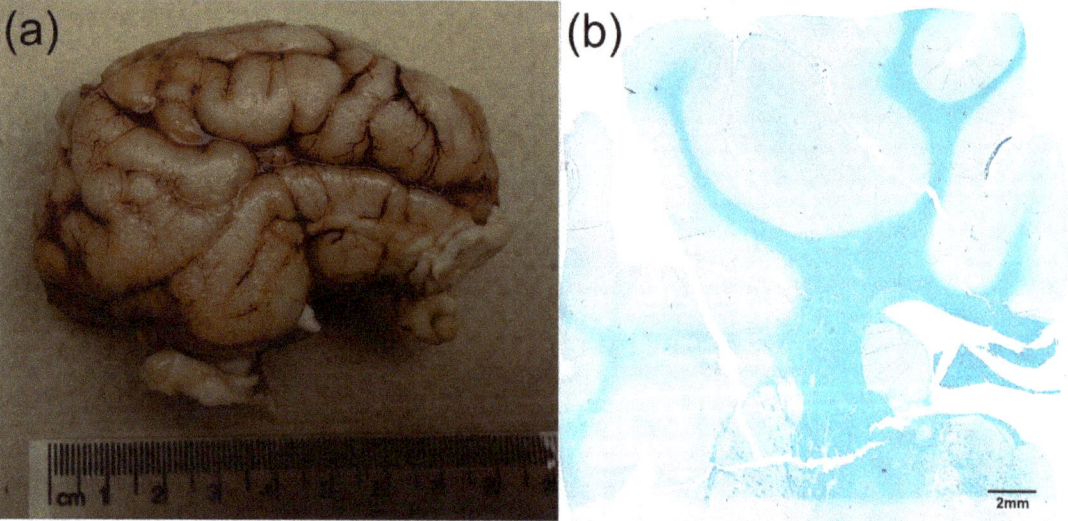

Figure 1. The gross anatomy of the pig brain shows similarities to human brains. (**a**) Photograph of an adult Yucatan minipig brain in sagittal view. This image shows the gross gyrencephalic anatomy of the pig cortex, depicting the presence of cerebral convolutions (gyri and sulci). (**b**) A 10× image of 5 μm thick brain slice stained for Luxol fast blue showing considerable white matter tracts in the Yucatan minipig cortex. Dark areas indicate the presence of myelin.

The pig brain undergoes a period of rapid growth and development during the perinatal period, which is comparable to human brain development during late gestation and early infancy [25,30,31]. Another important similarity to humans is the chronological development pattern of the neocortex; comparative studies demonstrated that the neurogenesis of a domestic pig is completed before term [29]. Additionally, imaging studies using Diffusion Tensor Imaging (DTI) have demonstrated that the myelination rate in the corpus callosum of pigs is similar to humans [25]. Pigs also exhibit many human-relevant behaviors, such as social structure, enabling the investigation of complex cognitive functions that are often impaired in human neurological disorders such as learning, memory, and social behavior [32–34]. Moreover, the pig's physiology and metabolism are closer to humans compared to smaller animal models, enabling a more accurate evaluation of pharmacokinetics and treatment responses [35,36]. Collectively, these factors make the pig a valuable model for studying human neurological conditions and injury, facilitating the translation of preclinical findings to clinical applications, and enhancing our understanding of these complex disorders.

2.4. Pig Cognition and Behavior

Behavioral assessment in preclinical research provides a bridge between basic neuroscience and clinical applications. The study of cognition and behavior is essential for understanding brain function and modeling human disorders. Many neurological and psychiatric disorders manifest as cognitive and behavioral impairments. By observing and quantifying behavioral responses in animal models, researchers can investigate various aspects of brain function, including sensory perception, motor control, learning and memory, social behavior, and emotional processes. Pigs possess a high level of cognitive complexity, exhibiting a wide range of behaviors and mental abilities that are relevant to studying human cognition. Pigs demonstrate advanced social behavior, problem-solving skills, spatial memory, and learning capabilities. In order to use pigs as a model animal for biobehavioral research, it is important to have reliable methods for assessing their behavior. Direct observation and scoring are commonly used, as well as video recording and computerized analysis, which enable researchers to monitor and measure behavioral parameters with high precision and accuracy. Automated systems, such as accelerometers and radio-frequency identification (RFID), are also being developed for monitoring pig behavior in a non-invasive manner. The advancement of technology continuously facilitates the development of sophisticated and precise behavioral assessment methods.

2.4.1. Spatial Memory and Maze Tests

Spatial memory deficits are observed in various neurological disorders. Maze tests can be used to assess spatial learning and memory in pigs. Pigs are trained to navigate through mazes, such as T-mazes or radial arm mazes, to locate rewards or escape routes. The T-maze apparatus consists of a central stem and two perpendicular arms, forming a T-shape. The stem is typically a long corridor or runway, while the arms are shorter and lead to distinct goal areas that customarily have a food reward. Similarly, a radial arm maze consists of a central hub and several radiating arms, typically eight, arranged in a circular pattern. Performance in these tests can reveal a pig's ability to remember and utilize spatial information. A T-maze was used to evaluate a pig's spatial cognition after brain injury, published by Kinder et al. (2019). They found that pigs with a brain injury required more time to make a decision and were less accurate when deciding which arm of the maze would contain a reward [37]. Similarly, Singh et al. (2019) found that pigs with hypoxia-ischemia (HI) performed more slowly in a T-maze compared to control pigs, suggesting that HI pigs have poorer working memory [38]. While less common, the eight-arm radial maze has also been used to assess spatial cognition in pigs. In a study published by Chen et al. (2021), neonatal pigs whose diets were supplemented with lactoferrin were able to reach learning objectives in the radial maze faster than the experimental controls [39]. These studies show that maze tasks are valuable assessment tools for their use with pigs.

The spatial hole board test is another commonly used assessment to test spatial learning and working memory in pigs [40,41]. The apparatus typically consists of a rectangular platform with evenly spaced holes or containers. During testing, food rewards or pellets are placed into some of the holes, while other holes are left empty. Pigs are tasked with retrieving the food in the most efficient way possible without returning to holes which have already been explored. Performance can be measured by the number of correct choices, the latency to find rewards, and the pattern of exploration. A recent study by Clouard et al. (2022) reported that pigs who were fed oligosaccharides spent more time in between visits to different baited buckets, which resulted in higher scores for memory and fewer errors. The researchers concluded that the pigs who were fed oligosaccharides demonstrated higher executive functioning than the controls who exhibited more sporadic, hyperactive behavior [42].

A spatial baited ball pit task was first published by Netzley et al. (2021), wherein food rewards were distributed evenly throughout a shallow pool and hidden under colorful plastic balls (Figure 2). The investigators found that over the course of several weeks, healthy pigs become faster and more accurate at retrieving the rewards [43].

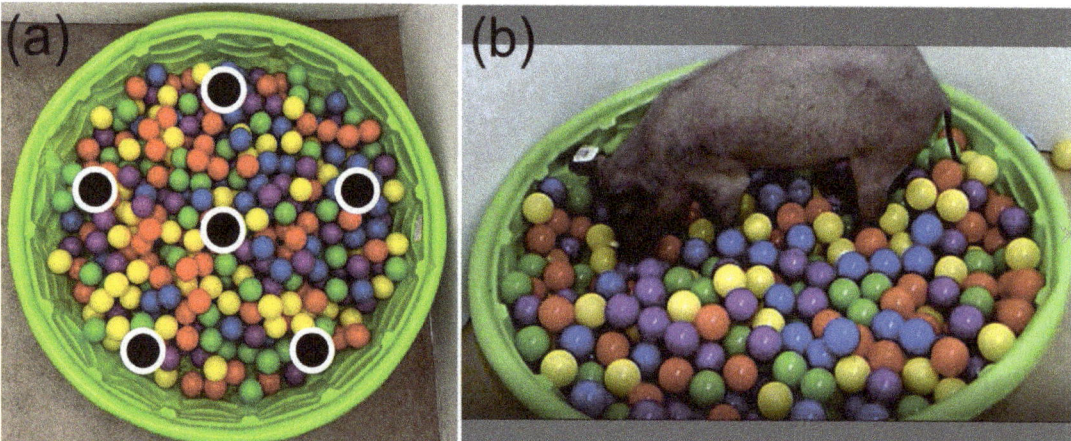

Figure 2. The baited ball pit—a novel spatial test to assess cognition and memory in pigs. (**a**) An overhead view of the apparatus and placement of food rewards (black circles with white outlines). (**b**) Photo showing a Yucatan minipig engaged in the ball pit task.

2.4.2. Object Recognition and Discrimination Tests

Object recognition tests can be used to assess the ability of pigs to discriminate between colors, shapes, or familiar and novel objects. These tasks test recognition memory, and often, the training for these tasks involves familiarizing the animal with a certain object and later presenting the trained object alongside one or more new or incorrect objects. For example, a three-choice color discrimination task was used to assess recognition memory in minipigs by Schramke et al. (2016). Pigs were presented with three colored boxes (blue, red, and yellow), each containing a food reward. Pigs were trained to recognize that only the blue box could be opened, while the red and yellow boxes were sealed. This task was given to both healthy pigs and pigs modeling Huntington's Disease, although no differences in task performance between groups were found [44]. Ao et al. (2022) developed a rack-mounted touchscreen device for pigs given a color discrimination task. Briefly, a touchscreen monitor was placed on an audio rack, and an automatic pellet dispenser was positioned across the room. Pigs were tasked with snout-touching a colored shape on the screen, and successful touches were reinforced with an audio cue while a fruit-flavored pellet was deposited from the dispenser [45].

2.4.3. Social Interaction and Aggression Tasks

Social interaction tests are used to evaluate a pig's social behavior and cognition. They often involve introducing a subject to an unfamiliar pig in order to observe social interactions between animals, dominance hierarchies, play behavior, and social recognition. A voluntary human approach test was conducted by Wegner et al. (2020), which revealed that pigs who approached a motionless human observer more quickly were more likely to engage in forceful means of contact, i.e., biting, when compared to pigs who were more reserved [46]. Aggression tests assess aggressive behavior and response to social challenges in pigs. In these tests, pigs are introduced to stimuli or provocations that elicit aggressive responses, such as competing for resources or territorial intrusions. One example of a dominance task was published by Schramke et al. (2016). Two pigs were led into a tunnel separated by a trap door. The trap door was removed, and the more dominant pig pushed past the less dominant one in order to reach a food reward given behind the opposing pig [44].

2.4.4. Vocalization Analysis

Vocalization analysis involves recording and analyzing the vocalizations emitted by pigs in different contexts. It provides insights into their communicative behaviors, emotional states, and responses to stimuli. Pigs are a vocal species, who communicate in a herd by calling to one another. A speech recognition paradigm for pigs was recently developed by Wu et al. (2022), utilizing a fusion network that combines both spectral and audio features to classify individual pig speech patterns [47]. Vocal expression has also been used to assess emotional states in pigs. In a study by Briefer et al. (2022), pig vocalizations were used to develop an automated vocal recognition system to monitor animal welfare in a farm setting [48].

2.4.5. Fear and Anxiety Tasks

Fear and anxiety tests assess an animal's responses to fear-inducing stimuli, such as open fields, sudden loud sounds, or even unfamiliar objects and locations. Observing behaviors like freezing, avoidance, or elevated stress markers can provide insights into porcine emotional states and stress responses. The open field test is a widely used behavioral test to assess exploratory behavior, locomotor activity, and anxiety-like responses in animals including pigs. The test is traditionally conducted in a square or rectangular arena, usually made of an open and brightly lit area, devoid of any specific cues or obstacles. The animal is placed in the center of the arena, and their behavior is recorded and analyzed. In pigs, typical behavioral parameters include the distance traveled, exploration of different zones such as the center or periphery of the arena, and specific behaviors such as rearing, sniffing, and defecation [49]. These measures provide insights into the animal's exploratory behavior, as well as anxiety levels. For example, greater distance traveled, and more time spent in the center of the arena are indicative of reduced levels of anxiety. On the other hand, freezing, defecating, and staying near the walls of the arena may indicate higher anxiety levels (Figure 3) [43,50]. In a study published by Haigh et al. (2020), it was determined that pigs who were bitten by other members of the herd exhibited more anxious behaviors in the open field compared to pigs who committed the biting acts [49]. Overall, the open field test is a well-established and widely used method to evaluate behavior in animals, including pigs, and has contributed significantly to our understanding of anxiety-related behaviors and exploratory tendencies.

Behavioral assessment is a fundamental component of preclinical neuroscience research, providing valuable information about the functioning of the nervous system and its relationship to behavior. It plays a crucial role in the development and evaluation of potential treatments for neurological and psychiatric disorders, ultimately contributing to the advancement of clinical neuroscience. Table 1 (below) lists several key behavioral assessments conducted in pigs and the various experimental conditions that have been evaluated.

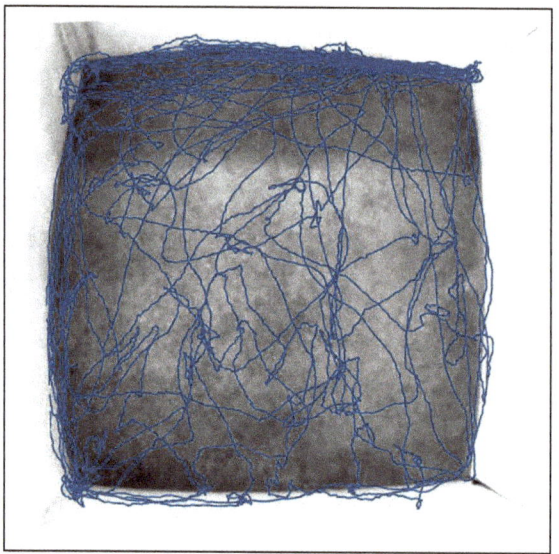

Figure 3. Example of pig locomotor activity in the open field. Coordinates were taken from the top of the head of a healthy Yucatan minipig over the course of 10 min via overhead video monitoring. Observations indicate that this individual spent much of their time rooting along the walls near the entrance to the chamber (top left corner). This can indicate some level of anxiety and the desire to escape the chamber.

Table 1. Summative table of highlighted behavioral tests used in pigs.

Type of Test	Authors	Experimental Condition
Maze Tasks	Kinder et al., 2019 [37] Singh et al., 2019 [38] Chen et al., 2021 [39]	Traumatic Brain Injury Hypoxia-Ischemia Diet
Spatial and Hole board	Clouard et al., 2022 [42]	Diet
	Netzley et al., 2021 [43]	Healthy
Object Discrimination	Schramke et al., 2016 [44]	Huntington's Disease
	Ao et al., 2022 [45]	Healthy
Socialization	Wegner et al., 2020 [46]	Healthy
	Schramke et al., 2016 [44]	Huntington's Disease
Vocalization	Wu et al., 2022 [47]	Healthy
	Briefer et al., 2022 [48]	Healthy
Open Field	Haigh et al., 2020 [49]	Tail biting

2.5. Physiological Assessments

Preclinical researchers utilize physiological assessments to gain a deeper understanding of the underlying biological mechanisms that drive changes in cognition and behavior. These assessment methods provide valuable data on various physiological parameters, such as heart rate, blood pressure, respiration rate, hormonal levels, and neural activity. Physiological assessments can be particularly important in preclinical research because they provide objective and quantitative measurements of biological processes that may not be directly observable through behavioral assessments alone. These assessments also help researchers investigate the effects of experimental manipulations on the function of different organ systems, including the cardiovascular, respiratory, endocrine, and nervous systems. For example, researchers can use electroencephalography (EEG) or functional magnetic resonance imaging (fMRI) to study brain activity patterns and neural responses. They can measure neurotransmitter levels or hormone concentrations to examine the impact of experimental interventions on biochemical signaling. Physiological assessments also enable researchers to monitor vital signs and physiological parameters during different experimental conditions, helping to ensure animal welfare and safety during the study.

2.5.1. Real-Time Physiological Monitoring

As technology advances, more accessible physiological monitoring devices have been developed. Pigs have been equipped with many of the technologies used to monitor health and wellness in humans. Wearable electrocardiograms have been used to monitor heart rate and blood pressure in pigs in a study published by Nachman et al. (2020). They found that the wearable device was able to consistently monitor these parameters in pigs with hemorrhagic shock despite unstable physiological conditions [51]. Other wearable devices have been used to monitor the health of pigs in research. Healthy minipigs were outfitted with human Fitbit® devices in Netzley et al. (2021) in order to track their activity during the course of a 12 h day, finding that the pigs were most active between the hours of 12 pm and 4 pm [43]. While wearable devices are desirable for human patients, pigs are curious and often chew anything they can reach, thus some studies may facilitate a need for implantable monitoring equipment. Martinez-Ramirez et al. (2022) demonstrated the capability of subdural implanted EEG monitoring systems for the long-term assessment of post-traumatic epilepsy in freely ambulating pigs for up to 13 months [52]. As a whole, pigs provide a fantastic model in which to test the validity of various novel devices or to use pre-established technologies for comparison to human data.

2.5.2. Neuroimaging

Neuroimaging plays a crucial role in translational research by providing valuable insights into the structure, function, and connectivity of the brain. Imaging studies can bridge the gap between preclinical studies and clinical research by facilitating the translation of findings from the laboratory to real-world applications. Neuroimaging techniques, such as magnetic resonance imaging (MRI), positron emission tomography (PET), and functional MRI (fMRI), allow researchers to non-invasively visualize and study the living brain. These techniques enable the investigation of brain abnormalities and alterations associated with various neurological and psychiatric disorders, providing crucial information for diagnosis, treatment planning, and monitoring responses to treatment. Additionally, neuroimaging provides a means to identify biomarkers, which can serve as objective measures of disease presence, progression, and treatment outcomes. By utilizing modern imaging techniques, researchers can better understand the underlying mechanisms behind brain disorders, evaluate the efficacy of interventions, and develop personalized treatment approaches.

The considerable biological similarities between pigs and humans make pigs a highly sought-after animal model for neuroimaging studies. Pig brains are anatomically similar to human brains (Figure 4), and the large body size of the pig allows researchers to utilize clinically available equipment and techniques to study various neurological conditions. The function of the pig brain is also remarkably similar to humans. Through resting-state fMRI (rs-fMRI), Simchick et al. (2019) demonstrated that pigs have homologous resting state networks similar to humans [53]. These networks include executive control, cerebellar, sensorimotor, visual, auditory, and default mode networks. These similarities can help researchers to better understand how different disease states may affect the functional connectivity of the brain. As such, Diffusion Tensor Imaging (DTI) was used to assess fractional anisotropy in pigs with hypoxic ischemia, as published by Lee et al. (2021). Specifically, they found that DTI could be used as a diagnostic tool for identifying pigs with more swollen astrocytes in the striatum [54]. Because of the considerable similarities that the pig brain shares with the human brain, pigs serve as an excellent resource for testing the efficacy of novel therapeutic interventions.

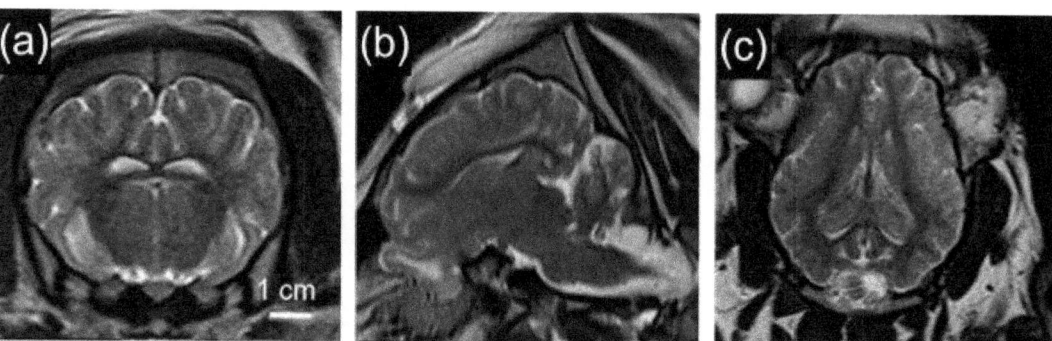

Figure 4. T2-weighted anatomical MR images of a minipig brain showing (**a**) coronal slice, (**b**) sagittal slice, and (**c**) axial slice. Gyri and sulci are clearly visible as well as regions of white matter.

2.6. Pigs as a Model Animal for Traumatic Brain Injury Research

In translational neuroscience, pigs are most commonly used for the study of traumatic brain injury (TBI). TBI is a major public health issue throughout the world, affecting millions of people annually and contributing to significant morbidity and mortality rates [55]. The Center for Disease Control and Prevention defines TBI as any disruption in brain function caused by an external force, such as a blow or jolt to the head. The severity of TBI can range from mild, such as a concussion, to severe, which often results in coma or death [56]. The pathophysiology of TBI is complex and involves both primary and secondary injury mechanisms, including mechanical damage to brain tissue, excitotoxicity, oxidative stress, inflammation, and cellular apoptosis [57,58]. Despite extensive research, effective treatments for TBI are still lacking, highlighting the need for improved understanding of the underlying mechanisms and the development of new therapeutic approaches [59]. There has been a growing concern among TBI researchers that a major reason for the poor clinical translation of treatments for TBI may be due to the animal models used [7]. As with most neuroscience research, TBI has traditionally been studied using rodent models. While the use of rats and mice has greatly increased our understanding of the cellular and molecular mechanisms behind neural injury, there are several key differences between rodent and human brains that can make it difficult to accurately predict the effects of potential treatments in humans based on rodent studies alone. As a result, there is increasing interest in using larger, more complex animal species, such as pigs, to better model the effects of TBI in humans and improve the translational potential of TBI research.

Studies That Use Pigs for TBI Research

The pig has been used to model various causes of TBI insults. One common method to induce TBI is the controlled cortical impact (CCI) [60]. CCI involves a craniotomy that is performed to expose the targeted region of the brain, usually the frontal or parietal cortex under general anesthesia. A pneumatic or electromagnetic impactor device is then used to deliver a controlled impact to the exposed brain tissue. The impactor is aligned and positioned to ensure accurate and consistent delivery of the impact. Researchers can adjust parameters such as impact velocity, depth, and dwell time to control the severity of the injury [61]. Following the impact, the craniotomy is typically covered with a protective material, and the scalp incision is sutured. While this is a well-characterized method for preclinical TBI in pigs [37,61–65], most human brain injuries are closed-head injuries [66]. Craniotomy-based models primarily mimic open-head injuries such as impalement or gunshot wounds, which may not fully capture the complexities and mechanisms of closed-head injuries, as the CCI is typically conducted with the pig's head secured in a stereotaxic frame, not allowing for any movement of the head during impact. Nevertheless, studies using the CCI to induce TBI have led to crucial understanding on primary and secondary injury mechanisms [67,68].

In contrast, rotational acceleration models in pigs involve the application of rotational forces to induce brain injury [69–72]. The procedure typically involves a specialized device that allows controlled rotational movement [69]. Under general anesthesia, the head is positioned securely to ensure accurate and consistent delivery of rotational forces. The rotational acceleration can be achieved using various methods, such as a custom-built device, a pendulum, or a rotational platform. The force and duration of rotation can be adjusted to control the severity of the injury. During rotation, the pig's head undergoes angular acceleration, leading to the deformation and shearing of brain tissue, which mimics the rotational forces experienced during closed-head TBIs in humans. The pig's relatively large brain size is needed to make scaled-up acceleration achievable. The rotational acceleration is a strong biomedical predictor in human TBI necessary to generate the characteristic manifestations used to diagnose severity: mild TBI with diffuse injury and no imaging abnormalities; moderate TBI with 30 min–24 h of unconsciousness; and severe TBI with over 24 h of unconsciousness. Therefore, rotational acceleration in pigs is vital for preclinical TBI studies to recreate the actual mechanisms and manifestations of human injury and bridge the translational gap in neurotrauma research.

Though less common than CCI or rotational acceleration models, pigs are also used to study the effects of blast injury to the brain [73–75]. The blast injury model involves exposing an anesthetized animal to a shockwave generated by an explosive or compressed gas. This model is specifically designed to simulate blast-related TBI, which occurs due to the impact of explosive forces on the brain such as those experienced by military personnel in active combat. Blast TBI is often associated with complex injury mechanisms, including primary blast waves, secondary injury from flying debris, and tertiary injury from body displacement. The blast model enables researchers to investigate the unique aspects of blast-induced brain injury, including the effects of shockwaves on brain tissue, neuroinflammation, and cognitive impairments (Table 2).

Table 2. Types of approaches used to model TBI in pigs.

Input Methods	Authors
Controlled Cortical Impact	Kinder et al., 2019 [37] Simchick et al., 2021 [62] Baker et al., 2019 [63] Manley et al., 2006 [64] Wang et al., 2023 [65]
Rotational Acceleration	Cullen et al., 2016 [69] Mayer et al., 2021 [71] Mayer et al., 2022 [70] O'Donnell et al., 2023 [72]
Blast Injury	Chen et al., 2017 [73] Kallakuri et al., 2017 [74] Cralley et al., 2022 [75]

Each of these TBI models has advantages and limitations, and the choice of injury model depends on the specific research questions and objectives. The different models play crucial roles in elucidating the pathophysiology of TBI, exploring therapeutic interventions and advancing our understanding of the mechanisms underlying brain injury.

Nevertheless, there are important considerations regarding the effects of TBI when conducting research on pigs. Specifically, there is a species-specific complication that is well known in the swine industry, porcine stress syndrome (PSS). PSS is a hereditary disorder that affects pigs, particularly Landrace and Yorkshire breeds [76]. This condition can have detrimental effects on imaging and behavioral data as PSS is characterized by a hypermetabolic response and is triggered by stressors such as handling or transport. Care must be taken to reduce stress in animals susceptible to PSS.

2.7. Considerations for Conducting Pig Research

Pigs can serve as an incredibly beneficial intermediary model species to bridge the gap between small animal research and clinical studies. While there are many benefits to using pigs in research, there are also several considerations that need to be addressed when designing studies that will use pigs.

Financial cost: The costs associated with using pigs for research can be much higher compared to small laboratory animals such as rodents. The initial costs to acquire pigs can vary depending on the specific breed, age, and supplier. Pigs bred specifically for research tend to be more expensive than those bred for agricultural purposes. Additionally, pig welfare is regulated under the United States Department of Agriculture, thus requiring specialized housing facilities with adequate space, ventilation, temperature control, and waste management systems. The overall size and difficulty of working with pigs can necessitate specially trained animal care staff to handle daily care such as feeding, health monitoring, and handling. Veterinary services such as routine health checkups, vaccinations, and treatment of any health issues should also be considered when budgeting for pig research. Further, conducting research with pigs often requires specialized supplies, equipment, and instrumentation. Often, clinical surgical tools and imaging equipment can be used; however, behavioral testing equipment generally must be specially engineered.

Lack of commercially available equipment: At present, there is limited availability of commercially available research supplies designed for working with pigs. Pigs are used in a wide range of research areas, and the various research applications require specialized supplies tailored to specific research objectives. The limited number of researchers who currently work with pigs and the diversity of research needs result in a lack of incentive for market suppliers to generate equipment for pig research. Additionally, pigs come in various sizes depending on breed and age. This size variability also makes it challenging to develop standardized research supplies that can accommodate the different sizes of the animals. Researchers often need to rely on adapting or modifying supplies used for other animal species, such as sheep or dogs. It is not unusual for research groups to commission customized equipment from vendors or to engage in an in-house fabrication of supplies.

Limited genetic tools: While pigs are becoming increasingly popular as a model organism in many fields of research, the genetic tools for working with pigs are limited compared to rodent species. Pigs have a larger and more complex genome than rats and mice, which can pose a challenge for genetic manipulation.

While these considerations and limitations exist, pigs offer unique advantages as a model organism for certain research questions, particularly in areas that require larger animal models with a closer resemblance to human physiology and behavior. Researchers must carefully weigh these factors and design experiments accordingly to maximize the benefits and minimize the limitations of using pigs for research purposes.

3. Discussion

Pigs have attracted much attention as a highly valuable model organism within the field of neuroscience. Pigs provide unique advantages and challenges for researchers interested in studying aspects of neural injury and disease that cannot be fully replicated in small animal models. Widely regarded for their similarities to humans in terms of neuroanatomy and neurodevelopment, pigs can help bridge the gap between small animal preclinical neuroscience and clinical applications. Pigs are particularly useful for studying cognition, behavior, and physiology in a variety of human health and disease states. Here, we have touched upon many of the commonly used behavioral tests, which have been conducted in pigs to demonstrate their intelligence and cognitive complexity. The range of relevant behaviors that pigs engage in can help researchers to identify more subtle differences in how a disease may affect an individual. Many easily replicable tests have been developed for use in pigs, and with the advancement of technology, more complex and integrative assessments are being conceptualized. Pigs are highly social animals who engage in their own forms of communication, social hierarchies, and emotional responses. The size of the pig allows for the usage of various wearable and implantable physiological monitoring devices, and neuroimaging studies can be conducted using the same equipment found in a clinical setting. Pigs are becoming one of the most sought-after model organisms for studying neurotrauma, as there are many aspects of brain injury that are unable to be replicated in a lissencephalic model species.

Author Contributions: All the authors analyzed the literature and wrote and read the manuscript. All authors have read and agreed to the published version of the manuscript.

Funding: Support for this work was provided by the National Institutes of Health, R01NS098231 (GP) and F99NS129171 (AN).

Institutional Review Board Statement: All experimental procedures were approved by the Michigan State University Institutional Animal Care and Use Committee and conducted in compliance with National Institutes of Health Animal Research Advisory Committee guidelines.

Data Availability Statement: All data supporting reported results are available upon request from the corresponding author.

Conflicts of Interest: The authors declare no conflict of interest.

References

1. Davies, C.; Hamilton, O.K.L.; Hooley, M.; Ritakari, T.E.; Stevenson, A.J.; Wheater, E.N.W. Translational neuroscience: The state of the nation (a Ph.D. student perspective). *Brain Commun.* **2020**, *2*, fcaa038. [CrossRef] [PubMed]
2. Horn, S.R.; Fisher, P.A.; Pfeifer, J.H.; Allen, N.B.; Berkman, E.T. Levers and barriers to success in the use of translational neuroscience for the prevention and treatment of mental health and promotion of well-being across the lifespan. *J. Abnorm. Psychol.* **2020**, *129*, 38–48. [CrossRef] [PubMed]
3. Faggion, C.M., Jr. Animal research as a basis for clinical trials. *Eur. J. Oral Sci.* **2015**, *123*, 61–64. [CrossRef] [PubMed]
4. Kobayashi, E.; Hishikawa, S.; Teratani, T.; Lefor, A.T. The pig as a model for translational research: Overview of porcine animal models at Jichi Medical University. *Transplant. Res.* **2012**, *1*, 8. [CrossRef]
5. Dai, J.X.; Ma, Y.B.; Le, N.Y.; Cao, J.; Wang, Y. Large animal models of traumatic brain injury. *Int. J. Neurosci.* **2018**, *128*, 243–254. [CrossRef]
6. Sorby-Adams, A.J.; Vink, R.; Turner, R.J. Large animal models of stroke and traumatic brain injury as translational tools. *Am. J. Physiol. Regul. Integr. Comp. Physiol.* **2018**, *315*, R165–R190. [CrossRef] [PubMed]
7. Vink, R. Large animal models of traumatic brain injury. *J. Neurosci. Res.* **2018**, *96*, 527–535. [CrossRef]
8. Gieling, E.T.; Nordquist, R.E.; van der Staay, F.J. Assessing learning and memory in pigs. *Anim. Cogn.* **2011**, *14*, 151–173. [CrossRef]
9. Gieling, E.T.; Schuurman, T.; Nordquist, R.E.; van der Staay, F.J. The pig as a model animal for studying cognition and neurobehavioral disorders. *Curr. Top. Behav. Neurosci.* **2011**, *7*, 359–383. [CrossRef]
10. Puccinelli, E.; Gervasi, P.G.; Longo, V. Xenobiotic metabolizing cytochrome P450 in pig, a promising animal model. *Curr. Drug Metab.* **2011**, *12*, 507–525. [CrossRef] [PubMed]
11. Flood, A.B.; Wood, V.A.; Schreiber, W.; Williams, B.B.; Gallez, B.; Swartz, H.M. Guidance to Transfer 'Bench-Ready' Medical Technology into Usual Clinical Practice: Case Study—Sensors and Spectrometer Used in EPR Oximetry. *Adv. Exp. Med. Biol.* **2018**, *1072*, 233–239. [CrossRef]
12. Huang, W.; Percie du Sert, N.; Vollert, J.; Rice, A.S.C. General Principles of Preclinical Study Design. *Handb. Exp. Pharmacol.* **2020**, *257*, 55–69. [CrossRef]
13. Bovenkerk, B.; Kaldewaij, F. The use of animal models in behavioural neuroscience research. *Curr. Top. Behav. Neurosci.* **2015**, *19*, 17–46. [CrossRef]
14. Ellenbroek, B.; Youn, J. Rodent models in neuroscience research: Is it a rat race? *Dis. Models Mech.* **2016**, *9*, 1079–1087. [CrossRef] [PubMed]
15. Shin, S.S.; Krishnan, V.; Stokes, W.; Robertson, C.; Celnik, P.; Chen, Y.; Song, X.; Lu, H.; Liu, P.; Pelled, G. Transcranial magnetic stimulation and environmental enrichment enhances cortical excitability and functional outcomes after traumatic brain injury. *Brain Stimul.* **2018**, *11*, 1306–1313. [CrossRef]
16. Peng, Z.; Zhang, C.; Yan, L.; Zhang, Y.; Yang, Z.; Wang, J.; Song, C. EPA is More Effective than DHA to Improve Depression-Like Behavior, Glia Cell Dysfunction and Hippcampal Apoptosis Signaling in a Chronic Stress-Induced Rat Model of Depression. *Int. J. Mol. Sci.* **2020**, *21*, 1769. [CrossRef] [PubMed]

17. Hughson, A.R.; Horvath, A.P.; Holl, K.; Palmer, A.A.; Solberg Woods, L.C.; Robinson, T.E.; Flagel, S.B. Incentive salience attribution, "sensation-seeking" and "novelty-seeking" are independent traits in a large sample of male and female heterogeneous stock rats. *Sci. Rep.* **2019**, *9*, 2351. [CrossRef]
18. Qiu, Y.; O'Neill, N.; Maffei, B.; Zourray, C.; Almacellas-Barbanoj, A.; Carpenter, J.C.; Jones, S.P.; Leite, M.; Turner, T.J.; Moreira, F.C.; et al. On-demand cell-autonomous gene therapy for brain circuit disorders. *Science* **2022**, *378*, 523–532. [CrossRef]
19. O'Collins, V.E.; Macleod, M.R.; Donnan, G.A.; Horky, L.L.; van der Worp, B.H.; Howells, D.W. 1,026 experimental treatments in acute stroke. *Ann. Neurol.* **2006**, *59*, 467–477. [CrossRef] [PubMed]
20. Stein, D.G.; Geddes, R.I.; Sribnick, E.A. Recent developments in clinical trials for the treatment of traumatic brain injury. *Handb. Clin. Neurol.* **2015**, *127*, 433–451. [CrossRef]
21. Todea, R.A.; Lu, P.J.; Fartaria, M.J.; Bonnier, G.; Du Pasquier, R.; Krueger, G.; Bach Cuadra, M.; Psychogios, M.N.; Kappos, L.; Kuhle, J.; et al. Evolution of Cortical and White Matter Lesion Load in Early-Stage Multiple Sclerosis: Correlation With Neuroaxonal Damage and Clinical Changes. *Front. Neurol.* **2020**, *11*, 973. [CrossRef]
22. Cloots, R.J.; Gervaise, H.M.; van Dommelen, J.A.; Geers, M.G. Biomechanics of traumatic brain injury: Influences of the morphologic heterogeneities of the cerebral cortex. *Ann. Biomed. Eng* **2008**, *36*, 1203–1215. [CrossRef]
23. Ventura-Antunes, L.; Mota, B.; Herculano-Houzel, S. Different scaling of white matter volume, cortical connectivity, and gyrification across rodent and primate brains. *Front. Neuroanat.* **2013**, *7*, 3. [CrossRef]
24. Hodge, R.D.; Bakken, T.E.; Miller, J.A.; Smith, K.A.; Barkan, E.R.; Graybuck, L.T.; Close, J.L.; Long, B.; Johansen, N.; Penn, O.; et al. Conserved cell types with divergent features in human versus mouse cortex. *Nature* **2019**, *573*, 61–68. [CrossRef]
25. Ryan, M.C.; Sherman, P.; Rowland, L.M.; Wijtenburg, S.A.; Acheson, A.; Fieremans, E.; Veraart, J.; Novikov, D.S.; Hong, L.E.; Sladky, J.; et al. Miniature pig model of human adolescent brain white matter development. *J. Neurosci. Methods* **2018**, *296*, 99–108. [CrossRef] [PubMed]
26. Henry, Y.; Sève, B.; Mounier, A.; Ganier, P. Growth performance and brain neurotransmitters in pigs as affected by tryptophan, protein, and sex. *J. Anim. Sci.* **1996**, *74*, 2700–2710. [CrossRef] [PubMed]
27. Dickerson, J.W.; Dobbing, J. Prenatal and postnatal growth and development of the central nervous system of the pig. *Proc. R. Soc. Lond. B Biol. Sci.* **1967**, *166*, 384–395. [PubMed]
28. Holm, I.E.; West, M.J. Hippocampus of the domestic pig: A stereological study of subdivisional volumes and neuron numbers. *Hippocampus* **1994**, *4*, 115–125. [CrossRef]
29. Jelsing, J.; Nielsen, R.; Olsen, A.K.; Grand, N.; Hemmingsen, R.; Pakkenberg, B. The postnatal development of neocortical neurons and glial cells in the Göttingen minipig and the domestic pig brain. *J. Exp. Biol.* **2006**, *209*, 1454–1462. [CrossRef]
30. Dobbing, J.; Sands, J. Comparative aspects of the brain growth spurt. *Early Hum. Dev.* **1979**, *3*, 79–83. [CrossRef]
31. Pond, W.G.; Boleman, S.L.; Fiorotto, M.L.; Ho, H.; Knabe, D.A.; Mersmann, H.J.; Savell, J.W.; Su, D.R. Perinatal ontogeny of brain growth in the domestic pig. *Proc. Soc. Exp. Biol. Med. Soc. Exp. Biol. Med.* **2000**, *223*, 102–108. [CrossRef]
32. Croney, C.; Adams, K.; Washington, C.; Stricklin, W. A note on visual, olfactory and spatial cue use in foraging behavior of pigs: Indirectly assessing cognitive abilities. *Appl. Anim. Behav. Sci.* **2003**, *83*, 303–308. [CrossRef]
33. de Jong, I.C.; Prelle, I.T.; van de Burgwal, J.A.; Lambooij, E.; Korte, S.M.; Blokhuis, H.J.; Koolhaas, J.M. Effects of environmental enrichment on behavioral responses to novelty, learning, and memory, and the circadian rhythm in cortisol in growing pigs. *Physiol. Behav.* **2000**, *68*, 571–578. [CrossRef] [PubMed]
34. Kornum, B.R.; Knudsen, G.M. Cognitive testing of pigs (Sus scrofa) in translational biobehavioral research. *Neurosci. Biobehav. Rev.* **2011**, *35*, 437–451. [CrossRef] [PubMed]
35. Howard, J.T.; Ashwell, M.S.; Baynes, R.E.; Brooks, J.D.; Yeatts, J.L.; Maltecca, C. Genetic Parameter Estimates for Metabolizing Two Common Pharmaceuticals in Swine. *Front. Genet.* **2018**, *9*, 40. [CrossRef]
36. Wolf, E.; Braun-Reichhart, C.; Streckel, E.; Renner, S. Genetically engineered pig models for diabetes research. *Transgenic Res.* **2014**, *23*, 27–38. [CrossRef]
37. Kinder, H.A.; Baker, E.W.; Howerth, E.W.; Duberstein, K.J.; West, F.D. Controlled Cortical Impact Leads to Cognitive and Motor Function Deficits that Correspond to Cellular Pathology in a Piglet Traumatic Brain Injury Model. *J. Neurotrauma* **2019**, *36*, 2810–2826. [CrossRef]
38. Singh, R.; Kulikowicz, E.; Santos, P.T.; Koehler, R.C.; Martin, L.J.; Lee, J.K. Spatial T-maze identifies cognitive deficits in piglets 1 month after hypoxia-ischemia in a model of hippocampal pyramidal neuron loss and interneuron attrition. *Behav. Brain Res.* **2019**, *369*, 111921. [CrossRef]
39. Chen, Y.; Wang, B.; Yang, C.; Shi, Y.; Dong, Z.; Troy, F.A., 2nd. Functional Correlates and Impact of Dietary Lactoferrin Intervention and its Concentration-dependence on Neurodevelopment and Cognition in Neonatal Piglets. *Mol. Nutr. Food Res.* **2021**, *65*, e2001099. [CrossRef]
40. Haagensen, A.M.; Klein, A.B.; Ettrup, A.; Matthews, L.R.; Sørensen, D.B. Cognitive performance of Göttingen minipigs is affected by diet in a spatial hole-board discrimination test. *PLoS ONE* **2013**, *8*, e79684. [CrossRef]
41. Grimberg-Henrici, C.G.; Vermaak, P.; Bolhuis, J.E.; Nordquist, R.E.; van der Staay, F.J. Effects of environmental enrichment on cognitive performance of pigs in a spatial holeboard discrimination task. *Anim. Cogn.* **2016**, *19*, 271–283. [CrossRef] [PubMed]
42. Clouard, C.; Reimert, I.; Fleming, S.A.; Koopmans, S.J.; Schuurman, T.; Hauser, J. Dietary sialylated oligosaccharides in early-life may promote cognitive flexibility during development in context of obesogenic dietary intake. *Nutr. Neurosci.* **2022**, *25*, 2461–2478. [CrossRef]

43. Netzley, A.H.; Hunt, R.D.; Franco-Arellano, J.; Arnold, N.; Vazquez, A.I.; Munoz, K.A.; Colbath, A.C.; Bush, T.R.; Pelled, G. Multimodal characterization of Yucatan minipig behavior and physiology through maturation. *Sci. Rep.* **2021**, *11*, 22688. [CrossRef] [PubMed]
44. Schramke, S.; Schuldenzucker, V.; Schubert, R.; Frank, F.; Wirsig, M.; Ott, S.; Motlik, J.; Fels, M.; Kemper, N.; Hölzner, E.; et al. Behavioral phenotyping of minipigs transgenic for the Huntington gene. *J. Neurosci. Methods* **2016**, *265*, 34–45. [CrossRef]
45. Ao, W.; Grace, M.; Floyd, C.L.; Vonder Haar, C. A Touchscreen Device for Behavioral Testing in Pigs. *Biomedicines* **2022**, *10*, 2612. [CrossRef]
46. Wegner, B.; Spiekermeier, I.; Nienhoff, H.; Große-Kleimann, J.; Rohn, K.; Meyer, H.; Plate, H.; Gerhardy, H.; Kreienbrock, L.; Beilage, E.G.; et al. Application of the voluntary human approach test on commercial pig fattening farms: A meaningful tool? *Porc. Health Manag.* **2020**, *6*, 19. [CrossRef]
47. Wu, X.; Zhou, S.; Chen, M.; Zhao, Y.; Wang, Y.; Zhao, X.; Li, D.; Pu, H. Combined spectral and speech features for pig speech recognition. *PLoS ONE* **2022**, *17*, e0276778. [CrossRef]
48. Briefer, E.F.; Sypherd, C.C.; Linhart, P.; Leliveld, L.M.C.; Padilla de la Torre, M.; Read, E.R.; Guérin, C.; Deiss, V.; Monestier, C.; Rasmussen, J.H.; et al. Classification of pig calls produced from birth to slaughter according to their emotional valence and context of production. *Sci. Rep.* **2022**, *12*, 3409. [CrossRef]
49. Haigh, A.; Chou, J.-Y.; O'Driscoll, K. Variations in the Behavior of Pigs During an Open Field and Novel Object Test. *Front. Vet. Sci.* **2020**, *7*, 607. [CrossRef]
50. Mathis, A.; Mamidanna, P.; Cury, K.M.; Abe, T.; Murthy, V.N.; Mathis, M.W.; Bethge, M. DeepLabCut: Markerless pose estimation of user-defined body parts with deep learning. *Nat. Neurosci.* **2018**, *21*, 1281–1289. [CrossRef]
51. Nachman, D.; Constantini, K.; Poris, G.; Wagnert-Avraham, L.; Gertz, S.D.; Littman, R.; Kabakov, E.; Eisenkraft, A.; Gepner, Y. Wireless, non-invasive, wearable device for continuous remote monitoring of hemodynamic parameters in a swine model of controlled hemorrhagic shock. *Sci. Rep.* **2020**, *10*, 17684. [CrossRef]
52. Martinez-Ramirez, L.; Slate, A.; Price, G.; Duhaime, A.C.; Staley, K.; Costine-Bartell, B.A. Robust, long-term video EEG monitoring in a porcine model of post-traumatic epilepsy. *eNeuro* **2022**, *9*, ENEURO.0025-22.2022. [CrossRef]
53. Simchick, G.; Shen, A.; Campbell, B.; Park, H.J.; West, F.D.; Zhao, Q. Pig Brains Have Homologous Resting-State Networks with Human Brains. *Brain Connect.* **2019**, *9*, 566–579. [CrossRef]
54. Lee, J.K.; Liu, D.; Jiang, D.; Kulikowicz, E.; Tekes, A.; Liu, P.; Qin, Q.; Koehler, R.C.; Aggarwal, M.; Zhang, J.; et al. Fractional anisotropy from diffusion tensor imaging correlates with acute astrocyte and myelin swelling in neonatal swine models of excitotoxic and hypoxic-ischemic brain injury. *J. Comp. Neurol.* **2021**, *529*, 2750–2770. [CrossRef]
55. Wang, K.K.; Yang, Z.; Zhu, T.; Shi, Y.; Rubenstein, R.; Tyndall, J.A.; Manley, G.T. An update on diagnostic and prognostic biomarkers for traumatic brain injury. *Expert Rev. Mol. Diagn.* **2018**, *18*, 165–180. [CrossRef]
56. Tenovuo, O.; Diaz-Arrastia, R.; Goldstein, L.E.; Sharp, D.J.; van der Naalt, J.; Zasler, N.D. Assessing the Severity of Traumatic Brain Injury-Time for a Change? *J. Clin. Med.* **2021**, *10*, 148. [CrossRef]
57. Kaur, P.; Sharma, S. Recent Advances in Pathophysiology of Traumatic Brain Injury. *Curr. Neuropharmacol.* **2018**, *16*, 1224–1238. [CrossRef]
58. Krishnamurthy, K.; Laskowitz, D.T. Cellular and Molecular Mechanisms of Secondary Neuronal Injury Following Traumatic Brain Injury. In *Translational Research in Traumatic Brain Injury*; Laskowitz, D., Grant, G., Eds.; CRC Press: Boca Raton, FL, USA; Taylor and Francis Group: Boca Raton, FL, USA, 2016.
59. Marklund, N.; Bellander, B.M.; Godbolt, A.K.; Levin, H.; McCrory, P.; Thelin, E.P. Treatments and rehabilitation in the acute and chronic state of traumatic brain injury. *J. Intern. Med.* **2019**, *285*, 608–623. [CrossRef]
60. Osier, N.D.; Korpon, J.R.; Dixon, C.E. Controlled Cortical Impact Model. In *Brain Neurotrauma: Molecular, Neuropsychological, and Rehabilitation Aspects*; Kobeissy, F.H., Ed.; CRC Press: Boca Raton, FL, USA, 2015.
61. Pareja, J.C.; Keeley, K.; Duhaime, A.C.; Dodge, C.P. Modeling Pediatric Brain Trauma: Piglet Model of Controlled Cortical Impact. *Methods Mol. Biol.* **2016**, *1462*, 345–356. [CrossRef]
62. Simchick, G.; Scheulin, K.M.; Sun, W.; Sneed, S.E.; Fagan, M.M.; Cheek, S.R.; West, F.D.; Zhao, Q. Detecting functional connectivity disruptions in a translational pediatric traumatic brain injury porcine model using resting-state and task-based fMRI. *Sci. Rep.* **2021**, *11*, 12406. [CrossRef]
63. Baker, E.W.; Kinder, H.A.; Hutcheson, J.M.; Duberstein, K.J.J.; Platt, S.R.; Howerth, E.W.; West, F.D. Controlled Cortical Impact Severity Results in Graded Cellular, Tissue, and Functional Responses in a Piglet Traumatic Brain Injury Model. *J. Neurotrauma* **2019**, *36*, 61–73. [CrossRef]
64. Manley, G.T.; Rosenthal, G.; Lam, M.; Morabito, D.; Yan, D.; Derugin, N.; Bollen, A.; Knudson, M.M.; Panter, S.S. Controlled cortical impact in swine: Pathophysiology and biomechanics. *J. Neurotrauma* **2006**, *23*, 128–139. [CrossRef]
65. Wang, J.; Shi, Y.; Cao, S.; Liu, X.; Martin, L.J.; Simoni, J.; Soltys, B.J.; Hsia, C.J.C.; Koehler, R.C. Polynitroxylated PEGylated hemoglobin protects pig brain neocortical gray and white matter after traumatic brain injury and hemorrhagic shock. *Front. Med. Technol.* **2023**, *5*, 1074643. [CrossRef]
66. Ginsburg, J.; Huff, J.S. *Closed Head Trauma*; StatPearls Publishing: Treasure Island, FL, USA, 2023.
67. Shin, S.S.; Gottschalk, A.C.; Mazandi, V.M.; Kilbaugh, T.J.; Hefti, M.M. Transcriptional Profiling in a Novel Swine Model of Traumatic Brain Injury. *Neurotrauma Rep.* **2022**, *3*, 178–184. [CrossRef]

68. Shin, S.S.; Chawla, S.; Jang, D.H.; Mazandi, V.M.; Weeks, M.K.; Kilbaugh, T.J. Imaging of White Matter Injury Correlates with Plasma and Tissue Biomarkers in Pediatric Porcine Model of Traumatic Brain Injury. *J. Neurotrauma* **2023**, *40*, 74–85. [CrossRef]
69. Cullen, D.K.; Harris, J.P.; Browne, K.D.; Wolf, J.A.; Duda, J.E.; Meaney, D.F.; Margulies, S.S.; Smith, D.H. A Porcine Model of Traumatic Brain Injury via Head Rotational Acceleration. *Methods Mol. Biol.* **2016**, *1462*, 289–324. [CrossRef]
70. Mayer, A.R.; Ling, J.M.; Patton, D.A.; Stephenson, D.D.; Dodd, A.B.; Dodd, R.J.; Rannou-Latella, J.G.; Smith, D.H.; Johnson, V.E.; Cullen, D.K.; et al. Non-Linear Device Head Coupling and Temporal Delays in Large Animal Acceleration Models of Traumatic Brain Injury. *Ann. Biomed. Eng.* **2022**, *50*, 728–739. [CrossRef]
71. Mayer, A.R.; Ling, J.M.; Dodd, A.B.; Rannou-Latella, J.G.; Stephenson, D.D.; Dodd, R.J.; Mehos, C.J.; Patton, D.A.; Cullen, D.K.; Johnson, V.E.; et al. Reproducibility and Characterization of Head Kinematics During a Large Animal Acceleration Model of Traumatic Brain Injury. *Front. Neurol.* **2021**, *12*, 658461. [CrossRef]
72. O'Donnell, J.C.; Browne, K.D.; Kvint, S.; Makaron, L.; Grovola, M.R.; Karandikar, S.; Kilbaugh, T.J.; Cullen, D.K.; Petrov, D. Multimodal Neuromonitoring and Neurocritical Care in Swine to Enhance Translational Relevance in Brain Trauma Research—Neurocritical Care in Swine. *Biomedicines* **2023**, *11*, 5.
73. Chen, C.; Zhou, C.; Cavanaugh, J.M.; Kallakuri, S.; Desai, A.; Zhang, L.; King, A.I. Quantitative electroencephalography in a swine model of blast-induced brain injury. *Brain Inj.* **2017**, *31*, 120–126. [CrossRef]
74. Kallakuri, S.; Desai, A.; Feng, K.; Tummala, S.; Saif, T.; Chen, C.; Zhang, L.; Cavanaugh, J.M.; King, A.I. Neuronal Injury and Glial Changes Are Hallmarks of Open Field Blast Exposure in Swine Frontal Lobe. *PLoS ONE* **2017**, *12*, e0169239. [CrossRef]
75. Cralley, A.L.; Moore, E.E.; Kissau, D.; Coleman, J.R.; Vigneshwar, N.; DeBot, M.; Schaid, T.R., Jr.; Moore, H.B.; Cohen, M.J.; Hansen, K.; et al. A combat casualty relevant dismounted complex blast injury model in swine. *J. Trauma Acute Care Surg.* **2022**, *93*, S110–S118. [CrossRef]
76. Wendt, M.; Bickhardt, K.; Herzog, A.; Fischer, A.; Martens, H.; Richter, T. Porcine stress syndrome and PSE meat: Clinical symptoms, pathogenesis, etiology and animal rights aspects. *Berl. Und Munch. Tierarztl. Wochenschr.* **2000**, *113*, 173–190.

Disclaimer/Publisher's Note: The statements, opinions and data contained in all publications are solely those of the individual author(s) and contributor(s) and not of MDPI and/or the editor(s). MDPI and/or the editor(s) disclaim responsibility for any injury to people or property resulting from any ideas, methods, instructions or products referred to in the content.

Review

Porcine Astrocytes and Their Relevance for Translational Neurotrauma Research

Erin M. Purvis [1,2,3], Natalia Fedorczak [1,2], Annette Prah [1,2], Daniel Han [1,2] and John C. O'Donnell [1,2,*]

1. Center for Neurotrauma, Neurodegeneration & Restoration, Corporal Michael J. Crescenz Veterans Affairs Medical Center, Philadelphia, PA 19104, USA; daniel.han@pennmedicine.upenn.edu (D.H.)
2. Center for Brain Injury & Repair, Department of Neurosurgery, Perelman School of Medicine, University of Pennsylvania, Philadelphia, PA 19104, USA
3. Department of Neuroscience, Perelman School of Medicine, University of Pennsylvania, Philadelphia, PA 19104, USA
* Correspondence: odj@pennmedicine.upenn.edu

Abstract: Astrocytes are essential to virtually all brain processes, from ion homeostasis to neurovascular coupling to metabolism, and even play an active role in signaling and plasticity. Astrocytic dysfunction can be devastating to neighboring neurons made inherently vulnerable by their polarized, excitable membranes. Therefore, correcting astrocyte dysfunction is an attractive therapeutic target to enhance neuroprotection and recovery following acquired brain injury. However, the translation of such therapeutic strategies is hindered by a knowledge base dependent almost entirely on rodent data. To facilitate additional astrocytic research in the translatable pig model, we present a review of astrocyte findings from pig studies of health and disease. We hope that this review can serve as a road map for intrepid pig researchers interested in studying astrocyte biology.

Keywords: swine; pig; porcine; astrocytes; glia; translational neurotrauma

Citation: Purvis, E.M.; Fedorczak, N.; Prah, A.; Han, D.; O'Donnell, J.C. Porcine Astrocytes and Their Relevance for Translational Neurotrauma Research. *Biomedicines* 2023, 11, 2388. https://doi.org/10.3390/biomedicines11092388

Academic Editor: Bruno Meloni

Received: 7 June 2023
Revised: 17 August 2023
Accepted: 22 August 2023
Published: 26 August 2023

Copyright: © 2023 by the authors. Licensee MDPI, Basel, Switzerland. This article is an open access article distributed under the terms and conditions of the Creative Commons Attribution (CC BY) license (https://creativecommons.org/licenses/by/4.0/).

1. Introduction

Astrocytes, the most abundant cell type in the brain, were once believed to play only a supporting role in brain function. However, research over the last few decades has revealed that they are active participants in nearly all facets of brain activity. They maintain ion gradient homeostasis essential for brain function, prevent edema, remove glutamate from the extracellular space, couple neuronal activity to changes in blood flow and glucose uptake, and engage directly in signaling events and plasticity mechanisms. As the hub between the vasculature and the synapse, they are also central to anabolic and catabolic metabolism in the brain, processing glucose for energy substrates (glycogen, lactate, etc.) or to synthesize essential molecules like glutamate, as well as providing antioxidant protection to neighboring neurons and performing the essential task of fixing NH_4 to allow incorporation of nitrogen into biological molecules. Neurotrauma and neurodegenerative diseases that impact these essential astrocytic functions can be devastating to the brain and the organism as a whole (for reviews, see [1–8]). Their central role in facilitating both healthy brain function and the mechanisms of neurotrauma and neurodegenerative disease, combined with their resilience relative to neighboring neurons, make astrocytes a very attractive therapeutic target [9–11]. Translating such therapies will require investigations in a large animal model like pigs to investigate the mechanisms and manifestations of human injuries more closely.

The overwhelming majority of our astrocyte knowledge stems from research in rodents due to their low cost and ease-of-use. However, recent genetic brain atlas comparisons between mice, pigs, and humans revealed greater variability between humans and mice relative to humans and pigs [12], and the distribution of splice variants of glutamate transporters (a signature family of astrocytic proteins) also shows a greater similarity between pigs and humans relative to mice and humans [13], suggesting that translation of knowledge even at the level of cell biology should involve a large animal model like the pig at some stage. In addition to the cellular similarities between humans and pigs (relative to rodents), pigs are high-fidelity models for the translational study of acquired brain injury due to their large gyrencephalic brains, structural similarities in limbic, subcortical, diencephalic, and brainstem regions, high white-to-gray matter ratio, and similar basal cistern geometry, among other features [14–22]. Due to their relatively large brain mass, pigs offer the unique opportunity to recreate rotational acceleration injury, the predominant mechanism of human traumatic brain injury (TBI), which in turn presents the unique opportunity to study TBI manifestations that cannot be recreated in rodents, such as coma and other disorders of consciousness [14,23]. Head rotational acceleration TBI in pigs also recreates acute crises such as apnea and increased intracranial pressure, which, coupled with their large size providing compatibility with clinical neuromonitoring equipment, makes them an ideal model for studying neurocritical care [24].

Beyond direct brain injury, we will also discuss studies of astrocytes in pigs that have experienced indirect brain injuries through mechanisms such as cardiac arrest, sepsis, and others. While we will have inevitably overlooked some research works, we believe that this neurotrauma-forward review provides a near-comprehensive assessment of studies that have assessed astrocytes in pigs. Such a review would be impossible in rodent literature, further emphasizing the relative lack of, and need for, astrocyte research in pigs.

2. Investigations of Astrocyte Structure and Function in the Porcine Brain

To date, there have been several studies that have investigated the structure and function of astrocytes in the pig brain. Porcine astrocytes have been investigated in vivo, in slice culture, histologically, and using cell culture techniques. Here, we review the details of these studies to highlight what is known about porcine astrocytes. We only discuss elements of these publications that are relevant to astrocytes in the porcine brain. Furthermore, as it can be difficult to find histological antibodies for pig specimens, we summarize all available antibody information from the reviewed papers in Supplementary Table S1.

2.1. Glutamate Transporters

The astrocytic sodium-dependent glutamate transporters GLAST (GLutamate ASpartate Transporter) and GLT-1 (GLutamate Transporter-1) are responsible for clearing glutamate from the extracellular space. Glutamate is the primary excitatory neurotransmitter in the mammalian brain and also, paradoxically, a potent neurotoxin. Therefore, one of the most obvious and important tasks astrocytic glutamate transporters perform is the constant clearance of glutamate from the extracellular space to prevent excitotoxic neuronal death. However, astrocytic glutamate transporters and associated calcium signals appear integral to many brain functions. Glutamate uptake into fine astrocytic processes results in local reversal of the sodium/calcium exchanger that positions astrocytic mitochondria near GLT-1 clusters that are servicing active synapses [25]. Aside from facilitating glutamate oxidation, local calcium signals in fine astrocytic processes appear between neighboring mitochondria, very rarely extending past a mitochondrion. However, when astrocytic mitochondria were damaged via transient oxygen/glucose deprivation in organotypic hippocampal slices (rat), most of these calcium signaling events extended beyond neighboring mitochondria [26]. These observations suggest that glutamate transporters may provide additional information integration via astrocytic calcium signaling between local synapses facilitated by mitochondria in fine astrocytic processes and that trauma can disrupt this communication. Astrocytic glutamate transporters are vitally important to brain function, but little is known about them outside of rodent models.

The Pow lab published a detailed article describing the expression patterns of the glutamate transporters GLAST, GLT-1alpha, and GLT-1v in the Large White/Landrace porcine brain [13]. They conducted immunohistochemistry for these three transporter proteins as well as GFAP in the coronal and sagittal sections of the porcine brain. They reported the highest level of immunoreactivity to GLAST in the cerebellum, with strong labeling in the Bergmann glial cells of the molecular layer and less labeling of astrocytes in the granular cell layer. The porcine hippocampus also expressed high levels of GLAST, with significant labeling in areas CA1, CA2, CA3, and the dentate regions. Additionally, there was GLAST labeling in the molecular layers of the fascia dentate and the hilar region of the dentate such that certain layers stained more strongly for GLAST than others. GLAST labeling was evident in all cortical gray matter layers. The motor cortex was evenly labeled for GLAST expression, while the frontal and temporal cortices had patchy labeling with some strongly immunoreactive areas interspersed with areas of weakly labeled tissue. These patchy areas were not contained within any specific cortical regions but rather seemed to be present in all cortical layers. GLAST co-labeled with GFAP in these patchy cortical regions. There was strong labeling of GLT-1alpha in the cortex, putamen, hippocampus, thalamus, and cerebellum. There was little labeling of GLT-1alpha in the forebrain white matter (Figure 1). In the cerebellum, GLT-1alpha was strongly labeled in Bergmann glial cells of the molecular layer, in astrocytes in the granular layer, and in the deep cerebellar nuclei. There was also significant GLT-1alpha labeling in hippocampal areas CA1, CA2, CA3, and the dentate gyrus, with the strongest dentate labeling occurring in the molecular layer. Immunolabeling of GLT-1alpha was strong but patchy in cortical areas such as the frontal cortex and temporal cortex. They also reported immunolabeling of the porcine brain for GLT-1v, but this labeling was seen in oligodendrocytes rather than astrocytes. GLT1-alpha expression was restricted to gray matter regions, whereas GLT-1v expression was restricted to white matter regions. They also reported that GLAST and GLT-1alpha expression sometimes, but not always, co-localized with GFAP staining. Overall, these results indicate that, as has been observed in other mammals, different brain regions have different glutamate transport properties and that GLAST and GLT-1apha glutamate transporters sometimes co-localized with GFAP in the porcine brain.

The Pow lab followed up on this with a publication describing the associated expression of the glutamate transporter GLAST and the astrocyte intermediate filament protein GFAP in the 1-day-old piglet brain [27]. In this study, they demonstrated co-immunoprecipitation of GFAP and GLAST in the cortex and cerebellum of the piglet brain. They immunoprecipitated total piglet brain lysate with anti-GFAP antibodies and detected GLAST, and immunoprecipitated total piglet brain lysate with anti-GLAST antibodies and detected GFAP. These results suggested an in vivo interaction between GFAP and GLAST in the piglet brain. They also performed immunohistochemistry to reveal that GFAP was strongly expressed by astrocytes in the gray and white matter of the piglet brain. GLAST was highly expressed in the cortical gray matter, hippocampus, thalamus, and hypothalamus. High magnification imaging revealed a close spatial association of these two markers, with GFAP expression in the core or cytoskeleton of the cell and GLAST expression in the plasma membrane. This publication also included data on changes in GFAP and GLAST expression following hypoxia in the piglet brain, which is discussed in the hypoxia–ischemia injury section below. These results, along with results gathered from experiments on the rat brain, led the authors to hypothesize that GFAP stabilizes astrocyte processes, which could help to anchor GLAST in the plasma membrane.

This group also demonstrated that the GLAST1c splice variant is expressed in the piglet brain [28]. RT-PCR and Western blot were used to identify the expression of GLAST1c in the pig cortex. They also used immunohistochemistry to demonstrate the presence of GLAST1c in Bergmann glial cells of the molecular layer, astrocytes of the granule cell layer, and cells in the white matter layer of the pig cerebellum.

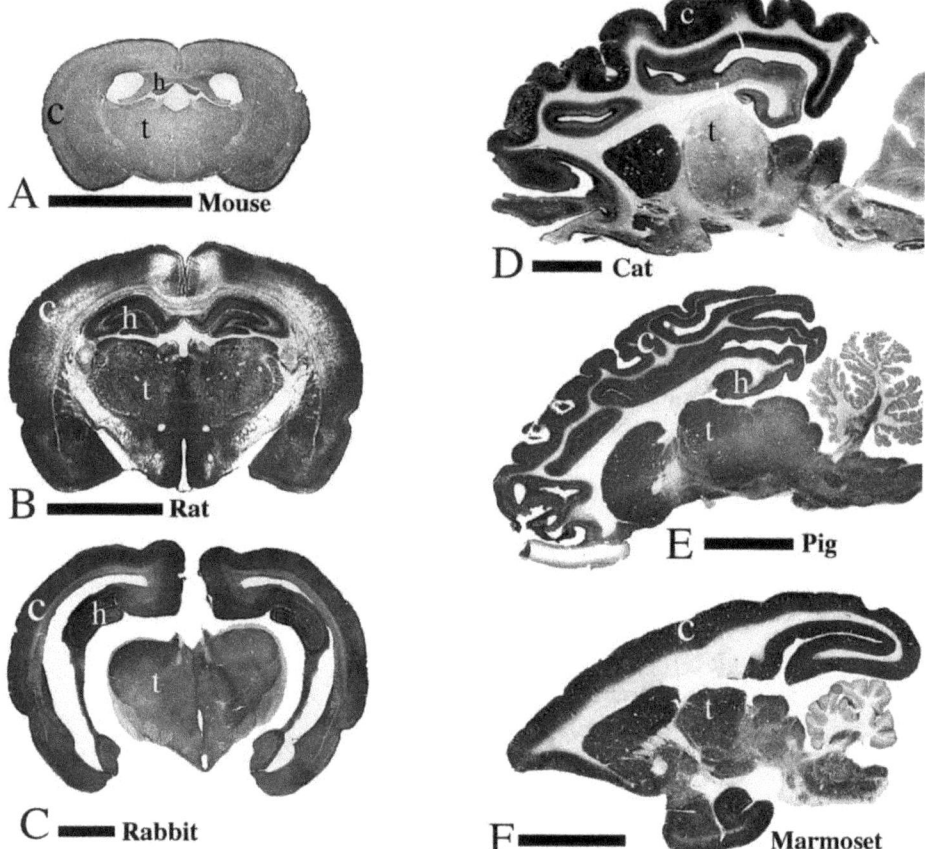

Figure 1. Cross-species comparison of astrocytic GLT-1alpha. Coronal brain sections from mouse (**A**), rat (**B**), and rabbit (**C**), and sagittal brain sections from cat (**D**), pig (**E**), and marmoset monkey (**F**), immunolabeled for GLT-1α. Cortex (c) and hippocampus (h) possessed high levels of GLT-1α in all species. Both rats (**B**) and pigs (**E**) exhibited strong labeling in the thalamus (t), but only pigs (**E**) and monkeys (**F**) exhibited strong labeling in deep cerebellar nuclei. Species such as cats (**D**), pigs (**E**), and monkeys (**F**) possess high amounts of white matter (like humans), which does not stain positive for GLT-1α. Rats, mice, and rabbits possess almost no white matter, but there does appear to be staining for GLT-1α in the paucity of white matter present. Scale bars = 5 mm in (**A**–**C**); 10 mm in (**D**–**F**). Figure reproduced with permission from Williams et al. 2005 [13].

2.2. GFAP (Glial Fibrillary Acidic Protein)

Throughout this review, it will become evident that histological staining of GFAP is by far the most common technique used to study astrocytes, primarily via morphological analyses or simply by measuring stain intensity. This section describes studies in pigs that focused on GFAP itself.

Blechinberg and colleagues published a report detailing the expression of the GFAP isoforms GFAPα, GFAPε, and GFAPκ in the adult and developing porcine brain [29]. They utilized real-time PCR (RT-PCR) to analyze and quantify the mRNA expression of these three GFAP isoforms in adult pig cortical tissue. They report that in the adult pig cortex, the expression of GFAPα is approximately 100-fold higher than the expression of GFAPκ and roughly 60-fold higher than the expression of GFAPε. The expression of GFAPε is almost double the expression of GFAPκ in the adult pig cortex. They also performed RT-PCR analysis of these three GFAP isoforms in fetal brain tissue comprised of the hippocampus, cortex, basal ganglia, cerebellum, and brain stem of embryonic day 40, 60, 80, 100, and 115 fetal pigs. They reported that mRNA levels of GFAPα, GFAPε, and GFAPκ increase across porcine development. The highest GFAP expression was observed in the brainstem and cerebellum brain regions, and the lowest GFAP expression was in the cortex. They reported the highest increase in GFAP expression occurred between e100 and e115 in the brainstem. The mRNA ratio of GFAPα/GFAPε is consistent across development, with GFAPα around 100-fold higher than GFAPε. The mRNA ratio of GFAPα/GFAPκ in the basal ganglia and brainstem increases on e115 compared to e60, whereas the GFAPα/GFAPκ ratio in the cerebellum increases between e60 and e80 and then decreases again. The GFAPκ/GFAPε mRNA ratio decreases from e60 to e115. Overall, the GFAPα/GFAPκ and GFAPε/GFAPκ ratios are higher in the adult cortex compared with the brain during fetal pig development. The results of this study indicate that GFAP is first expressed around e40 and is tightly regulated during fetal development. The authors speculate that changing mRNA ratios of the various GFAP isoforms throughout development could reflect cues for glial cell differentiation. Another group employing transcriptomic analyses to study glial differentiation in pigs reported that marker genes of astrocytes, including GFAP and AQP4, are differentially expressed in the male and female porcine brain during development between gestational days 45 and 90 [30].

2.3. Inflammatory Signaling

The Busija lab published two reports detailing prostaglandin synthesis in cultured porcine cerebral astrocytes [31,32]. The first report revealed that $PGF_{2\alpha}$ is the predominant prostaglandin produced by porcine astrocytes [31]. These cells were observed to produce minimal amounts of 6-keto-$PGF_{1\alpha}$, PGE_2, and LTC_4/D_4. Application of phorbol 12,13-dibutyrate (PDB), a protein kinase C (PKC) activator, onto astrocyte cultures led to increased levels of $PGF_{2\alpha}$. PDB application did not increase levels of 6-keto-$PGF_{1\alpha}$, PGE_2, or LTC_4/D_4. This PDB-induced increase in $PGF_{2\alpha}$ was prevented when cultures were treated with the drugs indomethacin, quinacrine (phospholipase A_2/PLA_2 inhibitor), or isoquinolinylsulfonylmethyl piperazine (PKC inhibitor) at the same time as PDB. Application of 4α-phorbol 12,13-didecanoate (PDD), which does not activate PKC, did not lead to increased levels of $PGF_{2\alpha}$ in porcine astrocyte cultures. These results indicate that the major prostaglandin produced by porcine cerebral astrocytes is $PGF_{2\alpha}$ and that PKC activation increases the production of $PGF_{2\alpha}$ through a mechanism that may involve PLA_2. In a separate study, they also demonstrated that the administration of interleukin 1α (IL-1α) to cultured porcine astrocytes rapidly increases their production of $PGF_{2\alpha}$ [32]. Using a previously validated protocol, $PGF_{2\alpha}$ was increased two-fold when 11 µg/mL of IL-1α was added and four-fold when 22 µg/mL of IL-1α to astrocyte media. To our knowledge, equivalent dosing in humans has not been determined. Levels of 6-keto-$PGF_{1\alpha}$ and PGE_2 did not change with IL-1α addition. These results confirm the results of their previous study indicating that $PGF_{2\alpha}$ is the major prostaglandin produced by porcine astrocytes.

The Zimmer lab conducted a study to investigate the cultured porcine astrocyte expression of major histocompatibility complex (MHC) antigens and their ability to induce the proliferation of human T-lymphocytes [33]. They reported that cultured astrocytes harvested from the fetal porcine brain were not autofluorescent, stained negative for CD18, and stained positive for GFAP, CD44, and the NCAM isoform of CD56. Cultured astrocytes also had upregulated levels of MHC class I antigens compared to freshly isolated cells, indicating that the process of culturing cells can lead to upregulated MHC antigens. Cultured astrocytes did not express MHC class II antigens. Cultured astrocytes also induced a proliferative response in human T lymphocytes, which provides relevant information to studies investigating neural xenotransplantation of porcine donor cells that contain an astrocytic population.

Ionescu and colleagues reported that astrocytes harvested from the adult porcine cortex express the markers GFAP, S100β, CD14, and interferon-γ receptor 2 (IFN-γ-R2) [34]. This group also examined the response of astrocytes to human SH-SY5Y neuroblastoma cells when exposed to proinflammatory mediators. When astrocytes were exposed to IFN-γ or a combination of lipopolysaccharide (LPS) and IFN-γ, they were cytotoxic to SH-SY5Y cells (as seen by increased levels of cell death). Exposure to LPS alone did not cause astrocyte toxicity toward SH-SY5Y cells. The authors did not detect increased levels of tumor necrosis factor-α (TNF-α) or nitric oxide (NO) in astrocyte culture media when astrocytes became cytotoxic toward SH-SY5Y cells, indicating that these neurotoxins were not mediating the astrocyte response to SH-SY5Y cells.

The Parfenova group published a detailed report investigating the antioxidant and cytoprotective effects of sulforaphane (SFN) in porcine cortical astrocyte cultures [35]. They demonstrate that adding the pro-inflammatory cytokine TNF-α or excitotoxic glutamate to astrocyte cultures rapidly led to increased production of reactive oxygen species (ROS). Independent application of tiron (a potent superoxide scavenger), apocynin (common Nox inhibitor), DPI (diphenylene iodonium; common Nox inhibitor), GKT137831 (novel Nox4 inhibitor), or SFN blocked ROS elevation in the presence of TNF-α or excessive glutamate. Application of DPI, GKT137831, or SFN to astrocyte cultures also inhibited NADPH oxidase activity under control conditions, TNF-α-induced inflammatory conditions, and excitotoxic glutamate conditions. Application of tiron or SFN to astrocyte cultures reduced DNA fragmentation and cell detachment, two indicators of apoptosis, under TNF-α-induced inflammatory conditions and excitotoxic glutamate conditions. These results indicate that the application of TNF-α or excitotoxic glutamate to astrocyte cultures activates Nox4 NADPH oxidase, which increases ROS, which in turn causes oxidative stress and apoptosis. The inhibitor compounds investigated in this study, including GKT137831 and SFN, exhibit antioxidant and cytoprotective properties toward astrocytes via inhibiting Nox4 activity.

2.4. Vascular Regulation

Barnes et al. investigated the presence of the cell surface peptidases aminopeptidase N (AP-N) and dipeptidyl peptidase IV (DPP-IV) in porcine striatal astrocytes [36]. They reported that AP-N, but not DPP-IV, is expressed by astrocytes in the piglet striatum. AP-N was seen in the cell membranes of striatal astrocytic endfeet, including in astrocytes associating with endothelial cells and pericytes. They demonstrated this with immunolabeling and transmission electron microscopy of primary cultures isolated from the postnatal-day-1 piglet striatum and brain sections from the postnatal-day-1 piglet striatum. The presence of AP-N in the cell membranes of striatal astrocytic endfeet suggests that astrocytes may actively participate in the enzymatic processing of peptides in the striatum. Additionally, the localization of AP-N in endothelial cells and pericytes suggests a potential role in the interaction between astrocytes and the neurovascular unit. This may enable striatal astrocytes to regulate the processing and transport of circulating peptide signaling throughout the brain.

Leffler and colleagues investigated the role of astrocytes in the dilation of porcine cerebral arteriole myocytes [36]. In porcine astrocyte cultures, inhibition of the enzyme heme oxygenase (HO) with chromium mesophorphyrin (CrMP) prior to being placed in contact with myocytes caused the elimination of glutamate-induced K_{Ca} channel activation in myocytes. Additionally, treatment of porcine cortical brain slices with the selective astrocyte toxin L-2-α-aminoadipic acid (L-AAA) blocked glutamate-induced dilation of arterioles. These results provide evidence that glutamate-induced dilation of monocyte arterioles is dependent on astrocyte-derived HO.

Authors from this same group then investigated the role of astrocytes in the mediation of Ca^{2+} signaling in arteriolar smooth muscle cells in live brain slices from the porcine cortex [37]. Following astrocyte injury by treating brain slices with L-AAA, glutamate-induced Ca^{2+} spark activation and reduction in intracellular Ca^{2+} concentration in arteriolar smooth muscle cells were both prevented. L-AAA treatment also increased glutamate-stimulated Ca^{2+} wave frequency in smooth muscle cells. This experiment demonstrates that astrocytes modulate glutamate-induced Ca^{2+} sparks, Ca^{2+} waves, and global intracellular Ca^{2+} levels in smooth muscle cells.

This group then investigated the mechanism of glutamate-induced carbon monoxide (CO) production in cultured porcine astrocytes [38]. They demonstrated that administration of glutamate, hemin (an exogenous HO substrate), and ionomycin (Ca^{2+} ionophore that held free intracellular calcium concentration constant) to astrocyte cultures each increased astrocyte CO production in a concentration-dependent manner. Administration of glutamate also increased intracellular calcium concentration in astrocyte cultures in a concentration-dependent manner. When astrocyte cultures were treated with the endoplasmic reticulum Ca^{2+}-ATPase blocker thapsigargin, which depleted intracellular Ca^{2+} stores, glutamate-induced Ca^{2+} signaling was blocked, and steady-state intracellular calcium levels were elevated, suggesting that glutamate stimulates Ca^{2+} release from the ER which in turn leads to intracellular calcium concentration. Glutamate did not increase astrocyte CO production when Ca^{2+} levels were held constant with ionomycin, suggesting that glutamate stimulates astrocyte CO production via Ca^{2+}. HO was inhibited by CrMP, which blocked glutamate-induced CO production in cultured astrocytes. Additionally, treatment of astrocytes with thapsigargin reduced basal CO levels and blocked glutamate-stimulated CO increase in astrocytes. Treatment of cultures with the calmodulin blocker calmidazolium also blocked glutamate-stimulated CO increase in astrocytes. Overall, these results suggest a mechanism by which glutamate causes the ER to release Ca^{2+}, which leads to elevated intracellular Ca^{2+} levels, in turn leading to Ca^{2+}-calmodulin-dependent HO activation and CO production in porcine cortical astrocytes. Astrocytic CO production then dilates cerebral arterioles.

This group also utilized Western blotting and immunohistochemistry to demonstrate that newborn porcine cortical astrocytes express the hydrogen sulfide-producing enzyme cystathionine β-synthase, which is important for cysteine synthesis [39].

2.5. Blood–Brain Barrier Modeling

Several different groups have published protocols outlining the isolation of cerebral astrocytes from the adult porcine brain [40–42]. These cells exhibit characteristic astrocyte morphology under phase microscopy, with multiple star-like processes extending from the cell body and uniformly expressed GFAP [41]. Cultured porcine astrocytes also express aquaporin 4 (AQP4), enhance neuronal survival, and demonstrate a dose-dependent loss of GFAP when exposed to sera from patients with Neuromyelitis optica, which is a disease characterized by astrocyte loss [42].

The Bobilya group and others have demonstrated that astrocytes can be cultured together with porcine brain capillary endothelial cells to create an in vitro blood–brain barrier model [40,43,44]. In a contact culture system, astrocytes form a confluent layer underneath the endothelial cells, and their presence in culture increases the transendothelial electrical resistance (TEER) up to nine times compared with culturing of endothelial cells alone [40]. In a non-contact culture system, astrocyte presence also contributes to increased TEER and barrier tightness [43,44] and influences endothelial cell expression of tight junction proteins [44]. The Moos lab has also demonstrated that porcine astrocytes can be cultured together with porcine brain endothelial cells and porcine pericytes in a triple culture model of the blood–brain barrier [45]. Here, they show that astrocyte presence in the tri-culture model contributes to high TEER and low permeability of endothelial cells.

2.6. Olfactory Bulb

The Osterberg lab published a detailed report about the structure of the adult porcine olfactory bulb (OB) [46]. Using GFAP immunohistochemistry, they reported that the porcine OB is densely populated by astrocytes, particularly in deep layers, including the granule cell layer and the mitral cell layer. The anterior olfactory nucleus (including the pars externa and pars principalis) was also populated with GFAP+ astrocytes, with the lowest astrocyte density seen in layer 2 of the pars principalis. The lateral olfactory tract was also found to be densely populated with GFAP+ astrocytes.

2.7. Optic Nerve and Retina

The Eppenberger group utilized GFAP immunohistochemistry and transmission electron microscopy to examine the astrocyte morphology in the porcine retina [47]. They demonstrated that GFAP staining is only located in the nerve fiber layer of the retina. Astrocytes extend an elaborate network of processes across the entire retina. Astrocytes are closely associated with blood vessels, particularly superficial blood vessels. Astrocytes were often seen wrapping their processes entirely around blood vessels, always with a higher quantity of GFAP+ processes located on the lateral/vitreal side and fewer on the scleral side of the vessels. The asymmetry with which astrocytes wrapped their processes around vessels was clearly visible with electron microscopy imaging. The authors describe astrocyte presence in the retina as forming a scaffold around blood vessels that bulge into the vitreous body.

Noda and colleagues described the expression of myocilin, the product of the gene MYOC/TIGR that is responsible for the pathogenesis of primary open-angle glaucoma in astrocytes of the optic nerve head [48]. Utilizing electron microscopy and GFAP and myocilin immunohistochemistry, they reported that myocilin was expressed in optic nerve astrocytes in the perinuclear region (outer nuclear membrane), pericentriolar region, glial filament, mitochondrial membrane, rough endoplasmic reticulum, and process endfeet near blood vessel walls. Myocilin was often found to be associated with microtubules and was also observed in astrocytes located in the lamina cribrosa region. The authors hypothesize that myocilin is an astrocyte membrane-associated cytoskeletal-linking protein that supports astrocyte cell shape and that dysregulation of myocilin in glaucoma may lead to structural alterations of the optic nerve head. This group published another report which showed that the amino acid sequence of myocilin expressed in porcine optic nerve head astrocytes is 82% homologous to human myocilin [49]. They also showed that optineurin, another glaucoma-associated gene, is expressed in porcine optic nerve head astrocytes and has an amino acid sequence that is 84% homologous with the human optineurin sequence.

Ripodas and colleagues reported that GFAP+ astrocytes in the inner layers of the porcine retina (nerve cell layer and ganglion cell layer) extend processes that make contact with blood vessels [50]. These astrocyte cell bodies and vasculature-contacting processes stain positive for vasoconstrictor endothelin-1 (ET-1), indicating that astrocytes may play a role in vascular regulation within the porcine retina.

Lee and colleagues reported the presence of heat shock protein 27 (HSP27) in the porcine retina [51]. In the adult (6 months) pig retina, they found expression of HSP27 on GFAP+ astrocytes in the ganglion cell layer and inner nuclear layer. In the newborn (postnatal day 1) pig retina, HSP27 expression was found on GFAP+ astrocytes in the ganglion cell layer.

Carreras and colleagues used immunohistochemistry and transmission electron microscopy to study the astrocyte cell–cell adhesions in the prelaminar region of the optic nerve head (ONH) [52]. They reported that the incomplete inner limiting membrane of Elschnig, which separates the ONH from the vitreous fluid of the eye, is covered with an interwoven expansion of GFAP+ astrocytes. Astrocytes closer to the vitreous stained more intensely for GFAP compared to astrocytes further away from the vitreous. However, the anterior surface of the ONH was only partially covered with astrocytes. Astrocytes at the surface of the vitreous exhibited cellular expansions that were conjoined by intercellular adherens junctions (zonulae adhesions) and occasionally gap junctions. These cells expressed calcium-dependent adhesion molecules neural cadherin (N-cadherin) in zonula adherens junctions and expressed neural cell adhesion molecules (N-CAM) in areas of cellular adhesions that were not junctional. These results indicate that astrocyte intercellular adhesions in the prelaminar region of the ONH are calcium-dependent.

Carreras and colleagues used GFAP immunohistochemistry, transmission electron microscopy, and perfusion with a fluorescent tracer to examine the pathways of fluid exchange between astrocytes within the prelaminar tissue of the porcine ONH [53]. They described the anterior interface of the ONH contacting the vitreous fluid is covered by a thick, uneven layer of astrocytes. Astrocytes expand in an interwoven fashion across this region and wrap around axons and blood vessels. The retina and optic nerve disc show extracellular spaces between the astrocyte networks that can be permeated by fluid. Astrocytes surrounding the meniscus of Kuhnt in the optic disc had large intracellular spaces and were organized together in a way the authors described as a "foamy" appearance. Astrocytes in the vitreous surface of the optic nerve are organized into an ordered "cobblestone-like" pattern. The distribution of dye across the optic disc was not uniform due to the structure of astrocyte processes, which had non-uniform, variably sized interconnected spaces between them. These extracellular spaces between astrocyte processes connected to form cavities which appeared to be preferred fluid flow routes to rid the prelaminar tissue of extra fluid.

Carreras and colleagues also published a report detailing the expression of glucose transporters in the ONH [54]. They described the ONH as having strong GFAP staining and columns of astrocytes alternating with columns of axonal bundles. GLUT1 was expressed in astrocyte endfeet, somas, and processes wrapping around axons in the ONH. They reported GLUT1 expression in the astrocyte endfeet of the nerve fiber layer, in the membranes of perivascular astrocytes of the optic disc, and in the astrocyte columns of the prelaminar regions.

Balaratnasingam and colleagues detailed astrocyte distribution in the lamina cribrosa, pre-laminar region, and post-laminar region of the porcine ONH [55–58]. Their results showed that astrocytes are the predominant glial cell in the optic nerve and that astrocytes in the ONH had radial, inter-digitating processes [56,57]. GFAP+ astrocytes in the nerve fiber layer followed the course and direction of axons, with their processes forming bundles that resembled retinal ganglion cell axons [55]. They found an absence of GFAP staining in tissue that contained laminar plates [57]. The area occupied by GFAP+ astrocytes was significantly greater in the pre-laminar region compared to the post-laminar region and the lamina cribrosa, indicating that the pre-laminar region may have the greatest metabolic demand out of these three areas [57,58]. In the pre-laminar region, GFAP staining was seen in areas where neuronal staining was present, in spaces around axonal bundles, and along the inner limiting membrane [56]. GFAP astrocytes that were present in the lamina cribrosa and post-laminar regions were also seen to closely associate with axonal bundles [56]. Additionally, higher regions of astrocyte density correlated to higher regions of axonal density in the ONH [58].

Kimball and colleagues reported on the co-expression of GFAP and the astrocyte aquaporin (AQP) channels AQP1, AQP4, and AQP9 in the ONH, retina, and myelinated optic nerve (MON) of the porcine eye [59]. They found AQP4 expression in the retinal nerve fiber layer, prelamina, and MON regions. Of particular note, they found that, similar to humans, there was no AQP4 expression in the lamina cribrosa region. There was AQP1 and AQP9 expression in the internal limiting membrane. There was no AQP1 or AQP9 expression in the lamina cribrosa or the MON. There was GFAP expression in the prelamina, lamina, lamina cribrosa, and MON. These results demonstrate that AQP4, which is normally expressed by healthy astrocytes, is not expressed in astrocytes of the lamina cribrosa of the porcine eye providing additional evidence supporting astrocytic porcine homology to humans.

Ederra and colleagues reported on the expression of the high-affinity nerve growth factor (NGF) receptor tyrosine kinase A (TrkA) in the porcine retina [60]. They performed GFAP and TrkA immunohistochemistry in the retinas of six adult porcine eyes and reported that the distribution of astrocytes in the porcine retina mimics that of the human retina. The highest quantity of GFAP+ astrocytes was found in the nerve fiber layer and ganglion cell layer of the retina, with fewer GFAP+ astrocytes found in the inner plexiform layer and inner nuclear layer. All astrocytes appeared to be closely associated with blood vessels.

Most GFAP+ astrocytes in the porcine retina expressed TrkA receptors. There was a subpopulation of GFAP+ astrocytes, some of which were seen in the inner nuclear layer, that did not express TrkA receptors. The results presented in this study demonstrate that most astrocytes in the porcine retina likely play a role in modulating NGF levels.

3. Previous Investigations of Porcine Astrocytes in Neurotrauma Research

Various research groups have conducted examinations of astrocyte response to injury in a range of porcine models. While undoubtedly not exhaustive, our literature review is extensive, and we summarize the number of studies in each porcine injury model that investigated astrocytes to provide a glimpse at where the focus has been thus far (Figure 2). These studies and their effects on astrocytes in the porcine brain are reviewed below and summarized in Supplementary Table S2. Most of the research articles discussed below contain details about astrocyte response to injury as well as injury response that is not astrocyte related. Here, we only review the details of these papers that are related to astrocytes specifically. We omit review of other non-astrocyte, injury-associated data that are described in these articles.

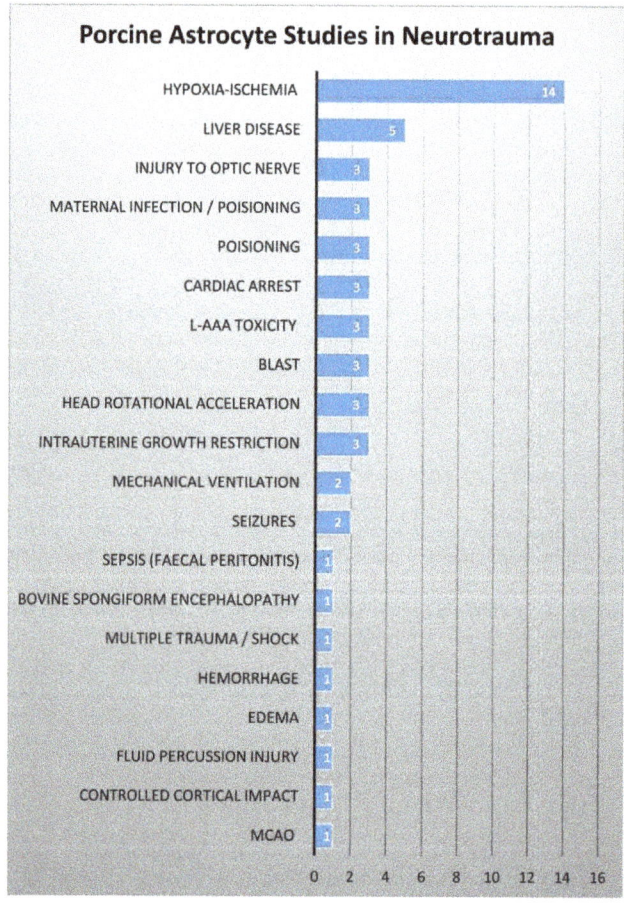

Figure 2. Porcine astrocyte studies in neurotrauma. Summarizing the number of astrocyte studies found in each pig injury model listed.

3.1. Hypoxia–Ischemia

Several research groups have investigated astrocyte response to injury in the porcine brain using models of hypoxia (oxygen availability deficiency) and/or ischemia (blood supply restriction). The Traystman lab utilized a piglet model of asphyxic cardiac arrest to examine striatal astrocyte response at 24-, 48-, and 96 h following hypoxic–ischemic (H-I) injury [61]. They reported astrocyte degeneration at early time points (24 h and 48 h) post-H-I, seen as astrocyte cytoplasmic swelling, distended processes, loss of GFAP staining, cell death, and fragmented DNA evidenced by co-labeling of TUNEL and GFAP. GFAP cell proliferation returned to normal levels at 96 h. They additionally reported changes in astrocytic GLT-1 expression following H-I injury. While astrocyte processes primarily stained for GLT-1 in control brains, astrocyte cell bodies (instead of processes) stained positive for GLT-1 at 24 h and 48 h, and astrocytes were virtually absent of GLT-1 staining at 96h following H-I. The authors hypothesize that rapid glutamate toxicity in the striatum following H-I could be responsible for these astrocytic changes. This group then followed up by investigating whether N-methyl-D-aspartate (NMDA) glutamatergic receptors are responsible for this striatal excitotoxicity following H-I [62]. They reported that astrocytes express elevated levels of the NMDA receptor subunit NR2B in the putamen at 24 h (but not at 3, 6, or 12 h) following H-I, but that elevated expression of this receptor subunit specifically did not correlate with striatal neuronal injury resulting from H-I.

Lee and colleagues examined whether H-I injury, with and without therapeutic hypothermia and rewarming, affected the expression of endoplasmic reticulum to nucleus signaling-1 protein (ERN1) expression in astrocytes [63]. ERN1 is a marker of unfolded protein response (UPR) activation which may contribute to cell death that occurs during the cellular stress caused by H-I injury, hypothermia, and rewarming (all independent cellular stressors). While endoplasmic reticulum stress-induced activation of UPR can be neuroprotective, it can also become maladaptive. They report that hypothermia and rewarming increase the quantity of ERN1+ astrocytes in the cerebral cortex and subcortical white matter of the motor gyrus and that this change occurs in the absence of H-I injury. They further report that H-I injury followed by hypothermia reduces the quantity of ERN1+ astrocytes. There was a correlation between ERN1+ astrocytes and astrocyte apoptosis in white matter after hypothermia and rewarming, but this correlation was not seen in animals also exposed to H-I. The authors conclude that hypothermia and rewarming cause astrocyte ER stress, as evidenced by UPR activation, leading to astrocyte apoptosis. They further conclude that H-I injury might interrupt this astrocyte ER stress response that is induced by hypothermia.

Lee and colleagues also examined changes in astrocyte morphology and the correlation of these changes to fractional anisotropy (FA) from diffusion tensor imaging (DTI) following H-I injury and excitotoxic injury caused by striatal quinolinic acid (QA) injection [64]. They examined astrocyte swelling as well as expression of astrocyte cytoskeletal (GFAP), glutamate reuptake (GLT-1), and water regulation (aquaporin 4/AQP4) markers, as well as the relationship of these markers to FA measurements during DTI. They reported that QA-induced excitotoxic injury causes swollen astrocytes that display AQP4+ aggregates, cytoplasmic swelling, and vacuoles, as well as degenerating astrocytes with fragmented processes in the putamen. H-I injury similarly causes swollen astrocytes in the caudate and degenerating astrocytes with fragmented processes in the putamen. QA and H-I injury both induce swollen GLT-1+ astrocytes in the putamen. Across both types of injury, swollen GFAP+ AQP4+ GLT-1+ astrocytes seen in the caudate and putamen correlate with lower FA measured during DTI. These results indicate that swelling of astrocytes that regulated water (AQP4+) and glutamate reuptake (GLT-1+) are associated with FA changes. The authors are unsure if astrocyte swelling directly or indirectly affects FA.

The Pow lab described the altered expression of GFAP and GLAST following hypoxia in the 1-day-old piglet brain [27]. They reported that GFAP expression was upregulated in gray matter areas, including the dentate gyrus of the hippocampus, some regions of the thalamus, and some outer cortical layers. GLAST expression following hypoxia remained high in these same regions and co-localized with GFAP. This colocalization simply served to confirm that the GLAST was, in fact, astrocytic and does not indicate protein–protein interactions. Astrocytes in these regions displayed typical morphology with co-localization of GFAP and GLAST in their distal processes. Following hypoxia, GFAP expression was minimal in area CA1 of the hippocampus. There was a significant loss of GLAST expression in area CA1 of the hippocampus and cortical layers 2–5. Subregions of the thalamus also lost GLAST expression following hypoxia. Astrocyte morphology was altered in regions damaged by hypoxia, with retracted processes and GFAP and GLAST expression restricted to the soma and proximal processes rather than expressed in the distal processes. These results support this group's hypothesis that GFAP and GLAST are co-localized such that GFAP helps to anchor GLAST in the astrocyte plasma membrane.

The Pow lab also examined the expression of the exon-9 skipping form of the glutamate-aspartate transporter EAAT1 (GLAST1b) following hypoxic injury to piglets [65]. They reported that expression of glutamate transporter 1 (GLT1a) decreases while expression of GLAST1b increases in the hippocampus (particularly the CA1 region) following H-I injury. They further report that some GFAP+ astrocytes co-localize with this GLAST1b following H-I but that most of the GLAST1b+ cells are MAP2+ neurons. The authors indicate that GLAST1b could be utilized as an important marker of neuronal damage following excitotoxic injuries. This group also reported no difference in GLAST1c expression as assessed by mRNA and protein expression in the piglet brain following hypoxia [28].

This same group then published a detailed article describing astrocyte structural changes following H-I injury in the newborn piglet brain [66]. By injecting Lucifer Yellow (LY) into fixed tissue and performing Golgi–Kopsch staining, they found that astrocytes from control subjects had fine, highly structured processes extending from astrocyte cell bodies and that multiple processes often extended from the cell to create complex arbors of processes. This contrasted astrocytes in damaged cortical gray matter regions following H-I, which had thicker processes, shorter processes, fewer processes, less complex branching structures, and often contained bulb-like swellings both along the processes and at the process terminals. Astrocytes in H-I injured animals had drastically fewer secondary and tertiary processes (i.e., processes extending from primary processes that extended directly from the soma) compared to the control. They performed Sholl analysis which revealed that astrocytes had far less branching complexity at 5–20 microns from the cell body compared to the control, a difference that was not found further out at 25–40 microns from the cell body. They also reported that astrocyte soma size significantly increased following H-I injury. These changes in astrocyte morphology occurred as early as 8 h following H-I injury and continued to become more abnormal at 72 h post-injury. This contrasted with neuronal injury, which was not observed at the 8 h time point. They reported that, in control subjects, astrocytes were of different sizes in different cortical layers and that H-I caused a decrease in astrocyte size across all cortical layers. They also performed D-aspartate uptake studies prior to fixation to examine changes in glutamate uptake in injured astrocytes. They reported that astrocytes in control brains uptake D-aspartate abundantly in their processes, whereas uptake was drastically reduced and restricted to uptake in astrocyte cell bodies at 8 and 72 h following H-I. These results collectively indicate that astrocytes in the cortex structurally and functionally change following H-I and that these astrocytic changes occur early and before neuronal changes. The authors speculate that astrocytes retract their processes quickly following injury, which damages neurons which in turn causes astrocyte proliferation and gliosis to re-establish connectivity within void brain tissue. The retraction of astrocyte processes may also lead to the observed phenomenon of glutamate transporters being closer to astrocyte cell bodies following H-I compared to their normal location along astrocyte processes. The reduced ability of astrocytes to uptake glutamate following H-I suggests that glutamate transporters may be reduced and/or distributed, and this indicates that glutamate-mediated excitotoxicity may play an important role in damage resulting from H-I injury. They also reported on changes in white matter astrocytes following H-I injury in newborn pig brains [67]. Utilizing GFAP immunolabeling, they reported that approximately 40% of cell bodies in subcortical white matter are GFAP+ astrocytes. They found reduced GFAP expression in the subcortical white matter following H-I, with the average area of GFAP+ astrocytes decreasing by 46% compared to the control. Utilizing Golgi–Kopsch staining, they also reported that white matter astrocytes were, on average, 34% smaller following H-I and that these cells had fewer processes that were shorter, thicker, and had abnormal swellings compared to control subcortical white matter astrocytes.

This group then published another report detailing changes in the presence of glutamine synthetase (GS), the glutamate detoxification enzyme that converts ammonia and glutamate into glutamine, in a newborn porcine model of H-I injury [68]. They reported changes in GS as soon as 1 h following H-I injury, with small patches of GS-devoid regions seen in regions of the cortex and CA1. At 24 and 72 h following H-I, they report extensive areas of GS loss in the cortex and CA1 (areas generally vulnerable to H-I injury) but not in the dentate gyrus or thalamus (areas not typically vulnerable to H-I injury). Astrocyte GS loss at these later time points overlapped with astrocyte GLAST loss. Like this group's previous work, these results suggest that astrocytes respond early to H-I injury, evidenced here by the early loss of GS that continues to become more pronounced over time. They argue that this early loss of GS likely leads to an accumulation of glutamate in astrocytes which would reduce their ability to uptake glutamate and, in turn, exacerbate the potential for glutamate toxicity following injury. Reduced GS may also lead to ammonia accumulation in the brain which would greatly disrupt ammonia homeostasis.

This group also investigated the presence of two phosphorylated GFAP proteins, p8GFAP and p13GFAP, following H-I injury in newborn piglets [69]. They reported that, even in control brains, pGFAP is expressed in astrocytes with normal morphology. Astrocytes that expressed pGFAP lacked the fine and bushy processes of normal astrocytes and instead had short processes and thickened varicose processes. Specifically, astrocytes expressed pGFAP in processes near the cell body, in abnormal terminal dilations on their processes, and in endfeet contacting blood vessels. They also reported upregulated pGFAP expression in brain regions that were injured following H-I. p8GFAP was upregulated in the cortex at 24 h following H-I and was upregulated in the cortex, basal ganglia, and thalamus at 72 h following H-I. p13GFAP expression did not change at 24 h following H-I but was upregulated in the cortex and basal ganglia at 72 h following H-I. They also reported that higher pGFAP expression correlated with higher histological injury. Overall, pGFAP expression appears in astrocytes with abnormal morphology and is upregulated in injured regions following H-I. The authors speculate that phosphorylation of GFAP, despite being an energy-dependent mechanism, could be part of a repair mechanism following injury.

Ruzafa and colleagues exposed newborn pigs to hypoxia and then examined astrocyte changes in the retina and superior colliculus at 4 h following injury [70]. They reported that astrocyte networks in the retina, which typically run parallel to retinal ganglion cell axons, appeared more disorganized following hypoxia. Although these differences in organization were not significant, they reported more laterally extending astrocyte processes and a higher degree of randomness of retinal astrocytes following hypoxia compared to control. They also reported that astrocytes in the superior colliculus exhibited hypertrophy, increased density, and increased astrocyte cytoskeletal area following hypoxia. These results indicate that astrocytes in the brain may be more susceptible to early damage from hypoxia compared to astrocytes in the retina. The authors speculate that this resilience of the retina to oxygen deprivation may be due to the presence of Müller glial cells, which exist in the retina but not in the brain.

Zheng and Wang investigated changes in lactate and glucose metabolism in the basal ganglia between 2 and 72 h following hypoxic–ischemic injury in piglets [71]. In this study, they utilized H&E staining to demonstrate that astrocytes in the basal ganglia become swollen at 6 h following H-I and that these swollen astrocytes further exhibit condensed nucleoli at 24 h and become degraded at 48 h following H-I. They also reported that expression of monocarboxylate transporter 4 (MCT-4), a lactate transporter that is primarily expressed by astrocytes, is increased in the basal ganglia at 12–24 h following H-I. These results indicate that astrocytes exhibit early morphological changes following H-I and that the lactate metabolism of astrocytes is altered following H-I.

Parfenova and colleagues investigated the role of astrocyte-produced carbon monoxide in both in vitro porcine astrocyte cultures and an in vivo model of neonatal asphyxia in the piglet brain [72]. They report that astrocyte cultures highly express heme oxygenase 2 (HO-2) and produce CO in the presence of the prooxidants glutamate and TNF-alpha. They also demonstrate that in vitro and in vitro porcine astrocytes drastically increase CO production (6–10-fold) when exposed to asphyxic conditions. Asphyxia also leads to increased astrocyte ROS production in vivo, which is further increased when subjects are pre-treated with the HO inhibitor SnPP and are decreased when subjects are pre-treated with the CO-releasing molecule A1 (CORM-A1) and bilirubin. Astrocyte cultures that were exposed to excitotoxic glutamate had increased ROS production and increased apoptosis, and this was augmented when astrocyte CO production was inhibited by SnPP. They also demonstrated that pial responses to the astrocyte-dependent vasodilators ADP and glutamate were compromised at 24 and 48 h post-asphyxia. Pial responses to ADP and glutamate were further inhibited when subjects were pretreated with SnPP prior to asphyxia. On the other hand, the reduced pial responses to ADP and glutamate resulting from asphyxia were prevented by pretreatment with CORM-A1. Overall, these results demonstrate that cortical astrocytes in the porcine brain respond to asphyxia by activating HO-2 and increasing the production of CO, which acts as an antioxidant and cytoprotective messenger against oxidative stress induced by asphyxia. These cytoprotective effects are increased with CO levels and augmented and reduced when HO is inhibited. These results indicate that CO donors could be an effective approach to treating prolonged asphyxia.

3.2. Middle Cerebral Artery Occlusion

Spellicy and colleagues recently published an article detailing their use of high-content image (HCI) analysis to examine changes in astrocyte morphology following middle cerebral artery occlusion (MCAO) in the adult Yucatan porcine brain [73]. Four weeks post-stroke, they performed GFAP immunohistochemistry and utilized HCI to examine 19 astrocyte morphological parameters in the perilesional area and ipsilateral hemisphere of stroke and non-stroke subjects. These methods provided a wealth of information on subtle changes in astrocyte morphology following injury. They reported larger, more extended, more ramified astrocytes in the perilesional area and ipsilateral hemisphere following stroke, compared to smaller, more rounded astrocytes in non-stroke subjects. Additionally, they reported significant increases in GFAP+ area shape perimeter, major axis length, and mean radius and a significant decrease in solidity in the ipsilateral hemisphere stroke subjects compared to non-stroke subjects. In the perilesional area, they reported significant increases in GFAP+ area, compactness, and major axis length and significant decreases in form factor, solidity, and orientation in stroke subjects compared to non-stroke subjects. Overall, their analysis indicated a higher GFAP+ area and more reactive astrocyte morphology following stroke. The semi-automated analyses employed in this article provided an in-depth analysis of changes in astrocyte morphology following MCAO.

3.3. Controlled Cortical Impact Brain Injury

Baker and colleagues developed a graded cortical control impact (CCI) brain injury model in 3-week-old piglets [74]. They described the effect of CCI on astrocytes by comparing GFAP reactivity in the perilesional area compared to the same area in the contralateral, uninjured hemisphere. They reported that increased GFAP reactivity in the perilesional area (compared to the contralateral side) correlates with increased CCI severity (increased impact velocity and impact depth). As CCI injury increased, the quantity and intensity of astrocytes in the perilesional area increased, indicating increased astrocyte proliferation (astrocytosis) and hypertrophy (astrogliosis) with more severe injury. The authors reported that augmented GFAP expression likely results from increased astrocyte size and proliferation following injury, as well as astrocyte migration toward the injury site.

3.4. Fluid Percussion Brain Injury

Lafrenaye and colleagues performed central fluid percussion injury (cFPI) in Yucatan mini pigs and collected blood samples at various time points up to 6 h post-injury [75]. They found steadily increasing blood serum levels of GFAP as time following injury increased, with significantly greater serum levels at 3 and 6 h post-injury compared to sham. Upon examining GFAP immunohistochemistry in the thalamus at 6 h post-injury, they reported that astrocyte GFAP intensity negatively correlated with cell area and cell roundness. These morphological changes are indicative of injured astrocytes. They also found more subtle differences in astrocyte morphology by examining GFAP+ cells in the thalamus using ultrastructural imaging. For example, GFAP+ astrocytic endfeet appeared more pronounced around vessels following cFPI injury compared to sham. Additionally, they discovered GFAP+ vesicles in the basement membrane and endothelial cell cytoplasm following cFPI injury. These findings indicate a possible mechanism for vesicular GFAP transport directly from astrocyte endfeet through endothelial cells lining blood vessel walls and into the blood. Additionally, these studies demonstrate that circulating serum biomarkers can be useful metrics to assess subtle changes in histopathology following diffuse brain injury.

3.5. Head Rotational Acceleration Injury

Currently, the only preclinical model that recreates the mechanisms and manifestations of human TBI relies on head rotational acceleration injury in pigs [14]. The core acceleration-induced forces exerted throughout the brain during human TBI are dependent on brain mass, which is why attempts to replicate acceleration-based injury in rodents with the CHIMERA model fail to reach scaled thresholds [76–80]. Pigs possess brains of sufficient mass to replicate human TBI forces by scaling up acceleration, and recreating this core mechanism of human TBI also recreates key manifestations of human TBI that do not appear in any other models, such as loss of consciousness [23,24,81]. Three previous studies utilizing this model reported from the Smith, Meaney, and Cullen labs include information on astrocyte response to rotational acceleration injury. In 1997, Smith and colleagues induced rotational acceleration injury in Hanford swine in the coronal plane and reported reactive astrocytosis in the molecular layer of the cerebral cortex, in subcortical white matter regions, in the corpus callosum, and in the CA1 and CA3 regions of the hippocampus [82]. Reactive astrocytosis was reported to be much greater in the hippocampus compared to other regions. Twenty years later, Johnson and colleagues investigated BBB disruption and associated astrocyte response at 6 h, 48 h, and 72 h following coronal head rotational acceleration in Hanford swine [83]. In addition to demonstrating BBB disruption via leaked serum proteins fibrinogen (FBG) and immunoglobulin (IgG), they report that these serum proteins are internalized by surrounding astrocytes in the brain parenchyma, as evidenced by GFAP-positive cells with astrocyte morphology surrounding FBG or IgG immunoreactivity (Figure 3). While the number of astrocytes internalizing these serum proteins was not quantified, striking images clearly demonstrate co-localization of GFAP with these serum proteins in the premotor cortex and parietal cortex at 72 h post-injury. The Cullen Lab then also conducted studies that included analysis of astrocyte response to head rotational acceleration injury. Grovola et al. analyzed astrocyte reactivity via GFAP expression following two severities of mild injury in the coronal plane [84]. They scored GFAP+ cell size and the density of GFAP+ cells in the hippocampus, periventricular white matter, inferior temporal gyrus, and cingulate gyrus and averaged these scores to provide a single score of reactivity for each subject. They reported no change in astrocyte size or density in either injury condition. Unfortunately, astrocytes have been mostly overlooked in this highly translational model, limited to GFAP-based morphological assessments after mild TBI, and deeper study is required.

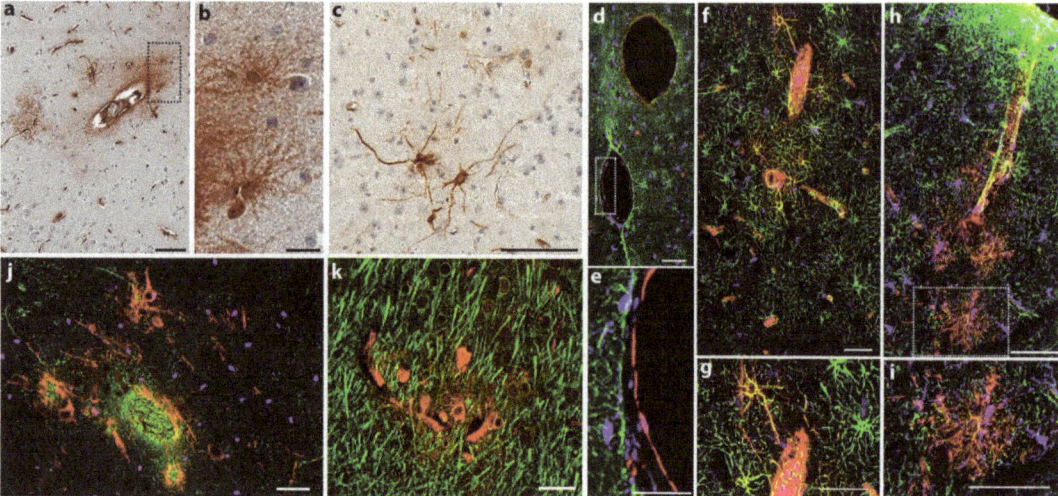

Figure 3. Internalization of bloodborne fibrinogen in astrocytes and neurons following mechanical permeabilization of the blood–brain barrier due to rotational acceleration injury in pigs. (**a**) Within the depth of the sulcus in the inferior temporal gyrus, cells with a glial morphology stain positive for fibrinogen at 48 h post-injury; (**b**) higher magnification of the call-out box. (**c**) In the caudate nucleus, cells with neuronal morphology were positive for fibrinogen 72 h post-injury. (**d**) In an uninjured brain, fibrinogen (red) is confined to the vessel lumen and absent from astrocytes (GFAP; green) or microglia (IBA-1; purple). (**e**) Higher magnification of box in (**d**). (**f**) Vessels in premotor cortex showing marked fibrinogen (red) extravasation 72 h post-experimental concussion. Co-localization with astrocytes (GFAP; green) is observed. Only minimal co-localization with microglia (IBA-1; purple) was observed in cells immediately adjacent to the vessel; (**g**) higher magnification of the call-out box. (**h**) In the parietal cortex, fibrinogen (red) extravasation is evident around penetrating surface vessels and co-localizes with astrocytes (GFAP; green) 72 h post-injury; (**i**) higher magnification of the call-out box. (**j**) In the frontal cortex, fibrinogen (red) is present within cells that have neuronal morphology and stain negative for IBA-1 (purple) and GFAP (green); (**k**) these fibrinogen-positive cells also stained positive for MAP-2 (green), offering confirmation of neuronal cell-type. Scale bars (**a**,**c**) 100 μm, (**b**,**e**) 25 μm, and (**d**,**f**–**k**) 50 μm. Figure reproduced with permission from Johnson et al. 2018 [83].

3.6. Explosive Blast Injury

To our knowledge, just three previous reports on the effects of explosive blast injury in the porcine brain have included an analysis of astrocyte response to this injury. In 2011, explosive blast injury was performed on Yorkshire swine to simulate a blast experienced in an open field (blast tube), tactile vehicle, or a building [85]. Seventy-two hours and two weeks following injury, GFAP+ cells were found to be greater in the hippocampus, corpus callosum, parasagittal cortex including cingulate gyrus, and superior, middle, and inferior frontal gyri. They reported increased cell density in the hilus and molecular layers of the hippocampus. Interestingly, despite the reported increase in GFAP+ astrocytes, the cells remained within distinct domains and did not display the phenotype typically characteristic of injured astrocytes. The authors speculate that the augmented astrocyte numbers combined with the normal astrocyte phenotype could indicate that astrogliosis resulted from proinflammatory factor release from activated microglia or transient blood–brain barrier permeability rather than resulting directly from neuronal injury. A few years later, this research group investigated the effects of a single, double, or triple blast on astrocyte proliferation and reactivity in Yucatan minipigs [86]. They found augmented astrocyte density in the dentate hilus and molecular layer of the hippocampus following blast injury (single, double, and triple injury) compared to sham. This astrocyte proliferation was found between 2 weeks and 6–8 months following blast exposure. Furthermore, this increase in astrocyte density was greater in animals exposed to two and three blasts compared to one blast. Examined astrocytes had a stellate appearance and non-overlapping domains. Additionally, animals exposed to three blasts also had astrocyte proliferation in the deep central white matter. The researchers speculate that astrocyte activation may be an early hippocampal response to injury. A different research group investigated the effects of open field blast exposure (medium and high blast overpressure) on astrocyte proliferation at 3 days following blast exposure [87]. They observed significantly higher quantities of astrocytes in frontal lobe regions in both the medium and high blast groups compared to sham animals, with most astrocytes seen in white matter. There were also more astrocytes in the high-blast group compared to the medium-blast group. The observed astrocytes had enlarged cell bodies and processes and more intense GFAP staining compared to sham, although these observations were not quantified. Overall, the above reports reveal that explosive blast injury causes astrocyte proliferation in all reported studies ranging from 3 days to 8 months following blast exposure.

3.7. Cerebral Edema via Water Intoxication

The Rosenthal group performed water intoxication in female swine as a mechanism to investigate cerebral edema leading to elevated intracranial pressure [88]. They took electron microscopy samples of brain tissue (harvested from over the left coronal suture 1 cm lateral to the midline) before and after serially inducing four levels of water intoxication in each subject. While no widespread brain astrocyte pathology was examined, transmission electron micrographs revealed substantial swelling of astrocyte endfeet in injured tissue when compared to tissue samples taken prior to injury. These swollen astrocyte endfeet were found around capillary endothelial cells.

3.8. Intracerebral Hemorrhage via Blood Injection

Zhou and colleagues induced intracerebral hemorrhage in adult male pigs by injecting blood into the frontal lobe to study the expression of the integrin-associated protein cluster of differentiation 47 (CD47) following injury [89]. CD47 plays a critical role in immune responses by providing inhibitory signals to phagocytic processes regulating inflammatory and immune responses. They reported that while CD47 was increased throughout the brain, this protein did not co-localize with astrocytes (defined as GFAP+ cells) in the perihematomal region 3 days following injury. These results indicate that astrocytes may not be directly involved in the increase in brain CD47 expression following injury.

3.9. Multiple Trauma and Hemorrhagic Shock

Vogt and colleagues studied the effects of multiple trauma associated with hemorrhagic shock (MT/HS) in male pigs [90]. They performed two different severities of MT/HS injury: MT with 45% blood loss and 90 min HS phase (T90) and MT with 50% blood loss and 120 min HS phase (T120). In separate groups for both injury severities, they induced hypothermia (TH90, TH120) to examine how this treatment might change the analyzed injury effects. They analyzed the levels of the calcium-binding astrocytic protein S100B at six time points up to 48.5 h following trauma induction and analyzed S100B immunohistochemistry at 48.5 h following trauma induction. They reported that S100B blood serum values were transiently increased in the T120 group compared to sham and that hypothermia did not influence serum S100B levels compared to normothermic groups across both injury conditions. Although they did not report any change between groups in S100B+ astrocyte cell count in the frontal lobe, they did note that S100B+ astrocytes in all four trauma groups had enlarged somata and elongated branches compared to astrocytes in control and sham groups. This lack of astrocyte proliferation following MT/HS injury caused authors to speculate that the temporary increase in S100B serum levels likely resulted from the trauma and/or shock but not directly from cerebral damage.

3.10. L-2-Alpha-Aminoadipic Acid (L-AAA) Toxicity

Leffler and colleagues performed a set of experiments examining the effects of glial toxin L-AAA on pig astrocytes in culture and in vivo [91]. They reported that treating cultured porcine astrocytes with 0.2–2 mM L-AAA for 2 h dose-dependently increased cell detachment, process retraction, loss of cell–cell contacts, and cytoskeletal changes. They further reported that in vivo administration of 2 mM L-AAA for 5 h significantly disrupted the confluent layer of GFAP+ superficial glia limitans, as evidenced by GFAP immunohistochemistry. In vivo L-AAA administration additionally eliminated pial arteriolar dilation to the astrocyte-dependent dilators ADP and glutamate, as well as eliminated glutamate-stimulated CO production at the cortical surface. These results suggest that astrocytes employ CO as a mechanism to induce glutamatergic vasodilation in the cerebral cortex. The authors hypothesize that glutamate activates astrocyte glutamate receptors to stimulate CO production, which then dilates pial arterioles. A few years later, these authors published another report further elucidating this mechanism [92]. They utilize a methodology to demonstrate that injuring astrocytes via the application of 2 mM L-AAA or the heme oxygenase (HO) inhibitor CrMP to the parietal cortex in vivo blocks both pial arteriolar dilation and ADP-induced CO production. These results further confirm their previous findings that glia limitans astrocytes utilize CO as a gaseous neurotransmitter to mediate ADP-induced pial arteriolar dilation. Further, these findings suggest that CO is produced by astrocytes via a HO-catalyzed reaction. This group continues their research to investigate whether cortical astrocyte ionotropic glutamate receptors (iGluRs), specifically NMDA- and AMPA/kainate-type receptors, mediate the glutamate-induced, astrocyte-dependent HO activation that in turn causes cerebral vasodilation [93]. They demonstrate that NMDAR agonists (NMDA and cic-ACPD) and AMPA receptor agonists (AMPA and kainate) cause pial arteriolar dilation in vivo and that the glial toxin L-AAA blocks this response. They further demonstrate that in vivo application of the NMDAR antagonist D-AP5 and the AMPA/kainite receptor antagonist DNQX block pial arteriolar dilation that results from AMPA, NMDA, or glutamate application. These results demonstrate that astrocytic iGluRs play an essential role in the dilation of pial arterioles in the piglet cortex.

3.11. Seizure

The Parfenova lab has also published two research articles which examine the effects of cerebral astrocyte response to neonatal seizure caused by bicuculline administration [94,95]. Both studies examine the effect of seizures on cerebral vascular response to the astrocyte-dependent vasodilator ADP and endothelium and astrocyte-dependent vasodilators glutamate, the AMPA receptor agonist L-quisqualic acid, and the HO substrate heme. In addition to examining general changes in the response of pial arterioles to these vasodilators after a seizure, they also investigate whether administration of CORM-1A (a carbon monoxide-releasing molecule) can improve vascular outcome following seizure. They report that cerebral vascular response to ADP, glutamate, quisqualic acid, and heme is significantly reduced following seizure compared to control groups [94]. Administration of CORM-A1 either 10 min prior to seizure induction or 20 min after seizure induction was able to prevent this lower cerebral vascular response to ADP, glutamate, and heme [94]. These studies suggest that the astrocyte components of the neurovascular unit are injured during a seizure, which could contribute to the dysregulation of blood flow in the brain. Moreover, these studies indicate that functional astrocytes are necessary for regular cerebral blood flow. Additionally, the positive effects of CORM-1 administration on this dysregulation could indicate that carbon monoxide administration in the brain could help alleviate some of this astrocyte damage before or after seizure induction. They further report no sex differences in these findings [95].

3.12. Bovine Spongiform Encephalopathy

Liberski and colleagues experimentally infected porcine with bovine spongiform encephalopathy (BSE) and performed transmission electron microscopy imaging to examine the ultrastructural pathology caused by this neurodegenerative disorder [96]. They reported that BSE causes astrocytosis and astrocyte processes to be in close conjunction with microglial cells. These ultrastructural findings parallel previous reports of cattle infected with BSE and humans infected with transmissible spongiform encephalopathies (TSEs).

3.13. Mechanical Ventilation

The Reynolds research group investigated brain injury caused by lung-protective mechanical ventilation in a porcine model [97]. Following 50 h of mechanical ventilation (MV), subjects had a higher percentage of GFAP-positive reactive astrocytes in the hippocampus compared to never-ventilated (NV) subjects. Astrocyte proliferation in the hippocampus indicates the hippocampus is damaged because of MV. Additionally, MV subjects had higher blood serum levels of GFAP compared to NV subjects, indicating astrocyte damage. The authors speculated that elevated serum GFAP levels could be caused either directly by lung injury or by a brain injury that occurred during MV. These researchers additionally reported that transvenous diaphragm neurostimulation (TTDN) during MV can reduce these negative effects [98]. Furthermore, neuroprotection was greater in subjects exposed to TTDN during every breath during MV compared to subjects exposed to TTDN every other breath during MV. The authors speculate that by mimicking the pulmonary stretch receptor response that occurs during spontaneous breathing, TTDN during MV may regulate dopamine release in the hippocampus to offer neuroprotection.

3.14. Cardiac Arrest and Cardiopulmonary Bypass (CPB)

Sharma et al. examined changes in GFAP expression and astrocyte morphology via transmission electron microscopy following 12 min of untreated cardiac arrest (ventricular fibrillation) in piglets [99]. They reported augmented GFAP reactivity in both the thalamus and cerebral cortex at 30 and 60 min following injury, with GFAP reactivity correlating well with neuronal damage and albumin leakage. Immunohistochemistry revealed an increase in star-shaped astrocytes around blood vessels and neurons, and electron microscopy revealed swollen astrocytes in the thalamus and cortex at 60 min following cardiac arrest. Overall, this study demonstrated astrocyte reactivity and morphological changes in diverse brain regions at early time points following cardiac arrest.

In 1989, Laursen and colleagues examined GFAP expression following total cardiopulmonary bypass (CPB) for two hours in pigs [100]. They did not observe any astrocytosis following CPB. However, they did observe perivascular swelling of astrocyte endfeet in white and gray matter following CPB that occurred at normothermia. This swelling of astrocytic endfeet was prevented when CPB was conducted at hypothermia. The authors speculate that astrocyte endfeet swelling could have been caused by a relative impairment of Na^+-K^+ exchange.

Around 30 years later, Stinnett and colleagues examined changes in fractional anisotropy (FA) obtained from diffusion tensor imaging and the correlation of these changes to changes in GS+GFAP+ astrocyte numbers following mild and severe CPB in piglets [101]. They reported acute astrogliosis following severe CPB and no differences in astrocyte quantities between sham, mild, and severe CPB injury by 4 weeks post-injury. They also reported that FA was not associated with astrocyte quantity in control subjects but that FA changes were positively correlated with astrocyte numbers in the acute postoperative period following CPB. Collectively, their results show that the quantity of white matter astrocytes changes acutely following CPB and that these changes can be captured with FA.

3.15. Sepsis

Papadopolous and colleagues utilized electron microscopy to explore astrocyte morphology changes in the cortex following sepsis induced by fecal peritonitis in adolescent pigs [102]. They reported that astrocyte endfeet were occasionally swollen in sham subjects. In contrast, astrocyte endfeet were frequently and intensely swollen in septic subjects. Astrocyte endfeet in septic subjects also presented ruptured perimicrovessel membranes.

The authors speculate that injury to astrocytic endfeet caused by sepsis could potentially impair astrocytic metabolic activity.

3.16. Hepatic Clamping and Hepatic Encephalopathy

Diemer and Tonnesen investigated glial changes following portocaval anastomosis (PCA) and either total or partial hepatic artery clamping/devascularization [103]. Utilizing H&E staining, they reported no astrocyte changes following PCA and total hepatic devascularization. They found increased density of astrocyte nuclei and increased astrocyte nuclei diameter in the frontal cortex and putamen following PCA and partial hepatic clamping for 30–60 min. Additionally, they found enlarged astrocytes with watery nuclei and peripheral nucleoli in the frontal cortex in the temporary hepatic clamping group, which is indicative of Alzheimer's type II astrocytes (AIIA) (Figure 4).

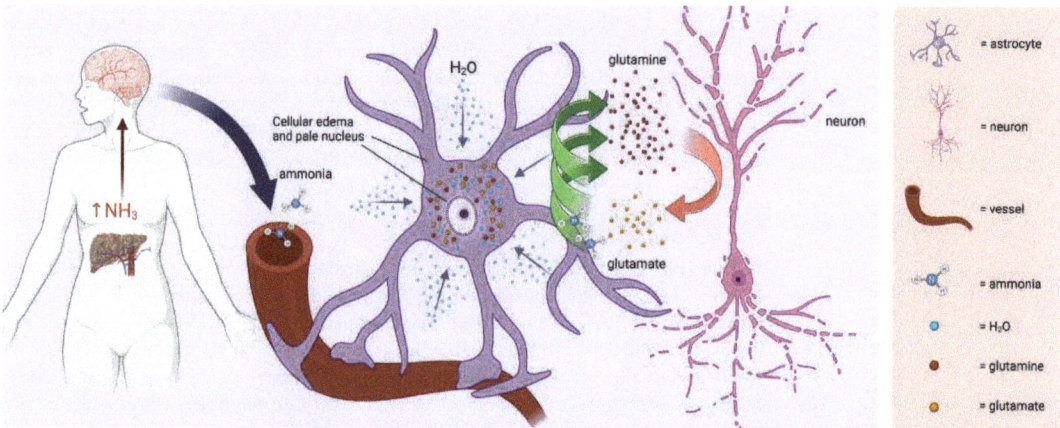

Figure 4. Alzheimer type II astrocytes (AIIAs). AIIAs are not associated with Alzheimer's disease but are similarly named due to their discovery by the same researcher. The AIIA phenotype can occur when hepatic dysfunction leads to the elevation of NH_3 in the bloodstream (blue arrow). Under normal conditions, glutamine from astrocytes is converted to glutamate in neurons, which is then released for synaptic signaling and take back up into astrocytes (pink arrow). The glutamine/osmolyte hypothesis suggests that elevated ammonia levels cause an increase in astrocytic glutamine synthesis (green arrows). The resulting increased intracellular glutamine concentration precipitates an osmotic disequilibrium, causing an influx of excess water molecules into the astrocyte. This leads to distinctive morphological changes, which include astrocytic swelling and a pronounced pale nucleus. Conversion of astrocytes to an AIIA phenotype can negatively affect neurons due to increased reactive oxygen species, metabolic insufficiency, and excitotoxicity due to loss of astrocytic glutamate clearance (created with BioRender.com; accessed on 15 August 2023).

Kristiansen and colleagues induced acute liver failure in pigs by placing an end-to-end portocaval shunt and ligating the hepatic arteries and analyzed the frontal lobe, pons, and cerebellum with electron microscopy imaging 8 h later [104]. They reported several ultrastructural astrocyte changes in these brain regions following ALF, including cytoplasmic swelling, increased electron density, condensation of the cytoplasm, clumping of nuclear chromatin, and cytoplasmic membrane dissolution. Electron microscopy imaging allowed this group to find several different areas of astrocyte injury following ALF that are indicative of necrotic cell death from this injury phenotype. Zeltser et al. created a model of diet-induced nonalcoholic fatty liver disease (NAFLD) in 13-day-old juvenile pigs by feeding them a high-fructose, high-fat (HFF) diet for 70 days [105]. They analyzed the GFAP intensity per area in the frontal cortex and reported increased astrogliosis in HFF subjects compared to subjects fed a control diet. They also demonstrated that GFAP intensity increased with the severity of liver disease, indicating that metabolic changes caused by NAFLD can lead to neurodegenerative changes, including astrogliosis.

Kanai and colleagues induced acute hepatic failure in male mini-pigs by administration of 0.05 mg/kg alpha-amanitin and 1 ug/kg lipopolysaccharide (LPS) in the splenic vein and investigated the effect of this injury on blood serum S100B levels with ELISA several hours after injury [106]. They reported that acute hepatic failure led to increased S100B serum protein levels and that S100B levels were particularly elevated in animals that died from induction of acute hepatic failure. They also reported that treatment of subjects with bioartificial liver (BAL) therapy for 4–6 h after hepatic failure induction decreased these plasma levels of S100B. This study reveals that acute hepatic failure induces astrocyte damage, as seen by elevated S100B serum levels, and that this astrocyte damage can be mitigated with BAL therapy.

The Cholich group induced hepatic encephalopathy by poisoning pigs with 5–10% *Senna occidentalis* (*S. occidentalis*) seeds [107]. Brain histology was studied within 12 h of the first observation of clinical symptoms, which occurred 7–11 days following poisoning. Utilizing H&E immunohistochemistry, they reported that hepatic encephalopathy subjects had AIIA in the cerebral cortex that exhibited nuclear pallor, chromatin margination, and swelling. GFAP immunohistochemistry revealed that white matter astrocytes in control subjects had long, branching processes that expressed GFAP, whereas white matter astrocytes in subjects treated with 10% S. occidentalis had reduced immunoreactivity to GFAP, cell shrinkage, and process retraction. Treated subjects had AIIA in gray matter regions that were GFAP and had overall reduced numbers of GFAP astrocytes and a reduced percentage of GFAP+ area in white matter compared to control. The authors speculated that these morphological and intermediate filament changes following poisoning with *S. occidentalis* may precede glial cell death resulting from poisoning.

3.17. Poisoning

Finnie and colleagues published a case report on the presence of AIIA in the brains of pigs that were subject to salt poisoning resulting from at least 2 days of water deprivation [108]. Utilizing H&E and GFAP histochemistry, they reported the presence of numerous AIIA randomly distributed in the cerebral cortical gray matter. These cells were minimal in subcortical white matter and were absent in the brains of control subjects. The morphology and appearance of these AIIA were described as swollen, clear, and watery. These cells also had scant chromatin, vacuolated nuclei, and reduced GFAP expression. They were sometimes aggregated in clusters of 2–3. The authors speculated that reduced GFAP staining in AIIA could be due to the instability of GFAP mRNA in these cells. Additionally, they hypothesized that the AIIA phenotype may be a transitionary phenotype between non-reactive astrocytes with normal GFAP expression and reactive astrocytes with augmented GFAP expression.

The Leifsson group examined changes in the astrocyte phenotype following the natural infection of pigs with the parasite *Taenia solium* (*T. solium*) [109]. They described normal astrocyte distribution in uninfected brains, with higher expression of astrocytes and astrocyte endfeet in the medulla compared to the cortex. There was an increase in astrocyte size, number, and GFAP expression in subjects infected with *T. solium*. This astrogliosis resulting from infection was more pronounced around cysticeri (larval cysts) found in the cortex compared to the medulla. They also noted that some lesions were surrounded by astrocyte endfeet forming a glial scar, while others were surrounded by blood vessels that lacked astrocyte endfeet (indicative of BBB loss). Overall, astrogliosis is prominent after *T. solium* neurocysticercosis.

Riet-Correra et al. examined the effects of ingesting 13% *Aeschynomene indica* (*A. indica*) seeds in adult pigs [110]. *A. indica* is a weed that grows abundantly in irrigated rice fields and can contaminate rice harvests. Utilizing H&E staining and transmission electron microscopy, they reported that astrocytes in the cerebellum were enlarged and swollen 24 h after ingestion, with endfoot processes often seen separating capillary endothelial cells and pericytes. They also reported astrocytosis in the cerebellum at 15 days following *A. indica* ingestion.

3.18. Intrauterine Growth Restriction

Intrauterine growth restriction (IUGR), caused by placental insufficiency, leads to fetal development in a chronic hypoxic environment that, in turn, leads to neurological disabilities. Wixey and colleagues published a report detailing the effects of IUGR on astrocyte density and morphology in postnatal-day 1 (p1) and 4 (p4) piglets [111]. They reported that GFAP+ astrocytes in the parietal white matter of normal growth (NG) brains possessed long, branching processes and small cell bodies. On the other hand, GFAP+ astrocytes in the parietal white matter of IUGR brains had morphology indicative of reactive astrocytes, including larger cell bodies and fewer, shorter, retracted processes. Analyzing GFAP+ areal density revealed that IUGR brains significantly increased GFAP+ density in intragyral white matter, subcortical white matter, and periventricular white matter compared with NG brains at both p1 and p4. Additionally, GFAP+ astrocytes co-localized with IL-1beta in periventricular white matter and with IL-18 and TNFalpha in the parietal cortex of IUGR brains. NG brains had a very minimal overlap of these inflammatory markers with GFAP+ astrocytes. Overall, these results reveal that chronic hypoxia during fetal development causes increased astrocyte density and astrocytes to possess a reactive morphology up to at least p4. This group then published two studies examining whether Ibuprofen treatment immediately after birth affected this inflammatory astrocytic morphology seen in IUGR brains [112,113]. They reported that treatment of Ibuprofen on postnatal days 1–3 alleviated the increase in GFAP+ density in intragyral white matter, subcortical white matter, and periventricular white matter that was seen in IUGR brains at p4 [112]. These results indicate that Ibuprofen treatment in the first few days after birth may hinder inflammatory mediators and thus reduce astrocyte reactivity in white matter regions. They also investigated the effects of IUGR and Ibuprofen treatment on the integrity of the neurovascular unit, including the interaction between astrocytic endfeet and blood vessels [113]. They reported that IUGR brains had significant loss of GFAP labeling around vasculature, increased hypertrophy of GFAP+ endfeet around vasculature, and decreased quantity of astrocyte endfeet interacting with the vasculature (overall loss of astrocyte coverage of vasculature). Ibuprofen treatment for three days following birth caused astrocyte interactions with vasculature to return to normal, with Ibuprofen subjects exhibiting normal astrocyte coverage of vessels and reduced endfeet hypertrophy. IUGR brains also contained GFAP+ astrocytes that took extraverted serum proteins albumin and IgG. Astrocytes co-labeling with these serum proteins tended to maintain their normal interactions with blood vessels but had very hypertrophic endfeet. Ibuprofen treatment reduced serum protein extraversion and co-labeling of astrocytes with serum proteins. IUGR brains also had increased expression of GFAP+ co-labeling with the tight junction protein claudin-1 (Cldn1) compared to NG brains, an increase that was also reduced with Ibuprofen treatment. Overall, these results indicate that the neurovascular unit is disrupted because of IUGR, including disrupted astrocyte interactions with vasculature and BBB disruption and leakage. Ibuprofen treatment decreased astrocyte reactivity and restored healthy astrocyte interactions with vasculature, helping to reduce BBB leakage. Anti-inflammatory signaling caused by Ibuprofen treatment appears to help restore BBB integrity in IUGR subjects. To our knowledge, this has not been investigated clinically.

3.19. Maternal Infection and Poisoning

Prenatal maternal infections can lead to fetal brain abnormalities. Antonson and colleagues investigated astrocyte changes in the Large White/Landrace porcine fetal hippocampus following maternal infection with porcine reproductive and respiratory syndrome virus (PRRSV) [114]. Pregnant female pigs were infected with PRRSV at gestational day (GD) 76, and fetuses were removed for analysis on GD111, 3 days before their expected delivery date. Using quantitative real-time PCR, they reported increased GFAP expression in the fetal hippocampus of PRRSV-infected mothers relative to fetuses of uninfected mothers. They also utilized GFAP immunohistochemistry to show that the relative integrated density of GFAP+ cells in the hilar region of the hippocampus was greater following maternal PRRSV infection compared to uninfected mothers. These results demonstrate upregulation of GFAP+ gliosis following late-gestation maternal PRRSV infection, indicating that maternal infection during pregnancy causes astrocyte responsiveness in developing fetuses.

Recently the Pankratova Lab published a study investigating the exposure of porcine fetuses to intra-amniotic injection of 1 mg liposaccharide (LPS), which can cause intrauterine inflammation at embryonic gestation day 103 (E103) prior to cesarean section delivery at E106 [115]. They analyzed GFAP immunoreactivity in the cortex, hippocampus, and periventricular white matter at day 1 (P1) and day 5 (P5) after birth. They reported no differences between control and LPS-treated subjects in GFAP immunoreactivity across these three brain regions at either P1 or P5. They did show an increase in GFAP immunoreactivity in the periventricular white matter from P1 to P5 in all subjects, reflecting a developmental increase in astrocytes during this time. GFAP did not increase from P1 to P5 across the cortex or hippocampus. The authors speculated that an increase in GFAP in the periventricular white matter specifically could be because astrocytes play an important role in myelination during this early time by supplying energy, trophic factors, and iron to oligodendrocytes and their precursors, among other supportive functions [116]. Additionally, their results showed that white matter astrocytes express overall higher levels of GFAP than gray matter astrocytes at both P1 and P5.

Brunse and colleagues investigated the effect of preterm birth, often associated with impaired neurodevelopment and necrotizing enterocolitis, on perivascular astrocyte coverage in the hippocampus and striatum of the porcine brain at 8 h and 5 days following birth [117]. Following the delivery of fetuses at preterm (106 days of gestation) and full-term (117 days of gestation), they conducted Western blotting and reported that GFAP protein levels in the hippocampus and striatum were similar between these groups at both 8 h and 5 days following birth. However, they noted that perivascular astrocyte coverage (determined by overlapping immunohistochemistry of GFAP and laminin) was three-fold higher in full-term compared to preterm pigs in the hippocampus. This difference was still found at 5 days following birth. Perivascular astrocyte coverage was similar between groups in the striatum. These results indicate that astrocyte endfeet coverage of the BBB is reduced in the hippocampus, but not the striatum, following preterm birth. This indicates that the hippocampus might be more vulnerable than the striatum to BBB disruption following preterm birth.

3.20. Injury to the Optic Nerve

In an earlier section, we described a publication that reported on the amino acid sequences of the glaucoma-associated genes myocilin and optineurin [49]. In this same publication, the authors examined the effects of different stressors on the level of these two genes in astrocytes isolated from the trabecular and prelaminar regions of the porcine optic nerve head. Expression of myocilin and optineurin was assessed by RT-PCR isolated from cultured cells following exposure to various stressors. Incubating porcine optic nerve head astrocytes under conditions of hydrostatic pressure (33 mg Hg above atmospheric pressure for between 12–72 h) or mechanical stretching (10% mechanical stretch over 24 h) caused no difference in myocilin or optineurin expression as analyzed via total RNA expression. Incubating porcine optic nerve head astrocytes under hypoxic conditions (7% O_2 and 5% CO_2 for 72 h) caused significantly decreased myocilin expression and no change in optineurin expression. Exposing porcine optic nerve head astrocytes to dexamethasone (500 nM added to culture media for 2 weeks) caused significantly decreased optineurin expression and significantly increased myocilin expression. The differing responses in expression levels of these two genes under different stressors indicate that myocilin and optineurin induce glaucoma via different mechanisms.

Balaratnasingam and colleagues examined astrocyte damage in the porcine optic nerve head following an acute increase in intraocular pressure (IOP) for 3, 6, 9, or 12 h [56]. There was no change in GFAP intensity following increased IOP for 3 h. After elevated IOP exposure for 6, 9, or 12 h, there was decreased GFAP expression in the pre-laminar, post-laminar, and lamina cribrosa regions. The percentage of ONH tissue that stained positive for GFAP decreased following 12 h of elevated IOP, a change that was not seen after shorter exposure times. They also reported that the architecture of astrocytes changed following increased IOP. Under normal conditions, GFAP+ astrocytes had a reticulated and skein organization across all regions of the ONH. Following increased IOP, astrocytes were disorganized and had round and ovoid changes to their structure. These changes were seen across all regions of the ONH (pre-laminar, post-laminar, and lamina cribrosa) but only in a subset of astrocytes. They observed patches of unaltered astrocytes interspersed with patches of morphologically altered astrocytes across these three regions following increased IOP. These morphological changes were seen in all groups. In subjects exposed to increased IOP for 12 h, the GFAP morphology changed drastically to elicit an amorphous appearance, having lost the structure of fine processes and gained nodular enlargements. In subjects exposed to elevated IOP for 3 h, morphological changes occurred in the lamina cribrosa. In subjects exposed to increased IOP for 6, 9, and 12 h, morphological changes were seen in the pre-laminar, post-laminar, and lamina cribrosa regions. The authors hypothesized that these morphological changes are due to astrocyte swelling and likely result from cell injury rather than cell reactivity.

This group also reported astrocyte changes following argon laser-induced axotomy of the porcine retinal ganglion cell axon [55]. A reduction in the intensity of GFAP staining was observed at the axotomized region as well as up to 2400 microns on the peripheral side of the axotomy location. Authors believe that reduced GFAP staining at the axotomized region is indicative of glial cell injury rather than reactive astrocytosis. They further hypothesize that reduced GFAP staining on the peripheral side of axotomy could be due to the enhanced coupling capacity of astrocytes by gap junction proteins or a reduced supply of neurotrophins necessary for astrocyte survival.

4. Conclusions

Historically, studies that assess astrocytes in the context of injury or disease suffer from an overreliance on GFAP and simplistic "reactive" scoring, and many of the above pig studies suffer from this oversimplification. Astrocyte reactivity is actually quite heterogeneous, and a great deal can be learned from a closer examination of expression patterns, morphological changes, and other properties [118,119]. However, we also reviewed several studies that went beyond GFAP to offer more insight into astrocyte biology. As we continue to pursue astrocyte research in pigs and other species, we must strive to collect and analyze meaningful data without depending on oversimplified GFAP scoring. These studies will be vital to therapeutic development in the field of neurotrauma. We have described the many reasons that astrocytes represent a very attractive therapeutic target for neuroprotection and recovery, provided the essential roles they play in neuronal survival as well as brain blood flow, metabolism, edema, and other functions that are affected by neurotrauma. Pigs are an essential translational bridge between rodents and humans in neurotrauma research. With the larger, more developed pig brain, we are able to recreate mechanisms and manifestations of human neurotrauma that are not possible in small animal models. Beyond size and anatomy, the brain cells of pigs also appear to be more similar to humans relative to rodents, further emphasizing the need for pig research [12]. We hope that in addition to providing a roadmap for previous pig astrocyte research and resources, this review will inspire researchers to expand the study of astrocytes in pigs in pursuit of a functioning translational neurotrauma pipeline.

Supplementary Materials: The following supporting information can be downloaded at: https://www.mdpi.com/article/10.3390/biomedicines11092388/s1, Table S1: Commercially Available Antibodies from the Reviewed Literature that have Reported Pig Reactivity; Table S2: Studies Including Analysis of Astrocytes under Pathological Conditions in Pigs.

Author Contributions: Conceptualization, J.C.O.; Methodology, E.M.P. and J.C.O.; Investigation, E.M.P.; Writing—Original Draft Preparation, E.M.P. and J.C.O.; Table Preparation, A.P. and N.F.; Figure Preparation, E.M.P., D.H. and J.C.O.; Writing—Review and Editing, E.M.P., A.P., N.F., D.H. and J.C.O.; Supervision, J.C.O.; Project Administration, J.C.O.; Funding Acquisition, E.M.P. and J.C.O. All authors have read and agreed to the published version of the manuscript.

Funding: Financial support was provided by the Department of Veterans Affairs [RR&D IK2-RX003376 (O'Donnell)] and the National Science Foundation Graduate Research Fellowship Program [DGE-1845298 (Purvis)]. Opinions, interpretations, conclusions, and recommendations are those of the authors and are not necessarily endorsed by the Department of Veterans Affairs or the University of Pennsylvania.

Institutional Review Board Statement: Not Applicable.

Informed Consent Statement: Not applicable.

Data Availability Statement: Not applicable.

Acknowledgments: We would like to thank D. Kacy Cullen for his support during the drafting of this review.

Conflicts of Interest: The authors declare no conflict of interest.

References

1. Nguyen, H.; Zerimech, S.; Baltan, S. Astrocyte Mitochondria in White-Matter Injury. *Neurochem. Res.* **2021**, *46*, 2696–2714. [CrossRef] [PubMed]
2. Chen, Y.; Swanson, R.A. Astrocytes and Brain Injury. *J. Cereb. Blood Flow Metab.* **2003**, *23*, 137–149. [CrossRef] [PubMed]
3. Rossi, D.J.; Brady, J.D.; Mohr, C. Astrocyte Metabolism and Signaling during Brain Ischemia. *Nat. Neurosci.* **2007**, *10*, 1377–1386. [CrossRef] [PubMed]
4. Barreto, G.E.; Sun, X.; Xu, L.; Giffard, R.G. Astrocyte Proliferation Following Stroke in the Mouse Depends on Distance from the Infarct. *PLoS ONE* **2011**, *6*, e27881. [CrossRef] [PubMed]
5. Lange, S.C.; Bak, L.K.; Waagepetersen, H.S.; Schousboe, A.; Norenberg, M.D. Primary Cultures of Astrocytes: Their Value in Understanding Astrocytes in Health and Disease. *Neurochem. Res.* **2012**, *37*, 2569–2588. [CrossRef] [PubMed]
6. Brambilla, L.; Martorana, F.; Rossi, D. Astrocyte Signaling and Neurodegeneration: New Insights into CNS Disorders. *Prion* **2013**, *7*, 28–36. [CrossRef]
7. Stary, C.M.; Giffard, R.G. Advances in Astrocyte-Targeted Approaches for Stroke Therapy: An Emerging Role for Mitochondria and MicroRNAS. *Neurochem. Res.* **2015**, *40*, 301–307. [CrossRef]
8. Sheldon, A.L.; Robinson, M.B. The Role of Glutamate Transporters in Neurodegenerative Diseases and Potential Opportunities for Intervention. *Neurochem. Int.* **2007**, *51*, 333–355. [CrossRef]
9. O'Donnell, J.C.; Swanson, R.L.; Wofford, K.L.; Grovola, M.R.; Purvis, E.M.; Petrov, D.; Cullen, D.K. Emerging Approaches for Regenerative Rehabilitation Following Traumatic Brain Injury. In *Regenerative Rehabilitation: From Basic Science to the Clinic*; Greising, S.M., Call, J.A., Eds.; Physiology in Health and Disease; Springer International Publishing: Cham, Switzerland, 2022; pp. 409–459, ISBN 978-3-030-95884-8.
10. Lee, H.-G.; Wheeler, M.A.; Quintana, F.J. Function and Therapeutic Value of Astrocytes in Neurological Diseases. *Nat. Rev. Drug Discov.* **2022**, *21*, 339–358. [CrossRef]
11. Brandebura, A.N.; Paumier, A.; Onur, T.S.; Allen, N.J. Astrocyte Contribution to Dysfunction, Risk and Progression in Neurodegenerative Disorders. *Nat. Rev. Neurosci.* **2023**, *24*, 23–39. [CrossRef]
12. Sjöstedt, E.; Zhong, W.; Fagerberg, L.; Karlsson, M.; Mitsios, N.; Adori, C.; Oksvold, P.; Edfors, F.; Limiszewska, A.; Hikmet, F.; et al. An Atlas of the Protein-Coding Genes in the Human, Pig, and Mouse Brain. *Science* **2020**, *367*, eaay5947. [CrossRef]
13. Williams, S.M.; Sullivan, R.K.P.; Scott, H.L.; Finkelstein, D.I.; Colditz, P.B.; Lingwood, B.E.; Dodd, P.R.; Pow, D.V. Glial Glutamate Transporter Expression Patterns in Brains from Multiple Mammalian Species. *Glia* **2005**, *49*, 520–541. [CrossRef] [PubMed]
14. Cullen, D.K.; Harris, J.P.; Browne, K.D.; Wolf, J.A.; Duda, J.E.; Meaney, D.F.; Margulies, S.S.; Smith, D.H. A Porcine Model of Traumatic Brain Injury via Head Rotational Acceleration. In *Injury Models of the Central Nervous System: Methods and Protocols*; Humana Press: Clifton, NJ, USA, 2016; Volume 1462, pp. 289–324. [CrossRef]
15. Bailey, E.L.; McCulloch, J.; Sudlow, C.; Wardlaw, J.M. Potential Animal Models of Lacunar Stroke: A Systematic Review. *Stroke J. Cereb. Circ.* **2009**, *40*, e451–e458. [CrossRef] [PubMed]

16. Howells, D.W.; Porritt, M.J.; Rewell, S.S.J.; O'Collins, V.; Sena, E.S.; van der Worp, H.B.; Traystman, R.J.; Macleod, M.R. Different Strokes for Different Folks: The Rich Diversity of Animal Models of Focal Cerebral Ischemia. *J. Cereb. Blood Flow Metab.* **2010**, *30*, 1412–1431. [CrossRef] [PubMed]
17. Zhang, K.; Sejnowski, T.J. A Universal Scaling Law between Gray Matter and White Matter of Cerebral Cortex. *Proc. Natl. Acad. Sci. USA* **2000**, *97*, 5621–5626. [CrossRef]
18. Lind, N.M.; Moustgaard, A.; Jelsing, J.; Vajta, G.; Cumming, P.; Hansen, A.K. The Use of Pigs in Neuroscience: Modeling Brain Disorders. *Neurosci. Biobehav. Rev.* **2007**, *31*, 728–751. [CrossRef] [PubMed]
19. Ostergaard, K.; Holm, I.E.; Zimmer, J. Tyrosine Hydroxylase and Acetylcholinesterase in the Domestic Pig Mesencephalon: An Immunocytochemical and Histochemical Study. *J. Comp. Neurol.* **1992**, *322*, 149–166. [CrossRef]
20. Wagner, K.R.; Xi, G.; Hua, Y.; Kleinholz, M.; de Courten-Myers, G.M.; Myers, R.E.; Broderick, J.P.; Brott, T.G. Lobar Intracerebral Hemorrhage Model in Pigs: Rapid Edema Development in Perihematomal White Matter. *Stroke* **1996**, *27*, 490–497. [CrossRef]
21. Hartings, J.A.; York, J.; Carroll, C.P.; Hinzman, J.M.; Mahoney, E.; Krueger, B.; Winkler, M.K.L.; Major, S.; Horst, V.; Jahnke, P.; et al. Subarachnoid Blood Acutely Induces Spreading Depolarizations and Early Cortical Infarction. *Brain J. Neurol.* **2017**, *140*, 2673–2690. [CrossRef]
22. Ma, Q.; Khatibi, N.H.; Chen, H.; Tang, J.; Zhang, J.H. History of Preclinical Models of Intracerebral Hemorrhage. *Acta Neurochir. Suppl.* **2011**, *111*, 3–8. [CrossRef]
23. O'Donnell, J.C.; Browne, K.D.; Kilbaugh, T.J.; Chen, H.I.; Whyte, J.; Cullen, D.K. Challenges and Demand for Modeling Disorders of Consciousness Following Traumatic Brain Injury. *Neurosci. Biobehav. Rev.* **2019**, *98*, 336–346. [CrossRef] [PubMed]
24. O'Donnell, J.C.; Browne, K.D.; Kvint, S.; Makaron, L.; Grovola, M.R.; Karandikar, S.; Kilbaugh, T.J.; Cullen, D.K.; Petrov, D. Multimodal Neuromonitoring and Neurocritical Care in Swine to Enhance Translational Relevance in Brain Trauma Research. *Biomedicines* **2023**, *11*, 1336. [CrossRef] [PubMed]
25. Jackson, J.G.; O'Donnell, J.C.; Takano, H.; Coulter, D.A.; Robinson, M.B. Neuronal Activity and Glutamate Uptake Decrease Mitochondrial Mobility in Astrocytes and Position Mitochondria Near Glutamate Transporters. *J. Neurosci.* **2014**, *34*, 1613–1624. [CrossRef] [PubMed]
26. O'Donnell, J.C.; Jackson, J.G.; Robinson, M.B. Transient Oxygen/Glucose Deprivation Causes a Delayed Loss of Mitochondria and Increases Spontaneous Calcium Signaling in Astrocytic Processes. *J. Neurosci.* **2016**, *36*, 7109–7127. [CrossRef]
27. Sullivan, S.M.; Lee, A.; Björkman, S.T.; Miller, S.M.; Sullivan, R.K.P.; Poronnik, P.; Colditz, P.B.; Pow, D.V. Cytoskeletal Anchoring of GLAST Determines Susceptibility to Brain Damage: An Identified Role for GFAP. *J. Biol. Chem.* **2007**, *282*, 29414–29423. [CrossRef] [PubMed]
28. Lee, A.; Anderson, A.R.; Beasley, S.J.; Barnett, N.L.; Poronnik, P.; Pow, D.V. A New Splice Variant of the Glutamate-Aspartate Transporter: Cloning and Immunolocalization of GLAST1c in Rat, Pig and Human Brains. *J. Chem. Neuroanat.* **2012**, *43*, 52–63. [CrossRef]
29. Blechingberg, J.; Holm, I.E.; Nielsen, K.B.; Jensen, T.H.; Jørgensen, A.L.; Nielsen, A.L. Identification and Characterization of GFAPkappa, a Novel Glial Fibrillary Acidic Protein Isoform. *Glia* **2007**, *55*, 497–507. [CrossRef]
30. Strawn, M.; Moraes, J.G.N.; Safranski, T.J.; Behura, S.K. Sexually Dimorphic Transcriptomic Changes of Developing Fetal Brain Reveal Signaling Pathways and Marker Genes of Brain Cells in Domestic Pigs. *Cells* **2021**, *10*, 2439. [CrossRef]
31. Thore, C.R.; Nam, M.; Busija, D. Phorbol Ester-Induced Prostaglandin Production in Piglet Cortical Astroglia. *Am. J. Physiol.* **1994**, *267*, R34–R37. [CrossRef]
32. Nam, M.J.; Thore, C.; Busija, D. Rapid Induction of Prostaglandin Synthesis in Piglet Astroglial Cells by Interleukin 1 Alpha. *Brain Res. Bull.* **1995**, *36*, 215–218. [CrossRef]
33. Brevig, T.; Kristensen, T.; Zimmer, J. Expression of Major Histocompatibility Complex Antigens and Induction of Human T-Lymphocyte Proliferation by Astrocytes and Macrophages from Porcine Fetal Brain. *Exp. Neurol.* **1999**, *159*, 474–483. [CrossRef]
34. Ionescu, V.A.; Villanueva, E.B.; Hashioka, S.; Bahniwal, M.; Klegeris, A. Cultured Adult Porcine Astrocytes and Microglia Express Functional Interferon-γ Receptors and Exhibit Toxicity towards SH-SY5Y Cells. *Brain Res. Bull.* **2011**, *84*, 244–251. [CrossRef] [PubMed]
35. Liu, J.; Chandaka, G.K.; Zhang, R.; Parfenova, H. Acute Antioxidant and Cytoprotective Effects of Sulforaphane in Brain Endothelial Cells and Astrocytes during Inflammation and Excitotoxicity. *Pharmacol. Res. Perspect.* **2020**, *8*, e00630. [CrossRef]
36. Li, A.; Xi, Q.; Umstot, E.S.; Bellner, L.; Schwartzman, M.L.; Jaggar, J.H.; Leffler, C.W. Astrocyte-Derived CO Is a Diffusible Messenger That Mediates Glutamate-Induced Cerebral Arteriolar Dilation by Activating Smooth Muscle Cell KCa Channels. *Circ. Res.* **2008**, *102*, 234–241. [CrossRef]
37. Xi, Q.; Umstot, E.; Zhao, G.; Narayanan, D.; Leffler, C.W.; Jaggar, J.H. Glutamate Regulates Ca2+ Signals in Smooth Muscle Cells of Newborn Piglet Brain Slice Arterioles through Astrocyte- and Heme Oxygenase-Dependent Mechanisms. *Am. J. Physiol. Heart Circ. Physiol.* **2010**, *298*, H562–H569. [CrossRef]
38. Xi, Q.; Tcheranova, D.; Basuroy, S.; Parfenova, H.; Jaggar, J.H.; Leffler, C.W. Glutamate-Induced Calcium Signals Stimulate CO Production in Piglet Astrocytes. *Am. J. Physiol. Heart Circ. Physiol.* **2011**, *301*, H428–H433. [CrossRef]
39. Leffler, C.W.; Parfenova, H.; Basuroy, S.; Jaggar, J.H.; Umstot, E.S.; Fedinec, A.L. Hydrogen Sulfide and Cerebral Microvascular Tone in Newborn Pigs. *Am. J. Physiol. Heart Circ. Physiol.* **2011**, *300*, H440–H447. [CrossRef]
40. Jeliazkova-Mecheva, V.V.; Bobilya, D.J. A Porcine Astrocyte/Endothelial Cell Co-Culture Model of the Blood-Brain Barrier. *Brain Res. Brain Res. Protoc.* **2003**, *12*, 91–98. [CrossRef] [PubMed]

41. Bobilya, D.J. Isolation and Cultivation of Porcine Astrocytes. In *Astrocytes: Methods and Protocols*; Humana Press: Clifton, NJ, USA, 2012; Volume 814, pp. 127–135. [CrossRef]
42. Tanti, G.K.; Srivastava, R.; Kalluri, S.R.; Nowak, C.; Hemmer, B. Isolation, Culture and Functional Characterization of Glia and Endothelial Cells From Adult Pig Brain. *Front. Cell Neurosci.* **2019**, *13*, 333. [CrossRef] [PubMed]
43. Nielsen, S.S.E.; Siupka, P.; Georgian, A.; Preston, J.E.; Tóth, A.E.; Yusof, S.R.; Abbott, N.J.; Nielsen, M.S. Improved Method for the Establishment of an In Vitro Blood-Brain Barrier Model Based on Porcine Brain Endothelial Cells. *J. Vis. Exp.* **2017**, *127*, e56277. [CrossRef]
44. Thomsen, M.S.; Humle, N.; Hede, E.; Moos, T.; Burkhart, A.; Thomsen, L.B. The Blood-Brain Barrier Studied in Vitro across Species. *PLoS ONE* **2021**, *16*, e0236770. [CrossRef] [PubMed]
45. Thomsen, L.B.; Burkhart, A.; Moos, T. A Triple Culture Model of the Blood-Brain Barrier Using Porcine Brain Endothelial Cells, Astrocytes and Pericytes. *PLoS ONE* **2015**, *10*, e0134765. [CrossRef]
46. Brunjes, P.C.; Feldman, S.; Osterberg, S.K. The Pig Olfactory Brain: A Primer. *Chem. Senses* **2016**, *41*, 415–425. [CrossRef]
47. Rungger-Brändle, E.; Messerli, J.M.; Niemeyer, G.; Eppenberger, H.M. Confocal Microscopy and Computer-Assisted Image Reconstruction of Astrocytes in the Mammalian Retina. *Eur. J. Neurosci.* **1993**, *5*, 1093–1106. [CrossRef] [PubMed]
48. Noda, S.; Mashima, Y.; Obazawa, M.; Kubota, R.; Oguchi, Y.; Kudoh, J.; Minoshima, S.; Shimizu, N. Myocilin Expression in the Astrocytes of the Optic Nerve Head. *Biochem. Biophys. Res. Commun.* **2000**, *276*, 1129–1135. [CrossRef] [PubMed]
49. Obazawa, M.; Mashima, Y.; Sanuki, N.; Noda, S.; Kudoh, J.; Shimizu, N.; Oguchi, Y.; Tanaka, Y.; Iwata, T. Analysis of Porcine Optineurin and Myocilin Expression in Trabecular Meshwork Cells and Astrocytes from Optic Nerve Head. *Investig. Ophthalmol. Vis. Sci.* **2004**, *45*, 2652–2659. [CrossRef]
50. Ripodas, A.; de Juan, J.A.; Roldán-Pallarés, M.; Bernal, R.; Moya, J.; Chao, M.; López, A.; Fernández-Cruz, A.; Fernández-Durango, R. Localisation of Endothelin-1 MRNA Expression and Immunoreactivity in the Retina and Optic Nerve from Human and Porcine Eye. Evidence for Endothelin-1 Expression in Astrocytes. *Brain Res.* **2001**, *912*, 137–143. [CrossRef]
51. Lee, J.; Kim, H.; Lee, J.-M.; Shin, T. Immunohistochemical Localization of Heat Shock Protein 27 in the Retina of Pigs. *Neurosci. Lett.* **2006**, *406*, 227–231. [CrossRef]
52. Carreras, F.J.; Porcel, D.; Alaminos, M.; Garzón, I. Cell-Cell Adhesion in the Prelaminar Region of the Optic Nerve Head: A Possible Target for Ionic Stress. *Ophthalmic Res.* **2009**, *42*, 106–111. [CrossRef]
53. Carreras, F.J.; Porcel, D.; Muñoz-Avila, J.I. Mapping the Surface Astrocytes of the Optic Disc: A Fluid-Conducting Role of the Astrocytic Covering of the Central Vessels. *Clin. Experiment. Ophthalmol.* **2010**, *38*, 300–308. [CrossRef]
54. Carreras, F.J.; Aranda, C.J.; Porcel, D.; Rodriguez-Hurtado, F.; Martínez-Agustin, O.; Zarzuelo, A. Expression of Glucose Transporters in the Prelaminar Region of the Optic-Nerve Head of the Pig as Determined by Immunolabeling and Tissue Culture. *PLoS ONE* **2015**, *10*, e0128516. [CrossRef]
55. Balaratnasingam, C.; Morgan, W.H.; Bass, L.; Kang, M.; Cringle, S.J.; Yu, D.-Y. Axotomy-Induced Cytoskeleton Changes in Unmyelinated Mammalian Central Nervous System Axons. *Neuroscience* **2011**, *177*, 269–282. [CrossRef] [PubMed]
56. Balaratnasingam, C.; Morgan, W.H.; Bass, L.; Ye, L.; McKnight, C.; Cringle, S.J.; Yu, D.-Y. Elevated Pressure Induced Astrocyte Damage in the Optic Nerve. *Brain Res.* **2008**, *1244*, 142–154. [CrossRef] [PubMed]
57. Balaratnasingam, C.; Kang, M.H.; Yu, P.; Chan, G.; Morgan, W.H.; Cringle, S.J.; Yu, D.-Y. Comparative Quantitative Study of Astrocytes and Capillary Distribution in Optic Nerve Laminar Regions. *Exp. Eye Res.* **2014**, *121*, 11–22. [CrossRef]
58. Chan, G.; Morgan, W.H.; Yu, D.-Y.; Balaratnasingam, C. Quantitative Analysis of Astrocyte and Axonal Density Relationships: Glia to Neuron Ratio in the Optic Nerve Laminar Regions. *Exp. Eye Res.* **2020**, *198*, 108152. [CrossRef]
59. Kimball, E.C.; Quillen, S.; Pease, M.E.; Keuthan, C.; Nagalingam, A.; Zack, D.J.; Johnson, T.V.; Quigley, H.A. Aquaporin 4 Is Not Present in Normal Porcine and Human Lamina Cribrosa. *PLoS ONE* **2022**, *17*, e0268541. [CrossRef]
60. Ruiz-Ederra, J.; Hitchcock, P.F.; Vecino, E. Two Classes of Astrocytes in the Adult Human and Pig Retina in Terms of Their Expression of High Affinity NGF Receptor (TrkA). *Neurosci. Lett.* **2003**, *337*, 127–130. [CrossRef]
61. Martin, L.J.; Brambrink, A.M.; Lehmann, C.; Portera-Cailliau, C.; Koehler, R.; Rothstein, J.; Traystman, R.J. Hypoxia-Ischemia Causes Abnormalities in Glutamate Transporters and Death of Astroglia and Neurons in Newborn Striatum. *Ann. Neurol.* **1997**, *42*, 335–348. [CrossRef] [PubMed]
62. Guerguerian, A.M.; Brambrink, A.M.; Traystman, R.J.; Huganir, R.L.; Martin, L.J. Altered Expression and Phosphorylation of N-Methyl-D-Aspartate Receptors in Piglet Striatum after Hypoxia-Ischemia. *Brain Res. Mol. Brain Res.* **2002**, *104*, 66–80. [CrossRef]
63. Lee, J.K.; Wang, B.; Reyes, M.; Armstrong, J.S.; Kulikowicz, E.; Santos, P.T.; Lee, J.-H.; Koehler, R.C.; Martin, L.J. Hypothermia and Rewarming Activate a Macroglial Unfolded Protein Response Independent of Hypoxic-Ischemic Brain Injury in Neonatal Piglets. *Dev. Neurosci.* **2016**, *38*, 277–294. [CrossRef]
64. Lee, J.K.; Liu, D.; Jiang, D.; Kulikowicz, E.; Tekes, A.; Liu, P.; Qin, Q.; Koehler, R.C.; Aggarwal, M.; Zhang, J.; et al. Fractional Anisotropy from Diffusion Tensor Imaging Correlates with Acute Astrocyte and Myelin Swelling in Neonatal Swine Models of Excitotoxic and Hypoxic-Ischemic Brain Injury. *J. Comp. Neurol.* **2021**, *529*, 2750–2770. [CrossRef]
65. Sullivan, S.M.; Macnab, L.T.; Björkman, S.T.; Colditz, P.B.; Pow, D.V. GLAST1b, the Exon-9 Skipping Form of the Glutamate-Aspartate Transporter EAAT1 Is a Sensitive Marker of Neuronal Dysfunction in the Hypoxic Brain. *Neuroscience* **2007**, *149*, 434–445. [CrossRef] [PubMed]

66. Sullivan, S.M.; Björkman, S.T.; Miller, S.M.; Colditz, P.B.; Pow, D.V. Structural Remodeling of Gray Matter Astrocytes in the Neonatal Pig Brain after Hypoxia/Ischemia. *Glia* **2010**, *58*, 181–194. [CrossRef] [PubMed]
67. Sullivan, S.M.; Björkman, S.T.; Miller, S.M.; Colditz, P.B.; Pow, D.V. Morphological Changes in White Matter Astrocytes in Response to Hypoxia/Ischemia in the Neonatal Pig. *Brain Res.* **2010**, *1319*, 164–174. [CrossRef] [PubMed]
68. Lee, A.; Lingwood, B.E.; Bjorkman, S.T.; Miller, S.M.; Poronnik, P.; Barnett, N.L.; Colditz, P.; Pow, D.V. Rapid Loss of Glutamine Synthetase from Astrocytes in Response to Hypoxia: Implications for Excitotoxicity. *J. Chem. Neuroanat.* **2010**, *39*, 211–220. [CrossRef] [PubMed]
69. Sullivan, S.M.; Sullivan, R.K.P.; Miller, S.M.; Ireland, Z.; Björkman, S.T.; Pow, D.V.; Colditz, P.B. Phosphorylation of GFAP Is Associated with Injury in the Neonatal Pig Hypoxic-Ischemic Brain. *Neurochem. Res.* **2012**, *37*, 2364–2378. [CrossRef]
70. Ruzafa, N.; Rey-Santano, C.; Mielgo, V.; Pereiro, X.; Vecino, E. Effect of Hypoxia on the Retina and Superior Colliculus of Neonatal Pigs. *PLoS ONE* **2017**, *12*, e0175301. [CrossRef]
71. Zheng, Y.; Wang, X.-M. Expression Changes in Lactate and Glucose Metabolism and Associated Transporters in Basal Ganglia Following Hypoxic-Ischemic Reperfusion Injury in Piglets. *Am. J. Neuroradiol.* **2018**, *39*, 569–576. [CrossRef]
72. Parfenova, H.; Pourcyrous, M.; Fedinec, A.L.; Liu, J.; Basuroy, S.; Leffler, C.W. Astrocyte-Produced Carbon Monoxide and the Carbon Monoxide Donor CORM-A1 Protect against Cerebrovascular Dysfunction Caused by Prolonged Neonatal Asphyxia. *Am. J. Physiol. Heart Circ. Physiol.* **2018**, *315*, H978–H988. [CrossRef]
73. Spellicy, S.E.; Scheulin, K.M.; Baker, E.W.; Jurgielewicz, B.J.; Kinder, H.A.; Waters, E.S.; Grimes, J.A.; Stice, S.L.; West, F.D. Semi-Automated Cell and Tissue Analyses Reveal Regionally Specific Morphological Alterations of Immune and Neural Cells in a Porcine Middle Cerebral Artery Occlusion Model of Stroke. *Front. Cell. Neurosci.* **2020**, *14*, 600441. [CrossRef]
74. Baker, E.W.; Kinder, H.A.; Hutcheson, J.M.; Duberstein, K.J.J.; Platt, S.R.; Howerth, E.W.; West, F.D. Controlled Cortical Impact Severity Results in Graded Cellular, Tissue, and Functional Responses in a Piglet Traumatic Brain Injury Model. *J. Neurotrauma* **2019**, *36*, 61–73. [CrossRef]
75. Lafrenaye, A.D.; Mondello, S.; Wang, K.K.; Yang, Z.; Povlishock, J.T.; Gorse, K.; Walker, S.; Hayes, R.L.; Kochanek, P.M. Circulating GFAP and Iba-1 Levels Are Associated with Pathophysiological Sequelae in the Thalamus in a Pig Model of Mild TBI. *Sci. Rep.* **2020**, *10*, 13369. [CrossRef]
76. Margulies, S.S.; Thibault, L.E.; Gennarelli, T.A. Physical Model Simulations of Brain Injury in the Primate. *J. Biomech.* **1990**, *23*, 823–836. [CrossRef] [PubMed]
77. Meaney, D.F.; Margulies, S.S.; Smith, D.H. Diffuse Axonal Injury. *J. Neurosurg.* **2001**, *95*, 1108–1110. [CrossRef]
78. Meaney, D.F.; Smith, D.H.; Shreiber, D.I.; Bain, A.C.; Miller, R.T.; Ross, D.T.; Gennarelli, T.A. Biomechanical Analysis of Experimental Diffuse Axonal Injury. *J. Neurotrauma* **1995**, *12*, 689–694. [CrossRef] [PubMed]
79. Namjoshi, D.R.; Cheng, W.H.; McInnes, K.A.; Martens, K.M.; Carr, M.; Wilkinson, A.; Fan, J.; Robert, J.; Hayat, A.; Cripton, P.A.; et al. Merging Pathology with Biomechanics Using CHIMERA (Closed-Head Impact Model of Engineered Rotational Acceleration): A Novel, Surgery-Free Model of Traumatic Brain Injury. *Mol. Neurodegener.* **2014**, *9*, 55. [CrossRef] [PubMed]
80. Sauerbeck, A.D.; Fanizzi, C.; Kim, J.H.; Gangolli, M.; Bayly, P.V.; Wellington, C.L.; Brody, D.L.; Kummer, T.T. ModCHIMERA: A Novel Murine Closed-Head Model of Moderate Traumatic Brain Injury. *Sci. Rep.* **2018**, *8*, 7677. [CrossRef] [PubMed]
81. Keating, C.E.; Cullen, D.K. Mechanosensation in Traumatic Brain Injury. *Neurobiol. Dis.* **2020**, *148*, 105210. [CrossRef] [PubMed]
82. Smith, D.H.; Chen, X.H.; Pierce, J.E.; Wolf, J.A.; Trojanowski, J.Q.; Graham, D.I.; McIntosh, T.K. Progressive Atrophy and Neuron Death for One Year Following Brain Trauma in the Rat. *J. Neurotrauma* **1997**, *14*, 715–727. [CrossRef]
83. Johnson, V.E.; Weber, M.T.; Xiao, R.; Cullen, D.K.; Meaney, D.F.; Stewart, W.; Smith, D.H. Mechanical Disruption of the Blood-Brain Barrier Following Experimental Concussion. *Acta Neuropathol.* **2018**, *135*, 711–726. [CrossRef]
84. Grovola, M.R.; Paleologos, N.; Brown, D.P.; Tran, N.; Wofford, K.L.; Harris, J.P.; Browne, K.D.; Shewokis, P.A.; Wolf, J.A.; Cullen, D.K.; et al. Diverse Changes in Microglia Morphology and Axonal Pathology during the Course of 1 Year after Mild Traumatic Brain Injury in Pigs. *Brain Pathol. Zur. Switz.* **2021**, *31*, e12953. [CrossRef]
85. de Lanerolle, N.C.; Bandak, F.; Kang, D.; Li, A.Y.; Du, F.; Swauger, P.; Parks, S.; Ling, G.; Kim, J.H. Characteristics of an Explosive Blast-Induced Brain Injury in an Experimental Model. *J. Neuropathol. Exp. Neurol.* **2011**, *70*, 1046–1057. [CrossRef]
86. Goodrich, J.A.; Kim, J.H.; Situ, R.; Taylor, W.; Westmoreland, T.; Du, F.; Parks, S.; Ling, G.; Hwang, J.Y.; Rapuano, A.; et al. Neuronal and Glial Changes in the Brain Resulting from Explosive Blast in an Experimental Model. *Acta Neuropathol. Commun.* **2016**, *4*, 124. [CrossRef] [PubMed]
87. Kallakuri, S.; Desai, A.; Feng, K.; Tummala, S.; Saif, T.; Chen, C.; Zhang, L.; Cavanaugh, J.M.; King, A.I. Neuronal Injury and Glial Changes Are Hallmarks of Open Field Blast Exposure in Swine Frontal Lobe. *PLoS ONE* **2017**, *12*, e0169239. [CrossRef]
88. Ramirez de Noriega, F.; Manley, G.T.; Moscovici, S.; Itshayek, E.; Tamir, I.; Fellig, Y.; Shkara, R.A.; Rosenthal, G. A Swine Model of Intracellular Cerebral Edema—Cerebral Physiology and Intracranial Compliance. *J. Clin. Neurosci.* **2018**, *58*, 192–199. [CrossRef] [PubMed]
89. Zhou, X.; Xie, Q.; Xi, G.; Keep, R.F.; Hua, Y. Brain CD47 Expression in a Swine Model of Intracerebral Hemorrhage. *Brain Res.* **2014**, *1574*, 70–76. [CrossRef] [PubMed]
90. Vogt, N.; Herden, C.; Roeb, J.; Roderfeld, M.; Eschbach, D.; Steinfeldt, T.; Wulf, H.; Ruchholtz, S.; Uhl, E.; Schöller, K. Cerebral Alterations Following Experimental Multiple Trauma and Hemorrhagic Shock. *Shock. Augusta Ga* **2018**, *49*, 164–173. [CrossRef]
91. Leffler, C.W.; Parfenova, H.; Fedinec, A.L.; Basuroy, S.; Tcheranova, D. Contributions of Astrocytes and CO to Pial Arteriolar Dilation to Glutamate in Newborn Pigs. *Am. J. Physiol. Heart Circ. Physiol.* **2006**, *291*, H2897–H2904. [CrossRef] [PubMed]

92. Kanu, A.; Leffler, C.W. Roles of Glia Limitans Astrocytes and Carbon Monoxide in Adenosine Diphosphate-Induced Pial Arteriolar Dilation in Newborn Pigs. *Stroke* 2009, *40*, 930–935. [CrossRef]
93. Parfenova, H.; Tcheranova, D.; Basuroy, S.; Fedinec, A.L.; Liu, J.; Leffler, C.W. Functional Role of Astrocyte Glutamate Receptors and Carbon Monoxide in Cerebral Vasodilation Response to Glutamate. *Am. J. Physiol. Heart Circ. Physiol.* 2012, *302*, H2257–H2266. [CrossRef]
94. Liu, J.; Fedinec, A.L.; Leffler, C.W.; Parfenova, H. Enteral Supplements of a Carbon Monoxide Donor CORM-A1 Protect against Cerebrovascular Dysfunction Caused by Neonatal Seizures. *J. Cereb. Blood Flow Metab.* 2015, *35*, 193–199. [CrossRef]
95. Liu, J.; Pourcyrous, M.; Fedinec, A.L.; Leffler, C.W.; Parfenova, H. Preventing Harmful Effects of Epileptic Seizures on Cerebrovascular Functions in Newborn Pigs: Does Sex Matter? *Pediatr. Res.* 2017, *82*, 881–887. [CrossRef]
96. Liberski, P.P.; Sikorska, B.; Wells, G.A.H.; Hawkins, S.A.C.; Dawson, M.; Simmons, M.M. Ultrastructural Findings in Pigs Experimentally Infected with Bovine Spongiform Encephalopathy Agent. *Folia Neuropathol.* 2012, *50*, 89–98. [PubMed]
97. Bassi, T.G.; Rohrs, E.C.; Fernandez, K.C.; Ornowska, M.; Nicholas, M.; Gani, M.; Evans, D.; Reynolds, S.C. Brain Injury after 50 h of Lung-Protective Mechanical Ventilation in a Preclinical Model. *Sci. Rep.* 2021, *11*, 5105. [CrossRef]
98. Bassi, T.G.; Rohrs, E.C.; Fernandez, K.C.; Ornowska, M.; Nicholas, M.; Gani, M.; Evans, D.; Reynolds, S.C. Transvenous Diaphragm Neurostimulation Mitigates Ventilation-Associated Brain Injury. *Am. J. Respir. Crit. Care Med.* 2021, *204*, 1391–1402. [CrossRef] [PubMed]
99. Sharma, H.S.; Miclescu, A.; Wiklund, L. Cardiac Arrest-Induced Regional Blood-Brain Barrier Breakdown, Edema Formation and Brain Pathology: A Light and Electron Microscopic Study on a New Model for Neurodegeneration and Neuroprotection in Porcine Brain. *J. Neural Transm.* 2011, *118*, 87–114. [CrossRef] [PubMed]
100. Laursen, H.; Waaben, J.; Gefke, K.; Husum, B.; Andersen, L.I.; Sørensen, H.R. Brain Histology, Blood-Brain Barrier and Brain Water after Normothermic and Hypothermic Cardiopulmonary Bypass in Pigs. *Eur. J. Cardio-Thorac. Surg.* 1989, *3*, 539–543. [CrossRef]
101. Stinnett, G.R.; Lin, S.; Korotcov, A.V.; Korotcova, L.; Morton, P.D.; Ramachandra, S.D.; Pham, A.; Kumar, S.; Agematsu, K.; Zurakowski, D.; et al. Microstructural Alterations and Oligodendrocyte Dysmaturation in White Matter After Cardiopulmonary Bypass in a Juvenile Porcine Model. *J. Am. Heart Assoc.* 2017, *6*, e005997. [CrossRef]
102. Papadopoulos, M.C.; Lamb, F.J.; Moss, R.F.; Davies, D.C.; Tighe, D.; Bennett, E.D. Faecal Peritonitis Causes Oedema and Neuronal Injury in Pig Cerebral Cortex. *Clin. Sci. Lond. Engl.* 1999, *96*, 461–466. [CrossRef]
103. Diemer, N.H.; Tonnesen, K. Glial Changes in Pigs with Porto-Caval Anastomosis and Temporary or Total Hepatic Artery Clamping. *Acta Pathol. Microbiol. Scand. Sect. A Pathol.* 1977, *85*, 721–730. [CrossRef]
104. Kristiansen, R.G.; Lindal, S.; Myreng, K.; Revhaug, A.; Ytrebø, L.M.; Rose, C.F. Neuropathological Changes in the Brain of Pigs with Acute Liver Failure. *Scand. J. Gastroenterol.* 2010, *45*, 935–943. [CrossRef] [PubMed]
105. Zeltser, N.; Meyer, I.; Hernandez, G.V.; Trahan, M.J.; Fanter, R.K.; Abo-Ismail, M.; Glanz, H.; Strand, C.R.; Burrin, D.G.; La Frano, M.R.; et al. Neurodegeneration in Juvenile Iberian Pigs with Diet-Induced Nonalcoholic Fatty Liver Disease. *Am. J. Physiol. Endocrinol. Metab.* 2020, *319*, E592–E606. [CrossRef]
106. Kanai, H.; Marushima, H.; Kimura, N.; Iwaki, T.; Saito, M.; Maehashi, H.; Shimizu, K.; Muto, M.; Masaki, T.; Ohkawa, K.; et al. Extracorporeal Bioartificial Liver Using the Radial-Flow Bioreactor in Treatment of Fatal Experimental Hepatic Encephalopathy. *Artif. Organs* 2007, *31*, 148–151. [CrossRef]
107. Chileski, G.S.; García, E.N.; Lértora, J.W.; Mussart, N.; Hernández, D.R.; Cholich, L.A. Hepatic Encephalopathy in Swine Experimentally Poisoned with Senna Occidentalis Seeds: Effects on Astrocytes. *Toxicon.* 2021, *201*, 86–91. [CrossRef]
108. Finnie, J.W.; Blumbergs, P.C.; Williamson, M.M. Alzheimer Type II Astrocytes in the Brains of Pigs with Salt Poisoning (Water Deprivation/Intoxication). *Aust. Vet. J.* 2010, *88*, 405–407. [CrossRef] [PubMed]
109. Sikasunge, C.S.; Johansen, M.V.; Phiri, I.K.; Willingham, A.L.; Leifsson, P.S. The Immune Response in *Taenia solium* Neurocysticercosis in Pigs Is Associated with Astrogliosis, Axonal Degeneration and Altered Blood-Brain Barrier Permeability. *Vet. Parasitol.* 2009, *160*, 242–250. [CrossRef] [PubMed]
110. Riet-Correa, F.; Timm, C.D.; Barros, S.S.; Summers, B.A. Symmetric Focal Degeneration in the Cerebellar and Vestibular Nuclei in Swine Caused by Ingestion of Aeschynomene Indica Seeds. *Vet. Pathol.* 2003, *40*, 311–316. [CrossRef] [PubMed]
111. Wixey, J.A.; Lee, K.M.; Miller, S.M.; Goasdoue, K.; Colditz, P.B.; Tracey Bjorkman, S.; Chand, K.K. Neuropathology in Intrauterine Growth Restricted Newborn Piglets Is Associated with Glial Activation and Proinflammatory Status in the Brain. *J. Neuroinflamm.* 2019, *16*, 5. [CrossRef] [PubMed]
112. Wixey, J.A.; Sukumar, K.R.; Pretorius, R.; Lee, K.M.; Colditz, P.B.; Bjorkman, S.T.; Chand, K.K. Ibuprofen Treatment Reduces the Neuroinflammatory Response and Associated Neuronal and White Matter Impairment in the Growth Restricted Newborn. *Front. Physiol.* 2019, *10*, 541. [CrossRef]
113. Chand, K.K.; Miller, S.M.; Cowin, G.J.; Mohanty, L.; Pienaar, J.; Colditz, P.B.; Bjorkman, S.T.; Wixey, J.A. Neurovascular Unit Alterations in the Growth-Restricted Newborn Are Improved Following Ibuprofen Treatment. *Mol. Neurobiol.* 2022, *59*, 1018–1040. [CrossRef]
114. Antonson, A.M.; Balakrishnan, B.; Radlowski, E.C.; Petr, G.; Johnson, R.W. Altered Hippocampal Gene Expression and Morphology in Fetal Piglets Following Maternal Respiratory Viral Infection. *Dev. Neurosci.* 2018, *40*, 104–119. [CrossRef] [PubMed]

115. Muk, T.; Stensballe, A.; Dmytriyeva, O.; Brunse, A.; Jiang, P.-P.; Thymann, T.; Sangild, P.T.; Pankratova, S. Differential Brain and Cerebrospinal Fluid Proteomic Responses to Acute Prenatal Endotoxin Exposure. *Mol. Neurobiol.* **2022**, *59*, 2204–2218. [CrossRef] [PubMed]
116. Li, J.; Zhang, L.; Chu, Y.; Namaka, M.; Deng, B.; Kong, J.; Bi, X. Astrocytes in Oligodendrocyte Lineage Development and White Matter Pathology. *Front. Cell. Neurosci.* **2016**, *10*, 119. [CrossRef] [PubMed]
117. Brunse, A.; Abbaspour, A.; Sangild, P.T. Brain Barrier Disruption and Region-Specific Neuronal Degeneration during Necrotizing Enterocolitis in Preterm Pigs. *Dev. Neurosci.* **2018**, *40*, 198–208. [CrossRef] [PubMed]
118. Sofroniew, M.V. Astrocyte Reactivity: Subtypes, States, and Functions in CNS Innate Immunity. *Trends Immunol.* **2020**, *41*, 758–770. [CrossRef] [PubMed]
119. Escartin, C.; Galea, E.; Lakatos, A.; O'Callaghan, J.P.; Petzold, G.C.; Serrano-Pozo, A.; Steinhäuser, C.; Volterra, A.; Carmignoto, G.; Agarwal, A.; et al. Reactive Astrocyte Nomenclature, Definitions, and Future Directions. *Nat. Neurosci.* **2021**, *24*, 312–325. [CrossRef] [PubMed]

Disclaimer/Publisher's Note: The statements, opinions and data contained in all publications are solely those of the individual author(s) and contributor(s) and not of MDPI and/or the editor(s). MDPI and/or the editor(s) disclaim responsibility for any injury to people or property resulting from any ideas, methods, instructions or products referred to in the content.

Review

Porcine Models of Spinal Cord Injury

Connor A. Wathen [1,2], Yohannes G. Ghenbot [1,2], Ali K. Ozturk [1], D. Kacy Cullen [1,2,3], John C. O'Donnell [1,2] and Dmitriy Petrov [1,*]

[1] Center for Brain Injury & Repair, Department of Neurosurgery, Perelman School of Medicine, University of Pennsylvania, Philadelphia, PA 19104, USA; connor.wathen@pennmedicine.upenn.edu (C.A.W.); yohannes.ghenbot@pennmedicine.upenn.edu (Y.G.G.); ali.ozturk@pennmedicine.upenn.edu (A.K.O.); dkacy@pennmedicine.upenn.edu (D.K.C.); odj@pennmedicine.upenn.edu (J.C.O.)

[2] Center for Neurotrauma, Neurodegeneration & Restoration, Corporal Michael J. Crescenz VA Medical Center, Philadelphia, PA 19104, USA

[3] Department of Bioengineering, School of Engineering and Applied Science, University of Pennsylvania, Philadelphia, PA 19104, USA

* Correspondence: dmitriy.petrov@pennmedicine.upenn.edu; Tel.: +1-215-294-9494

Abstract: Large animal models of spinal cord injury may be useful tools in facilitating the development of translational therapies for spinal cord injury (SCI). Porcine models of SCI are of particular interest due to significant anatomic and physiologic similarities to humans. The similar size and functional organization of the porcine spinal cord, for instance, may facilitate more accurate evaluation of axonal regeneration across long distances that more closely resemble the realities of clinical SCI. Furthermore, the porcine cardiovascular system closely resembles that of humans, including at the level of the spinal cord vascular supply. These anatomic and physiologic similarities to humans not only enable more representative SCI models with the ability to accurately evaluate the translational potential of novel therapies, especially biologics, they also facilitate the collection of physiologic data to assess response to therapy in a setting similar to those used in the clinical management of SCI. This review summarizes the current landscape of porcine spinal cord injury research, including the available models, outcome measures, and the strengths, limitations, and alternatives to porcine models. As the number of investigational SCI therapies grow, porcine SCI models provide an attractive platform for the evaluation of promising treatments prior to clinical translation.

Keywords: spinal cord injury; porcine models; regeneration; translational neurotrauma; hemisection; contusion; balloon compression technique

1. Introduction

Traumatic spinal cord injury (SCI) is a devastating event that leads to varying loss of sensorimotor, autonomic, sphincter, and sexual dysfunction. SCI is a major driver of disability worldwide, afflicting over 20 million individuals [1]. Motor vehicle collisions (MVC), falls, and violence are the most common causes of SCI [2], which disproportionately affect young individuals aged 15 to 29. Injury in this age group accounts for the majority of disability adjusted life years in the SCI population [3]. In recent years, however, there has been an increasing incidence of SCI amongst older adults, potentially driven by global increases in life expectancy [4]. On average, the annual cost of injury is 14.5 billion dollars, which positively correlates with the severity of injury [2].

Spinal cord topography changes along its rostrocaudal and ventrodorsal axes, making the location of injury a major determinant of SCI subtype and severity [5]. For example, high cervical spine injuries may lead to tetraplegia, autonomic dysfunction, and loss of respiratory drive requiring lifelong mechanical ventilation, while thoracic cord injury spares respiratory and upper extremity function [6,7]. Both cervical and thoracic spine injuries are suprasacral injuries and may result in neurogenic detrusor overactivity that has disabling social consequences and increases the risk of urinary tract infections [8]. Given

the disability that results from these devastating injuries, substantial efforts have been made to mitigate secondary injury following SCI [9].

Secondary injury stems from a variety of sources including host response to mechanically damaged tissue and pathologic states of spinal cord blood flow that lead to impaired tissue metabolism [10]. Direct cellular injury that occurs from the primary insult also triggers a cascade of damaging events including glutamate-mediated excitotoxicity and oxidative stress [11]. These effects combine to create a hostile environment for recovery, which is made worse by the limited regenerative capacity of the central nervous system. Advances in first responder, hospital, and rehabilitation care have led to dramatic improvements in SCI outcomes [12]. However, meaningful recovery of the sensorimotor and autonomic function that injured patients desire have remained elusive and are an active area of preclinical investigation [9–11].

Animal models have been developed using different species and mechanisms of injury to characterize pathogenesis of SCI and test new therapies. Unfortunately, many promising preclinical treatments have failed to show efficacy in clinical trials [13]. For example, pharmacologic therapies such as high-dose methylprednisolone and GM-1 ganglioside are not recommended by current guidelines since randomized clinical trials have failed to show sufficient benefit [14]. One potential reason for translational failure is the inadequacy of widely used rodent models of SCI, which account for 72% of SCI models [13]. Porcine models of SCI account for 1.5% of animal SCI studies and are a promising alternative species that may hold translational relevance due to comparable spinal anatomy and cardiovascular physiology. In this article, we review the current state of porcine models of SCI.

2. SCI Pathophysiology

The majority of SCI results from sudden, traumatic impact on the spine causing fractured or dislocated vertebrae. The impacts on the spinal cord can be broken down into the primary injury, which occurs due to direct mechanical forces at the time of injury, and the secondary injury, which is the result of downstream effects including ischemic injury, excitotoxicity, and inflammatory damage [11].

The primary injury can be classified based on mechanism of trauma, including (1) impact plus persistent compression, (2) impact alone with only transient compression, (3) distraction, and (4) laceration/transection [15]. Impact leads to destabilization of the spinal column as a result of bony and ligamentous injuries. Bone and ligamentous injuries can cause further direct injury to the spinal cord through fragments compressing the cord or ligamentous instability causing shearing or stretching injuries. Distraction injuries occur when two adjacent vertebrae are pulled apart, and laceration and transection injuries can occur through missile injuries, severe dislocations, or sharp bone fragment dislocations. Intraparenchymal, subdural, or epidural hematomas may also form and further compound injury from bony compression [11].

The primary injury then sets off a cascade of pathophysiologic mechanisms leading to further morbidity and mortality in SCI patients. Numerous secondary injury processes are initiated following the primary insult. In addition to the direct damage to the cord parenchyma, the compressive forces on the spinal cord can compromise spinal cord perfusion leading to tissue ischemia and further cell death [16,17]. Disruption of the brain spinal cord barrier facilitates the formation of cytotoxic and vasogenic edema in addition to an influx of peripheral inflammatory cells. A complex interaction of these processes result in apoptosis, necrosis, axonal degeneration, gliotic and fibrotic scar formation, and demyelination, which contribute to the persistent neurological deficits experienced by patients with SCI [18].

3. Broad Overview of SCI Models

Development of new SCI therapeutics is dependent on effective preclinical models that can be used to develop and evaluate new treatments. The first widely publicized SCI model was developed by Alfred Allen at the University of Pennsylvania [19]. In 1911, he published the first report on a canine model of SCI via controlled weight drop on the exposed spinal cord. Since his early experiments, numerous methods have been developed to induce experimental SCI across several species [13]. Experimental methods are broadly categorized as contusion, compression, and transection injuries. Contusion models are the most frequently studied, followed by transection and compression models [13]. Although cervical SCI has the highest clinical prevalence, most SCI models employ thoracic injuries (81%), a practice that is frequently attributed to ethical and resource concerns with cervical SCI-associated morbidity [13]. Rodent models of SCI are the most frequently cited species in the published literature, accounting for 88.4% of studies (72.4% rat, 16% mouse) [13].

Large animal models of SCI are less frequently reported in the literature. The first porcine model of SCI was developed in 1996, 85 years after Allen's canine SCI report [20]. Prior to development of porcine and nonhuman primate models, small animal models of SCI using rats and mice predominated due to low cost, small size, a well-characterized genome, and established functional assessment tools. However, clear anatomic and physiologic differences between rodents and humans present major limitations when generalizing insights from the rodent SCI literature to the human SCI population. Differences in the functional organization of the rodent spinal cord, vascular supply to the spinal cord, size of the spinal column, and the well-described potential for spontaneous recovery following SCI in rodents limit translatability of these models [21]. In a review of animal SCI models by Shari-Alhoseini et al., only 1.5% of over 2000 SCI studies reported use of porcine models with another 1.5% using nonhuman primate (NHP) models [13]. In comparison, 72.4% of studies used rats, 6% mice, 2.4% rabbit, 2.3% dog, 2.2% cat, and 0.4% goat, sheep, or bovine models [13]. Despite their infrequent use in the published literature, large animal models remain important in the study of SCI and development and preclinical testing of novel treatment strategies.

Comparative studies of spinal cord structure and function across species have found a significant degree of homology between the porcine and human spinal cord and column anatomy when compared to rodents [22–24]. A more accurate recapitulation of the human spinal cord is of great translational importance. Ongoing strategies for reanimation of limbs following SCI include implanting tissue engineered axonal constructs to bypass the region of SCI [25], neuromodulation to convert cortical motor intent into action via implantable spinal cord electrodes [26], and fiberoptic monitoring of spinal cord blood flow and oxygenation for closed-loop blood pressure augmentation using epidural probes [27]. Translation of these implantable technologies is limited by scalability when using small animal models. Additionally, the lateral location of the corticospinal tract in pigs is more similar to the corticospinal tract (CST) in humans, whereas rodent CST is divided into dorsal, ventral, and lateral components [24]. An increased level of similarity in the functional organization and size of the relevant anatomic tracts is essential in determining the efficacy of new therapies [28].

4. Porcine Models of SCI

Similar to rodent models of SCI, porcine spinal cord injuries are experimentally produced by contusion, compression, selective spinal tractotomy, or transection of the spinal cord. These injuries produce varying degrees of axotomy and axonal regeneration [29]. Neuronal death, vascular damage, and connective tissue scarring are differentially impacted based on mechanism of injury, allowing experimental modeling of the varied presentations of human SCIs. Below, we discuss porcine models of SCI, with summaries provided in Table 1.

Table 1. Porcine spinal cord injury models—strengths and limitations.

	Porcine Spinal Cord Injury Models—Strengths and Limitations		
	Strengths	Limitations	Clinical Translation
Contusion only [30]	• Titratable force applied can produce graded injury that has been shown to correlate with clinical and histopathologic changes. • Cavitation, axonal loss, and syrinx formation mimic sequela of human SCI.	• Requires laminectomy, decompressing the spinal cord before injury. • Lateral cord displacement during impaction may produce variable injury patterns. • Invasive, laminectomy +/− instrumentation requires large surgical exposures and more postoperative care.	• Model for central cord where spinal cord is contused during extension from disc–osteophyte complex or compressed from tricompartmental stenosis.
Compression [31]	• Force and duration of compression are titratable that can create graded changes in clinical exam. • Cavitation, axonal loss, and syrinx formation mimic sequela of human SCI. • Procedure can be performed with minimally invasive technique using balloon compression.	• Compression without contusion has little clinical translation in traumatic spinal cord injury.	• Model for pathology that slowly develops such as degenerative stenosis.
Contusion–compression [30]	• Best mimics real-life traumatic SCI. • Titratable force applied can produce graded injury that has been shown to correlate with clinical and histopathologic changes.	• Mimics initial force against cord and time under compression prior to surgical intervention. • Lateral cord displacement during impaction may produce variable injury patterns. • Invasive, laminectomy +/− instrumentation requires large surgical exposures and more post operative care.	• Model for fracture dislocation injury and central cord with ongoing degenerative stenosis.
Transection [32]	• Highly selective injury. • Produces clinical and histologic changes similar to hemisection of the human spinal cord.	• Rare pathology. • Non-titratable injury.	• Stab wounds.
Ischemic [27,33,34]	• Duration and degree of ischemia can be titrated to grade both clinical injury and neuronal cell death on histology. • Ischemia can be induced in a minimally invasive fashion through aortic balloon occlusion. • Devices used to measure SCBF can be used in traumatic SCI applications to characterize and treat pathologic changes in SCBF to reduce secondary injury.	• Prolonged ischemic time may damage nonneural tissue such as the limbs and kidneys, producing widespread damage and increasing animal care needs.	• Iatrogenic injury during aortic surgery. • Anterior spinal artery compression from fracture fragments. • Blood pressure augmentation following spinal cord injury.
Penetrating [20]	• Useful for the study of pathophysiologic mechanisms of injury that can provide insights for penetrating injuries in humans, which disproportionately affect young healthy individuals.	• Ethical issues. • Difficult reproducibility.	• Military and civilian SCI from direct and blast-related injury.

4.1. Contusion Models

Contusion models of SCI remain the most widely utilized in porcine models. In a recent systematic review, 70% of published studies of porcine SCI used contusion-based injuries, 43% weight drop with subsequent compression, 16% weight drop alone, and 11% using modified computer-controlled impactors similar to controlled cortical impaction in traumatic brain injury [35]. Contusion-induced injuries first require a laminectomy at the level of injury to expose the dura and underlying spinal cord [30]. To secure the device platform to the animal, pedicle screws can be placed to mount the device. After ensuring appropriate positioning of the guide rail and impactor overtop the laminectomy defect, the impactor is dropped from a predetermined height to then impact the cord. After injury, the impactor may be left in place and additional weight may be added to simulate ongoing spinal cord compression after the initial injury. The severity of injury may be adjusted by changing the mass of the impactor, the height from which it is dropped, and the duration of compression after initial impact [36].

To reduce the variability in size and severity of lesions produced by the weight drop method, Zuchner et al. developed a spring-loaded impactor device equipped with a load sensor to more accurately estimate the force of impact applied during a given injury [37]. Although the authors encountered a degree of variability in the injuries produced with this device, efforts to mitigate that variability through optimizing positioning of the animals during surgery, rigidly fixating the spine to minimize dissipation of force through extension of the spine in response to the impact, and standardization of the treated levels via preoperative X-ray substantially reduced this variation [37].

Kuluz et al. also described the use of a controlled cortical impactor (CCI) device to induce SCI in piglets [38]. In this study, a 6 mm impactor tip was used to injure the spinal cord of 3–5-week-old piglets in which the average spinal cord diameter was 5.5 to 6.5 mm. Complete and incomplete injuries could be selectively and reliably obtained by varying the depth of impact and pressure generated.

4.2. Compression Models

Spinal cord compression injuries account for the next most commonly cited porcine SCI model with balloon compression models being cited in 5% of studies and surgical clip application in 6% [35]. The compression model of SCI was first introduced in the 1950s by Tarlov in a canine model [39]. Tarlov's device consisted of a hydraulic device that inflated a bulb-shaped balloon with either water or iodinated contrast that was inserted into the spinal canal after a laminectomy was performed at the site of injury. Acute or chronic compression injuries were induced by variation in the time course of balloon inflation.

Foditsch et al. described a similar method of inducing SCI in minipigs via minimally invasive techniques. A needle is introduced percutaneously into the lumbar epidural space which facilitates placement of a guidewire and serial dilations before insertion of a kyphoplasty balloon that is threaded into the thoracic epidural space before inflation. This minimally invasive, percutaneous method has the advantage of inducing less pain to the animal. The technique also avoids the need for upfront laminectomy in contusion and transection methods, which increases translatability as injury does not occur to a decompressed spinal cord in clinical practice.

In addition to balloon compression, variations of the rodent clip compression SCI have also been studied in porcine models. In the clip application model, a laminectomy is performed at the level to be injured, followed by the placement of a calibrated clip, such as an aneurysm clip, around the cord. Given the size of the porcine spinal cord, these models have been adapted with the use of specially developed devices capable of providing precisely controlled compressive forces. Injury severity can be modulated by using clips that are calibrated to deliver a different amount of force or by adjusting the length of time the clip is left in place on the spinal cord. Zurita et al. described a technique in which a durotomy is performed prior to application of two Heifetz's clips directly onto the spinal cord [40]. Following the surgical procedure, the authors reported the formation of a reproducible necrotic centromedullary lesion 2 weeks following injury. Similarly, Kowalski et al. showed reproducible injuries after epidural application of Heifetz's clips for 30 min [41]. While the clip application method is notable for its ease of implementation and consistent delivery of a calibrated force, the force applied by the clips produces a predominantly laterally directed compression, unlike the typical dorsal or ventral compressive forces generated in clinical SCI.

4.3. Transection Models

Unlike rodent models, in which hemisection and complete transection injuries are commonly utilized, such methods are reported at a significantly lower rate in porcine models [36]. The damage caused by transection of the spinal cord in animal models is representative of lacerating spinal cord injury in humans, which may account for up to 20% of SCI [42]. Similar to the weight drop method, a laminectomy is first required to expose the dura and underlying spinal cord in order to perform hemisection or complete transection of the porcine spinal cord.

Many animal SCI models produce complete paraplegia. However, given the large number of human injuries meeting criteria for incomplete SCI (American Spinal Injury Association Grades B–E), hemisection of the spinal cord offers an injury mechanism that can mimic the deficits associated with incomplete SCI [13]. Prior work in other animal models has demonstrated that hemisection of the cord causes a reliable pattern of neural damage with significant neuronal loss in the spinal cord tissue adjacent to the lesion and chronic motor disability [43]. Hemisection of the spinal cord can produce injury phenotypes similar to Brown–Séquard syndrome in human patients, but also hold the potential to induce monoparesis of the upper extremity, which maybe a useful tool for studying cervical lesions while minimizing the morbidity associated with more severe, tetraplegia-inducing complete cervical injuries [32]. Injury phenotypes for hemisection of the cervical spinal cord included severe and chronic paresis of the ipsilateral forelimb with substantial recovery of the hindlimb, consistent with the motor syndrome observed in humans with asymmetric SCI [32].

Another spinal cord transection model that has been described is a transection caudal to the last sacral spinal cord segment [44]. In this model, a spinal cord injury was induced in the sacrocaudal spinal cord of Yucatan minipigs to cause paralysis of the tail while sparing pelvic limb, rectal, and bladder function. Dorsal laminectomy of the seventh lumbar and first two sacral vertebrae is performed, and the spinal cord is then transected at the junction of the last sacral and first caudal spinal cord segment using tenotomy scissors. This spinal cord transection model has been utilized for cellular transplantation research and offers a novel method for investigating the effect of cellular transplantation on axonal regeneration and functional recovery. Transection of the sacral spinal cord provides the benefit of producing paralysis of the tail only, allowing for a more humane model than those leading to severe thoracolumbar SCIs in pigs. In addition, this reduces the practical nursing challenges associated with inducing pelvic limb and bladder paralysis in a large mammal. However, as motor impairment and bowel and bladder dysfunction are the most frequently cited drivers of disability in patients, the sacrocaudal model may prove inadequate in evaluating the efficacy of potential therapeutics targeted towards those functional domains.

In a study designed to investigate the early postinjury response of sympathetic nerve activity following high cervical injuries, Ruggiero et al. described a complete transection at the C1 level in a non-survival injury model [45]. Given the significant anesthetic and physiologic support required to maintain the animals following this type of injury, the widespread applicability of this model remains limited to studies of the hyper-acute period after injury, which may ultimately limit frequent utilization of this technique. Overall, as large animal studies are utilized as a more advanced step on the path to translation, the significant incongruencies between clinical SCI pathophysiology and that induced by cord transection models may explain why few studies have been described in the porcine SCI literature.

4.4. Ischemic Models

Spinal cord injury from ischemia may occur following fractures that compress the anterior spinal artery causing anterior cord syndrome, aortic surgery, and thromboembolic phenomena during endovascular procedures. Comparable vascular anatomy between humans and pigs have made pigs an ideal candidate to study spinal cord ischemia in the area of aortic surgery research [35].

Recently, a porcine model of pure spinal cord ischemia (i.e., no compression) was used to test SCBF monitoring devices [27]. Busch et al. performed a lumbar laminectomy with placement of a spinal cord blood flow (SCBF) monitoring probe in the epidural space that utilized diffuse correlation spectroscopy (DCS) to measure changes in SCBF. This noninvasive optical technique measures fluctuations in near infrared light (NIR) due to red blood cell (RBC) motion. Ischemic injury was produced by placing a REBOA balloon catheter into the aorta via femoral artery cannulation. Sensors on the probe were found to have a sensitivity of 0.87 and specificity of 0.91 for detecting a 25% decrement in SCBF below the level of aortic occlusion [27]. Monitoring of SCBF is also important in contusion models of SCI as spinal cord ischemia is a driver of secondary injury. For example, prior porcine models of SCI that have used laser doppler flowmetry (LDF) to measure SCBF following moderate trauma from weight drop measured a 47.5–61.1% reduction in SCBF following injury [46].

Interestingly, when Busch et al. compared DCS to LDF in the porcine model of spinal cord ischemia, reperfusion after releasing aortic occlusion measured using DCS transiently increased above baseline values, whereas LDF returned to baseline. This may reflect differences in sampling, as LDF can only measure surface microcirculation, while DCS penetrates deeper microcirculation [27]. The ability to monitor deeper tissues may be especially important in incomplete spinal cord injuries such as central cord syndrome where damage primarily occurs in the central grey matter with white matter sparing, resulting in segmental sensorimotor dysfunction without long tract injury.

Thus, ischemic models of SCI can be used to prevent ischemia during vascular procedures, investigate ideal pressors to optimize spinal cord blood flow [47], and serve as a platform for testing devices that allow targeted hyperdynamic therapy following traumatic SCI. Indeed, current clinical guidelines following traumatic SCI recommend maintaining mean arterial pressure (MAP) between 85 and 90 mm Hg for 5–7 days following SCI to mitigate secondary injury from hypoperfusion, but little is known on how this strategy affects SCBF across individuals [48].

4.5. Other Models

Penetrating SCI accounts for only ~5% of traumatic SCI. However, such injuries disproportionately affect young male patients, which leads to significant disability costs due to lost productivity and larger duration of long-term care needs. SCI resulting from gunshot wounds likely display different pathophysiologic characteristics than blunt injuries as a result of the unique processes of ballistic injury and cavitation. Given their large size, pigs provide a much more suitable model system to evaluate pathophysiologic mechanisms following penetrating SCI [20]. Few studies employing porcine models of penetrating trauma have been published, which reflects the relative rarity of this entity. Given the relative infrequency of penetrating SCI and the significant ethical concerns of such models, their larger utility remains in doubt.

5. Porcine Strains Used in SCI Models

There are several wildtype porcine strains used in medical research. While domestic farm pigs were one of the first porcine strains to be used, their large size presents significant challenges in the study of SCI, as adults frequently achieve weights of greater than 200 kg [49,50]. This has led to exploration of minipig strains as more favorable for biomedical research purposes. Minipigs offer significant advantages for biomedical research given their more manageable size and gentle disposition, which simplify the logistics of animal care while providing a more accurate representation of human spine biomechanics [51,52]. The Yucatan porcine strain is most widely used within SCI research, followed by the Gottingen mini, Vietnamese potbellied mini, and Yorkshire pigs [35].

The Yucatan minipig is a hairless and docile pig that reaches a weight of 70 kg (154 lbs) [52]. The Yucatan pig is an inbred strain, leading to less genetic variability [50]. This is important when considering genetic manipulations in SCI research and responses to new drug therapies. A genetically defined population is advantageous in animal research when compared to outbred strains that have more genetic variability [53].

In contrast, the Gottingen minipig offers an even smaller alternative, reaching up to 38 kg (83 lbs) in weight [50]. Size differences in porcine strains are important, as researchers must balance replicating human height and weight with research expenditures (e.g., housing size, food, and handling).

As the majority of porcine SCI injury models require laminectomy to expose the cord prior to contusion, transection, or compression, the effect of the laminectomy on subsequent spinal instability must be evaluated and addressed due to potential confounding effect of iatrogenic SCI secondary to mechanical instability. These size considerations make domestic farm pigs most suitable for non-survival or short-term studies, as demonstrated by Ruggerio et al [45].

As with any animal model, the use of porcine strains for biomedical research should be conducted in accordance with legal and ethical principles. In addition, researchers should be careful to follow strain-specific considerations such as acclimation periods and individual housing following surgical procedures [54].

6. Strengths of Porcine Models

Due to the similarity of porcine spinal cord morphology and physiology to humans, porcine models offer significant advantages for preclinical therapeutic trials for human disease. Pig spinal cords are similar in size, dimensions, vertebral body height, and circulatory system to humans [22,55,56]. Toossi et al. recently published a comparative study examining the anatomy of the lumbosacral spina cord in humans and domestics pigs, in addition to Sprague–Dawley rats, rhesus macaques, and cats [57]. Relative to the other species studied, there were significantly greater similarities between several metrics between humans and pigs including the length of the lumbosacral enlargement, cross-sectional area of the spinal cord, and morphology of the central gray matter. This varies across porcine strains, as minipigs have a spinal cord $\sim\frac{1}{2}$ the diameter of humans [58]. As clinical SCI can sometimes span multiple vertebral levels, the increased length of the porcine spinal cord allows for the study of axonal regeneration across large distances more commensurate with humans, an essential factor in determining the translational potential of bioengineered SCI therapies [28]. Additionally, the position of the porcine corticospinal tract is lateral to the central gray matter, similar to humans [24]. Such anatomic considerations make porcine models better suited to study the effects of implantable therapies on motor recovery compared to rodent models. With respect to sensory pathways, nociceptive neurons traverse the ventrolateral spinal cord in pigs [59]. The location of these fibers correspond with the ventrolateral positions of the anterior and lateral spinothalamic tracts in humans, which are the primary tracts responsible for conveying nociceptive input to the thalamus. Especially in porcine models of SCI, in which the larger caliber of the spinal cord can magnify the regional differences in pathology based upon the injury model employed, a more accurate arrangement of critical functional pathways is helpful in establishing the translational relevance of both the injury model itself and any therapies investigated within a given model.

The vascular anatomy of the porcine spinal cord also shares significant similarities to humans [56]. Vascular compression and damage lead to tissue ischemia and subsequent reperfusion injury, which are important components in the acute phase of SCI. In the chronic phase, subsequent angiogenesis occurs and plays a further role in the chronic remodeling that occurs after SCI. As injury patterns may vary significantly depending upon differences in vascular supply to the cord as well as the mechanism of injury employed in a given study, this homology is an important factor in the interpretation of the effects of tissue ischemia following experimental SCI. In addition to the similarities in spinal cord vascular anatomy, the circulatory system of the pig also more closely approximates the human than many other large animal models. The well-characterized cardiovascular system of the pig also makes it an attractive model in the study of SCI-induced spinal shock in the acute phase of SCI, and long-term cardiovascular dysfunction that results in the chronic phase of SCI [60].

Recovery of voluntary bladder control is cited as a high priority for spinal cord-injured patients, and pigs have comparable lower urinary tract anatomy and physiology. Anatomic studies in pigs have shown comparable features including slit-like urethral openings and similar urethral epithelial lining, although differences in smooth and striated detrusor muscle arrangement exist [61]. Importantly, humans and pigs share similar voiding activity as both species have relaxed bladders during filling stages prior to detrusor contraction. Awake urodynamic studies in Yucatan minipigs undergoing contusion–compression injury were recently demonstrated by Keung et al., which demonstrated similar pathophysiologic changes in detrusor activity following spinal cord injury [62]. Compared to healthy cohorts, spinal cord-injured pigs displayed detrusor contractions during filling stages, which mimics neurogenic detrusor overactivity seen in spinal cord-injured humans [62]. Importantly, pigs did not display recovery of spontaneous voiding weeks after injury that can be seen in rodent models.

Several studies have investigated the neurophysiologic changes following both complete and incomplete SCI in pigs [38,63–65]. Numerous outcome measures have been described in pigs including histologic; electrophysiologic; CSF sampling; microdialysis within spinal cord parenchyma; spinal cord pressure and perfusion; multiple imaging modalities including MRI, CT, and ultrasound; as well as behavioral outcomes, especially motor function scoring [35]. The wide array of available neurophysiologic outcomes and significant similarities between human and porcine SCI have thus enabled the rigorous study of both clinically accepted and investigational therapies in attempts to better understand the mechanisms through which these treatments may affect recovery after SCI. For example, Zurita et al. showed that functional locomotor recovery in pigs after transplantation of bone marrow stromal cells (BMSC) into an experimental SCI lesion was paralleled by the recovery of somatosensory-evoked potentials, reduction in lesion size upon MRI, and by the formation of tissue-containing axon bundles mixed with differentiated BMSCs expressing a variety of markers including those of both neuronal and glial lineages [63]. Hu et al. described the different neurophysiologic profiles of pigs who sustained complete versus incomplete SCI induced by balloon compression. In their study, motor-evoked potentials (MEPs), SSEPs, and novel "spine-to-spine-evoked spinal cord potentials" (SP-EPs), which were generated by direct stimulation of the spinal cord and subsequent measuring of potentials at more distal segments located both above and below the injured level [65]. They found that in pigs that received complete SCI lesions, MEPs, SSEPs, and SP-EPs were all completely diminished without recovery. In animals who received incomplete lesions, however, while MEPs and SP-Eps were significantly impaired, SSEPs remained unchanged throughout the operative procedure. Additionally, in the incomplete group, SP-EPs recovered partially in some subjects, although not back to their preinjury baseline. These findings are consistent with findings in human electrophysiologic research in which SSEPs may be preserved even in patients with significant neurologic deficits, while MEPs and D-waves (directly recorded spinal cord impulses, similar to the SP-EPs described by HU et al.) are much more sensitive and specific predictors of subsequent neurologic deficits [66].

In addition, pigs are the only large omnivore in which complex transgenic manipulations have been successfully performed [67]. Transgenic manipulation of porcine lines opens the door for a wide variety of experimental designs that require reliable reporter gene expression, immunocompromised phenotypes, or manipulation of other genes of interest.

7. Limitations of Porcine Models

One important limitation of ungulate models, such as porcine SCI models, is the inability to study upper extremity deficits and recovery, a major driver of disability and frequently cited desire for recovery by SCI patients [68]. Consideration must be made for more suitable models, particularly nonhuman primates, for studies aiming to evaluate upper extremity recovery with a high degree of translational potential.

As a result of the less frequent utilization of porcine SCI models, there is less standardization in quantification of behavioral outcomes [35]. While the Porcine Thoracic Injury Behavior Scale (PTIBS) is commonly used [30], numerous others are frequently cited, including the Porcine Neurological Motor (PNM) Score [31], Miami Porcine Walking Scale [69], Tarlov Scale [63], Individual Limb Motor Scale [32], and Quadruped Position Global Scale [32]. Comparisons across studies of porcine models of SCI are thus impaired by the variety of scoring systems described in the literature. Of the frequently used scoring systems, the PTIBS offers advantages over other available scoring systems. The 10-point scale allows for detailed hindlimb function scoring while also grouping scores into gross locomotor function—hindlimb dragging (score 1–3), stepping (score 4–6), impaired walking (score 7–9), and normal function (score 10) [30]. Additionally, the PTIBS has good inter- and intraobserver reliability.

The detailed 14-point PNM grading system evaluates movement of the tail and movements across the bilateral hip, knee, and ankle joints in the hind limbs [31]. Prior reports using this scale have not documented inter- and intrarater reliability and have required experienced scorers to complete the task. This limits generalizability of the PNM scale [31,70]. The original 5-point Tarlov scale was created in the 1980s to evaluate hindlimb function in rats following SCI and has been used across species with modifications into a short 4-point and longer 6- and 10-point scales [40,71]. Modifications to the scale make it difficult to perform direct comparisons between SCI studies.

In addition to the variation in motor outcome scales reported in the literature, evaluation of sensory function and pain are highly limited in porcine models relative to other animals [72]. Chronic neuropathic pain is an important factor negatively influencing quality of life after SCI and a thorough understanding of how any proposed therapies will modulate that pain is critical in assessing the translational potential of any new treatment. Assessment of pain in swine is limited and past porcine models of SCI have used vocalization in response to pressure applied to the back as a crude measure of allodynia. To maximize the data obtained from porcine studies of SCI, development and standardization evaluations across multiple functional domains is essential.

8. Other Large Animal Models

NHP represent the primary alternative to swine for large animal SCI investigations. One systematic review of tissue engineering approaches to SCI showed NHP models were used in 3.2% of identified studies [28]. Another, larger systematic review, evaluating all published animal models of SCI, not just those limited to tissue engineered therapies, similarly found a low frequency of NHP studies, with only 1.5% of over 2000 publications reporting their use. In comparison, 77.4% used rat models and 1.9% used porcine models. NHP models of SCI larger mirror the techniques used in swine and rats. These include contusion via weight drop or mechanical impaction, various transection methods, and compression with balloons or clips.

The benefits of NHP models include decreased genetic interspecies differences relative to humans [73], larger size and more similar neuroanatomic organization of the spinal cord relative to rodents [74], and the potential to evaluate more advanced motor behaviors including bipedal locomotion and hand dexterity [75].

However, NHPs are subject to similar barriers to adoption as other large animal models [76]. Similar to pigs, NHPs are more expensive and require specialized housing and veterinary care. Furthermore, NHPs are subject to a greater degree of ethical concerns given their higher degree of phylogenetic similarity to humans. To our knowledge, while these considerations do not preclude their use in SCI research, it is important to ensure that the selection of such models is done only after the careful exclusion of other available animal models.

Historically, canines served as one of the first SCI model systems as reported by Allen's seminal work detailing his device for facilitating contusive SCI via weight drop [19]. Tarlov also developed the balloon compression model in canines [39]. More recently, Fukuda et al. refined this by detailing methods for a laminectomy-free balloon compression injury in mixed breed dogs by insertion of a balloon catheter through the intervertebral foramen [77]. As with pigs, the larger size of canines provide a more direct comparison to human SCI and facilitates more detailed histologic examinations. In addition, canines are more docile, which facilitates examination of neurologic function. Consequently, canine models have been cited more frequently in the SCI literature relative to porcine models: 2.2% vs. 1.5% of published studies, respectively.

9. Conclusions

While large animal models have been used since Allen published the first report of experimental SCI in canines, they have largely been supplanted by rodent models. However, a renewed interest in large animal models has been growing as a potential tool to further validate new SCI therapeutics in light of prior translational failures. Perhaps the only translational success in animal research following spinal cord injury is hyperdynamic therapy to improve spinal cord perfusion, as reduced spinal cord blow flow has been demonstrated response to injury across species [78,79].

Porcine models of SCI are of particular interest due to significant anatomic and physiologic similarities to humans. The similar size and functional organization of the porcine spinal cord, for instance, may facilitate more accurate evaluation of axonal regeneration across long distances that more closely resemble the realities of clinical SCI. Furthermore, the porcine cardiovascular system closely resembles that of humans, including at the level of the spinal cord vascular supply. These anatomic and physiologic similarities to humans not only enable more representative SCI models with the ability to accurately evaluate the translational potential of novel therapies—especially biomaterial and cell-based therapies—they also facilitate the collection of physiologic data to assess response to therapy in a setting similar to those used in the clinical management of SCI. The collection of such data can significantly aid in both translation of novel therapies, but also may provide insights into optimizing clinical treatment strategies. Furthermore, relative to NHP models of SCI, the primary large animal model alternative used to test promising small molecule and immune therapies targeting secondary injury, porcine models are less expensive and have significantly less ethical objections [8].

However, the benefits of large animal models with respect to improving translational success remain theoretical. No empirical data exist to support the claim that therapies that prove efficacious in swine are more likely to show benefit in clinical trials relative to therapies evaluated in rodent models alone. While extrapolations may be made from data in other fields where clinical trial outcomes demonstrated a significantly higher degree of concordance with data derived from porcine studies relative to rodent studies, the highly complex and heterogeneous pathophysiologic mechanisms that occur after SCI may present a larger hurdle to overcome. Given this uncertainty, careful consideration must be taken in determining the role of these models in the development pipeline for SCI therapeutics. The situations in which it is appropriate to employ porcine models of SCI must also be clarified. Due to their increased associated costs, reserving these models for the evaluation of therapies that have shown efficacy in lower-order animal models is prudent. While there may be specific circumstances unique to a given study that may drive the selection of one strain, method of injury, or outcome measure over another, efforts to standardize these measures should be undertaken to facilitate cross-study comparisons. Such efforts have the potential to significantly increase the value and utility of investigations utilizing porcine models.

Author Contributions: Writing—original draft preparation, C.A.W.; writing—review and editing, Y.G.G., A.K.O., D.K.C., J.C.O. and D.P. All authors have read and agreed to the published version of the manuscript.

Funding: Financial support was provided by the National Institutes of Health (T32-NS043126 (Ghenbot); R01-NS117757 (Cullen)), the Department of Veterans Affairs (RR&D IK2-RX003376 (O'Donnell); BLR&D I01-BX003748 (Cullen)), the University Research Foundation at the University of Pennsylvania (Ozturk), and the Department of Neurosurgery, Perelman School of Medicine, University of Pennsylvania (Wathen, Petrov). Opinions, interpretations, conclusions, and recommendations are those of the authors and are not necessarily endorsed by the National Institutes of Health, the Department of Veterans Affairs, or the University of Pennsylvania.

Institutional Review Board Statement: Not applicable.

Informed Consent Statement: Not applicable.

Data Availability Statement: Not applicable.

Conflicts of Interest: The authors declare no conflict of interest.

Abbreviations

Cortical impactor (CCI), corticospinal tract (CST), diffuse correlation spectroscopy (DCS), laser doppler flowmetry (LDF), mean arterial pressure (MAP), near-infrared light (NIR), nonhuman primate (NHP), red blood cell (RBC), spinal cord blood flow (SCBF), spinal cord injury (SCI).

References

1. Ding, W.; Hu, S.; Wang, P.M.; Kang, H.; Peng, R.; Dong, Y.; Li, F. Spinal Cord Injury: The Global Incidence, Prevalence, and Disability FROM the Global Burden of Disease Study 2019. *Spine* **2022**, *47*, 1532–1540. [CrossRef]
2. Chen, Y.; Tang, Y.; Vogel, L.; DeVivo, M. Causes of Spinal Cord Injury. *Top. Spinal Cord Inj. Rehabil.* **2013**, *19*, 1–8. [CrossRef]
3. Hall, O.T.; McGrath, R.P.; Peterson, M.D.; Chadd, E.H.; DeVivo, M.J.; Heinemann, A.W.; Kalpakjian, C.Z. The Burden of Traumatic Spinal Cord Injury in the United States: Disability-Adjusted Life Years. *Arch. Phys. Med. Rehabil.* **2018**, *100*, 95–100. [CrossRef]
4. Thompson, C.; Mutch, J.; Parent, S.; Mac-Thiong, J.-M. The changing demographics of traumatic spinal cord injury: An 11-year study of 831 patients. *J. Spinal Cord Med.* **2013**, *38*, 214–223. [CrossRef]
5. Amidei, C.B.; Salmaso, L.; Bellio, S.; Saia, M. Epidemiology of traumatic spinal cord injury: A large population-based study. *Spinal Cord* **2022**, *60*, 812–819. [CrossRef]
6. Chen, Y.; He, Y.; DeVivo, M.J. Changing Demographics and Injury Profile of New Traumatic Spinal Cord Injuries in the United States, 1972–2014. *Arch. Phys. Med. Rehabil.* **2016**, *97*, 1610–1619. [CrossRef]
7. DeVivo, M.J. Epidemiology of traumatic spinal cord injury: Trends and future implications. *Spinal Cord* **2012**, *50*, 365–372. [CrossRef]
8. Dodd, W.; Motwani, K.; Small, C.; Pierre, K.; Patel, D.; Malnik, S.; Lucke-Wold, B.; Porche, K. Spinal cord injury and neurogenic lower urinary tract dysfunction: What do we know and where are we going? *J. Men's Health* **2022**, *18*, 24. [CrossRef]
9. Fehlings, M.G.; Tetreault, L.A.; Wilson, J.R.; Kwon, B.K.; Burns, A.S.; Martin, A.R.; Hawryluk, G.; Harrop, J.S. A Clinical Practice Guideline for the Management of Acute Spinal Cord Injury: Introduction, Rationale, and Scope. *Glob. Spine J.* **2017**, *7*, 84S–94S. [CrossRef]
10. Bradbury, E.J.; Burnside, E.R. Moving beyond the glial scar for spinal cord repair. *Nat. Commun.* **2019**, *10*, 1–15. [CrossRef]
11. Anjum, A.; Yazid, M.D.; Daud, M.F.; Idris, J.; Ng, A.M.H.; Naicker, A.S.; Ismail, O.H.R.; Kumar, R.K.A.; Lokanathan, Y. Spinal Cord Injury: Pathophysiology, Multimolecular Interactions, and Underlying Recovery Mechanisms. *Int. J. Mol. Sci.* **2020**, *21*, 7533. [CrossRef]
12. Donovan, W.H. Spinal Cord Injury-Past, Present, and Future. *J. Spinal Cord Med.* **2007**, *30*, 85–100. [CrossRef]
13. Sharif-Alhoseini, M.; Khormali, M.; Rezaei, M.; Safdarian, M.; Hajighadery, A.; Khalatbari, M.M.; Meknatkhah, S.; Rezvan, M.; Chalangari, M.; Derakhshan, P.; et al. Animal models of spinal cord injury: A systematic review. *Spinal Cord* **2017**, *55*, 714–721. [CrossRef]
14. Hextrum, S.; Bennett, S. A Critical Examination of Subgroup Analyses: The National Acute Spinal Cord Injury Studies and Beyond. *Front. Neurol.* **2018**, *9*, 11. [CrossRef]
15. Alizadeh, A.; Dyck, S.M.; Karimi-Abdolrezaee, S. Traumatic spinal cord injury: An overview of pathophysiology, models and acute injury mechanisms. *Front. Neurol.* **2019**, *10*, 282. [CrossRef]
16. Dimitrijevic, M.R.; Danner, S.M.; Mayr, W. Neurocontrol of Movement in Humans with Spinal Cord Injury. *Artif. Organs* **2015**, *39*, 823–833. [CrossRef]

17. Turtle, J.D.; Henwood, M.K.; Strain, M.M.; Huang, Y.-J.; Miranda, R.C.; Grau, J.W. Engaging pain fibers after a spinal cord injury fosters hemorrhage and expands the area of secondary injury. *Exp. Neurol.* **2018**, *311*, 115–124. [CrossRef]
18. Tran, A.P.; Warren, P.M.; Silver, J. The Biology of Regeneration Failure and Success After Spinal Cord Injury. *Physiol. Rev.* **2018**, *98*, 881–917. [CrossRef]
19. Allen, A.R. Surgery of Experimental Lesion of Spinal Cord Equivalent to Crush Injury of Fracture Dislocation of Spinal Column. *J. Am. Med. Assoc.* **1911**, *LVII*, 878–880. [CrossRef]
20. Wang, D.; Wang, Z.; Yin, X.; Li, Y.; Nu, Z.; Wang, X.; Wang, B.; Yang, Y.; Hu, W. Histologic and Ultrastructural Changes of the Spinal Cord after High Velocity Missile Injury to the Back. *J. Trauma Acute Care Surg.* **1996**, *40* (Suppl. S3), 90S–93S. [CrossRef]
21. Kjell, J.; Olson, L. Rat models of spinal cord injury: From pathology to potential therapies. *Dis. Model. Mech.* **2016**, *9*, 1125–1137. [CrossRef]
22. Busscher, I.; Ploegmakers, J.J.W.; Verkerke, G.J.; Veldhuizen, A.G. Comparative anatomical dimensions of the complete human and porcine spine. *Eur. Spine J.* **2010**, *19*, 1104–1114. [CrossRef]
23. Schomberg, D.T.; Miranpuri, G.S.; Chopra, A.; Patel, K.; Meudt, J.J.; Tellez, A.; Resnick, D.K.; Shanmuganayagam, D.; Kwon, B.K.; Streijger, F.; et al. Translational Relevance of Swine Models of Spinal Cord Injury. *J. Neurotrauma* **2017**, *34*, 541–551. [CrossRef]
24. Leonard, A.V.; Menendez, J.Y.; Pat, B.M.; Hadley, M.N.; Floyd, C.L. Localization of the corticospinal tract within the porcine spinal cord: Implications for experimental modeling of traumatic spinal cord injury. *Neurosci. Lett.* **2017**, *648*, 1–7. [CrossRef]
25. Sullivan, P.Z.; AlBayar, A.; Burrell, J.C.; Browne, K.D.; Arena, J.; Johnson, V.; Smith, D.H.; Cullen, D.K.; Ozturk, A.K. Implantation of Engineered Axon Tracts to Bridge Spinal Cord Injury Beyond the Glial Scar in Rats. *Tissue Eng. Part A* **2021**, *27*, 1264–1274. [CrossRef]
26. Capogrosso, M.; Milekovic, T.; Borton, D.; Wagner, F.; Moraud, E.M.; Mignardot, J.-B.; Buse, N.; Gandar, J.; Barraud, Q.; Xing, D.; et al. A brain–spine interface alleviating gait deficits after spinal cord injury in primates. *Nature* **2016**, *539*, 284–288. [CrossRef]
27. Busch, D.R.; Lin, W.; Goh, C.C.; Gao, F.; Larson, N.; Wahl, J.; Bilfinger, T.V.; Yodh, A.G.; Floyd, T.F. Towards rapid intraoperative axial localization of spinal cord ischemia with epidural diffuse correlation monitoring. *PLoS ONE* **2021**, *16*, e0251271. [CrossRef]
28. Li, J.J.; Liu, H.; Zhu, Y.; Yan, L.; Liu, R.; Wang, G.; Wang, B.; Zhao, B. Animal Models for Treating Spinal Cord Injury Using Biomaterials-Based Tissue Engineering Strategies. *Tissue Eng. Part B Rev.* **2022**, *28*, 79–100. [CrossRef]
29. Chen, K.; Liu, J.; Assinck, P.; Bhatnagar, T.; Streijger, F.; Zhu, Q.; Dvorak, M.F.; Kwon, B.K.; Tetzlaff, W.; Oxland, T.R.; et al. Differential Histopathological and Behavioral Outcomes Eight Weeks after Rat Spinal Cord Injury by Contusion, Dislocation, and Distraction Mechanisms. *J. Neurotrauma* **2016**, *33*, 1667–1684. [CrossRef]
30. Lee, J.H.T.; Jones, C.F.; Okon, E.B.; Anderson, L.; Tigchelaar, S.; Kooner, P.; Godbey, T.; Chua, B.; Gray, G.; Hildebrandt, R.; et al. A Novel Porcine Model of Traumatic Thoracic Spinal Cord Injury. *J. Neurotrauma* **2013**, *30*, 142–159. [CrossRef]
31. Navarro, R.; Juhas, S.; Keshavarzi, S.; Juhasova, J.; Motlik, J.; Johe, K.; Marsala, S.; Scadeng, M.; Lazar, P.; Tomori, Z.; et al. Chronic Spinal Compression Model in Minipigs: A Systematic Behavioral, Qualitative, and Quantitative Neuropathological Study. *J. Neurotrauma* **2012**, *29*, 499–513. [CrossRef]
32. Del-Cerro, P.; Barriga-Martín, A.; Vara, H.; Romero-Muñoz, L.M.; Rodríguez-De-Lope, Á.; Collazos-Castro, J.E. Neuropathological and Motor Impairments after Incomplete Cervical Spinal Cord Injury in Pigs. *J. Neurotrauma* **2021**, *38*, 2956–2977. [CrossRef]
33. Papakostas, J.C.; Matsagas, M.I.; Toumpoulis, I.K.; Malamou-Mitsi, V.D.; Pappa, L.S.; Gkrepi, C.; Anagnostopoulos, C.E.; Kappas, A.M. Evolution of Spinal Cord Injury in a Porcine Model of Prolonged Aortic Occlusion. *J. Surg. Res.* **2006**, *133*, 159–166. [CrossRef]
34. Nielsen, E.W.; Miller, Y.; Brekke, O.-L.; Grond, J.; Duong, A.H.; Fure, H.; Ludviksen, J.K.; Pettersen, K.; Reubsaet, L.; Solberg, R.; et al. A Novel Porcine Model of Ischemia-Reperfusion Injury After Cross-Clamping the Thoracic Aorta Revealed Substantial Cardiopulmonary, Thromboinflammatory and Biochemical Changes without Effect of C1-Inhibitor Treatment. *Front. Immunol.* **2022**, *13*, 852119. [CrossRef]
35. Weber-Levine, C.; Hersh, A.M.; Jiang, K.; Routkevitch, D.; Tsehay, Y.; Perdomo-Pantoja, A.; Judy, B.F.; Kerensky, M.; Liu, A.; Adams, M.; et al. Porcine Model of Spinal Cord Injury: A Systematic Review. *Neurotrauma Rep.* **2022**, *3*, 352–368. [CrossRef]
36. Kim, K.-T.; Streijger, F.; Manouchehri, N.; So, K.; Shortt, K.; Okon, E.B.; Tigchelaar, S.; Cripton, P.; Kwon, B.K. Review of the UBC Porcine Model of Traumatic Spinal Cord Injury. *J. Korean Neurosurg. Soc.* **2018**, *61*, 539–547. [CrossRef]
37. Züchner, M.; Lervik, A.; Kondratskaya, E.; Bettembourg, V.; Zhang, L.; Haga, H.A.; Boulland, J.-L. Development of a Multimodal Apparatus to Generate Biomechanically Reproducible Spinal Cord Injuries in Large Animals. *Front. Neurol.* **2019**, *10*, e00223. [CrossRef]
38. Kuluz, J.; Samdani, A.; Benglis, D.; Gonzalez-Brito, M.; Solano, J.P.; Ramirez, M.A.; Luqman, A.; Santos, R.D.L.; Hutchinson, D.; Nares, M.; et al. Pediatric Spinal Cord Injury in Infant Piglets: Description of a New Large Animal Model and Review of the Literature. *J. Spinal Cord Med.* **2010**, *33*, 43–57. [CrossRef]
39. Tarlov, I.M.; Klinger, H.; Vitale, S. Spinal Cord Compression Studies: I Experimental Techniques to Produce Acute and Gradual Compression. *AMA Arch. Neurol. Psychiatry* **1953**, *70*, 813. [CrossRef]
40. Zurita, M.; Aguayo, C.; Bonilla, C.; Otero, L.; Rico, M.; Rodríguez, A.; Vaquero, J. The pig model of chronic paraplegia: A challenge for experimental studies in spinal cord injury. *Prog. Neurobiol.* **2012**, *97*, 288–303. [CrossRef]
41. Kowalski, K.E.; Kowalski, T.; DiMarco, A.F. Safety assessment of epidural wire electrodes for cough production in a chronic pig model of spinal cord injury. *J. Neurosci. Methods* **2016**, *268*, 98–105. [CrossRef]

42. Bunge, R.P.; Puckett, W.R.; Hiester, E.D. Observations on the pathology of several types of human spinal cord injury, with emphasis on the astrocyte response to penetrating injuries. *Adv. Neurol.* **1997**, *72*, 305–315.
43. Wilson, S.; Nagel, S.J.; Frizon, L.A.; Fredericks, D.C.; DeVries-Watson, N.A.; Gillies, G.T.; Howard, M.A. The Hemisection Approach in Large Animal Models of Spinal Cord Injury: Overview of Methods and Applications. *J. Investig. Surg.* **2018**, *33*, 240–251. [CrossRef]
44. Lim, J.-H.; Piedrahita, J.A.; Jackson, L.; Ghashghaei, T.; Olby, N.J.; Khatri, M.; O'brien, T.D.; Chattha, K.S.; Saif, L.J.; Duberstein, K.J.; et al. Development of a Model of Sacrocaudal Spinal Cord Injury in Cloned Yucatan MiniPigs for Cellular Transplantation Research. *Cell. Reprogramming* **2010**, *12*, 689–697. [CrossRef]
45. Ruggiero, D.; Sica, A.; Anwar, M.; Frasier, I.; Gootman, N.; Gootman, P. Induction of c-fos gene expression by spinal cord transection in Sus scrofa. *Brain Res.* **1997**, *759*, 301–305. [CrossRef]
46. Martirosyan, N.L.; Kalani, M.Y.S.; Bichard, W.D.; Baaj, A.A.; Gonzalez, L.F.; Preul, M.C.; Theodore, N. Cerebrospinal Fluid Drainage and Induced Hypertension Improve Spinal Cord Perfusion After Acute Spinal Cord Injury in Pigs. *Neurosurgery* **2015**, *76*, 461–469. [CrossRef]
47. Altaf, F.; E Griesdale, D.; Belanger, L.; Ritchie, L.; Markez, J.; Ailon, T.; Boyd, M.C.; Paquette, S.; Fisher, C.G.; Street, J.; et al. The differential effects of norepinephrine and dopamine on cerebrospinal fluid pressure and spinal cord perfusion pressure after acute human spinal cord injury. *Spinal Cord* **2016**, *55*, 33–38. [CrossRef]
48. Walters, B.C.; Hadley, M.N.; Hurlbert, R.J.; Aarabi, B.; Dhall, S.S.; Gelb, D.E.; Harrigan, M.R.; Rozelle, C.J.; Ryken, T.C.; Theodore, N. Guidelines for the Management of Acute Cervical Spine and Spinal Cord Injuries. *Neurosurgery* **2013**, *60*, 82–91. [CrossRef]
49. Swindle, M.M.; Smith, A.C.; Laber-Laird, K.; Dungan, L. Swine in Biomedical Research: Management and Models. *ILAR J.* **1994**, *36*, 1–5. [CrossRef]
50. Gutierrez, K.; Dicks, N.; Glanzner, W.G.; Agellon, L.B.; Bordignon, V. Efficacy of the porcine species in biomedical research. *Front. Genet.* **2015**, *6*, 293. [CrossRef]
51. Bollen, P.J.A.; Hansen, A.K.; Alstrup, A.K.O. *The Laboratory Swine*, 2nd ed.; CRC Press: Boca Raton, FL, USA, 2010.
52. Panepinto, L.M.; Phillips, R.W.; Wheeler, L.R.; Will, D.H. The Yucatan minature pig as a laboratory animal. *Lab. Anim. Sci.* **1978**, *28*, 308–313.
53. Festing, M.F.W. Evidence Should Trump Intuition by Preferring Inbred Strains to Outbred Stocks in Preclinical Research. *ILAR J.* **2014**, *55*, 399–404. [CrossRef]
54. Smith, A.C.; Swindle, M.M. Preparation of swine for the laboratory. *ILAR J.* **2006**, *47*, 358–363. [CrossRef]
55. Sheng, S.-R.; Wang, X.-Y.; Xu, H.-Z.; Zhu, G.-Q.; Zhou, Y.-F. Anatomy of large animal spines and its comparison to the human spine: A systematic review. *Eur. Spine J.* **2009**, *19*, 46–56. [CrossRef]
56. Strauch, J.T.; Lauten, A.; Zhang, N.; Wahlers, T.; Griepp, R.B. Anatomy of Spinal Cord Blood Supply in the Pig. *Ann. Thorac. Surg.* **2007**, *83*, 2130–2134. [CrossRef]
57. Toossi, A.; Bergin, B.; Marefatallah, M.; Parhizi, B.; Tyreman, N.; Everaert, D.G.; Rezaei, S.; Seres, P.; Gatenby, J.C.; Perlmutter, S.I.; et al. Comparative neuroanatomy of the lumbosacral spinal cord of the rat, cat, pig, monkey, and human. *Sci. Rep.* **2021**, *11*, 1–15. [CrossRef]
58. Engelke, E.C.; Post, C.; Pfarrer, C.D.; Sager, M.; Waibl, H.R. Radiographic Morphometry of the Lumbar Spine in Munich Miniature Pigs. *J. Am. Assoc. Lab. Anim. Sci.* **2016**, *55*, 336–345.
59. Breazile, J.E.; Kitchell, R.L. A study of fiber systems within the spinal cord of the domestic pig that subserve pain. *J. Comp. Neurol.* **1968**, *133*, 373–381. [CrossRef]
60. West, C.R.; Poormasjedi-Meibod, M.; Manouchehri, N.; Williams, A.M.; Erskine, E.L.; Webster, M.; Fisk, S.; Morrison, C.; Short, K.; So, K.; et al. A porcine model for studying the cardiovascular consequences of high-thoracic spinal cord injury. *J. Physiol.* **2020**, *598*, 929–942. [CrossRef]
61. Dass, N.; McMurray, G.; Greenland, J.E.; Brading, A.F. Morphological aspects of the female pig bladder neck and urethra: Quantitative analysis using computer assisted 3-dimensional reconstructions. *J. Urol.* **2001**, *165*, 1294–1299. [CrossRef]
62. Keung, M.S.; Streijger, F.; Herrity, A.; Ethridge, J.; Dougherty, S.M.; Aslan, S.; Webster, M.; Fisk, S.; Deegan, E.; Tessier-Cloutier, B.; et al. Characterization of Lower Urinary Tract Dysfunction after Thoracic Spinal Cord Injury in Yucatan Minipigs. *J. Neurotrauma* **2021**, *38*, 1306–1326. [CrossRef]
63. Zurita, M.; Vaquero, J.; Bonilla, C.; Santos, M.; De Haro, J.; Oya, S.; Aguayo, C. Functional Recovery of Chronic Paraplegic Pigs After Autologous Transplantation of Bone Marrow Stromal Cells. *Transplantation* **2008**, *86*, 845–853. [CrossRef]
64. Fadeev, F.; Eremeev, A.; Bashirov, F.; Shevchenko, R.; Izmailov, A.; Markosyan, V.; Sokolov, M.; Kalistratova, J.; Khalitova, A.; Garifulin, R.; et al. Combined Supra- and Sub-Lesional Epidural Electrical Stimulation for Restoration of the Motor Functions after Spinal Cord Injury in Mini Pigs. *Brain Sci.* **2020**, *10*, 744. [CrossRef]
65. Hu, C.-K.; Chen, M.-H.; Wang, Y.-H.; Sun, J.-S.; Wu, C.-Y. Integration of multiple prognostic predictors in a porcine spinal cord injury model: A further step closer to reality. *Front. Neurol.* **2023**, *14*, 1136267. [CrossRef]
66. Ghadirpour, R.; Nasi, D.; Iaccarino, C.; Romano, A.; Motti, L.; Sabadini, R.; Valzania, F.; Servadei, F. Intraoperative neurophysiological monitoring for intradural extramedullary spinal tumors: Predictive value and relevance of D-wave amplitude on surgical outcome during a 10-year experience. *J. Neurosurg. Spine* **2019**, *30*, 259–267. [CrossRef]

67. Aigner, B.; Renner, S.; Kessler, B.; Klymiuk, N.; Kurome, M.; Wünsch, A.; Wolf, E. Transgenic pigs as models for translational biomedical research. *J. Mol. Med.* **2010**, *88*, 653–664. [CrossRef]
68. Simpson, L.A.; Eng, J.J.; Hsieh, J.T.; Re, D.L.W.A.T.S.C.I.; Gaudet, A.D.; Fonken, L.K.; Ayala, M.T.; Dangelo, H.M.; Smith, E.J.; Bateman, E.M.; et al. The Health and Life Priorities of Individuals with Spinal Cord Injury: A Systematic Review. *J. Neurotrauma* **2012**, *29*, 1548–1555. [CrossRef]
69. Santamaria, A.J.; Benavides, F.D.; Padgett, K.R.; Guada, L.G.; Nunez-Gomez, Y.; Solano, J.P.; Guest, J.D. Dichotomous Locomotor Recoveries Are Predicted by Acute Changes in Segmental Blood Flow after Thoracic Spinal Contusion Injuries in Pigs. *J. Neurotrauma* **2019**, *36*, 1399–1415. [CrossRef]
70. Foditsch, E.E.; Miclaus, G.; Patras, I.; Hutu, I.; Roider, K.; Bauer, S.; Janetschek, G.; Aigner, L.; Zimmermann, R. A new technique for minimal invasive complete spinal cord injury in minipigs. *Acta Neurochir.* **2018**, *160*, 459–465. [CrossRef]
71. Huang, L.; Lin, X.; Tang, Y.; Yang, R.; Li, A.-H.; Ye, J.-C.; Chen, K.; Wang, P.; Shen, H.-Y. Quantitative assessment of spinal cord perfusion by using contrast-enhanced ultrasound in a porcine model with acute spinal cord contusion. *Spinal Cord* **2012**, *51*, 196–201. [CrossRef]
72. Yuan, T.; Li, J.; Shen, L.; Zhang, W.; Wang, T.; Xu, Y.; Zhu, J.; Huang, Y.; Ma, C. Assessment of Itch and Pain in Animal Models and Human Subjects. In *Translational Research in Pain and Itch*; Ma, C., Huang, Y., Eds.; Springer: Dordrecht, The Netherlands, 2016; pp. 1–22.
73. Nardone, R.; Florea, C.; Höller, Y.; Brigo, F.; Versace, V.; Lochner, P.; Golaszewski, S.; Trinka, E. Rodent, large animal and non-human primate models of spinal cord injury. *Zoology* **2017**, *123*, 101–114. [CrossRef]
74. Friedli, L.; Rosenzweig, E.S.; Barraud, Q.; Schubert, M.; Dominici, N.; Awai, L.; Nielson, J.L.; Musienko, P.; Nout-Lomas, Y.; Zhong, H.; et al. Pronounced species divergence in corticospinal tract reorganization and functional recovery after lateralized spinal cord injury favors primates. *Sci. Transl. Med.* **2015**, *7*, 302ra134. [CrossRef]
75. Courtine, G.; Bunge, M.B.; Fawcett, J.W.; Grossman, R.G.; Kaas, J.H.; Lemon, R.; Maier, I.; Martin, J.; Nudo, R.J.; Ramon-Cueto, A.; et al. Can experiments in nonhuman primates expedite the translation of treatments for spinal cord injury in humans? *Nat. Med.* **2007**, *13*, 561–566. [CrossRef]
76. Kwon, B.K.; Streijger, F.; Hill, C.E.; Anderson, A.J.; Bacon, M.; Beattie, M.S.; Blesch, A.; Bradbury, E.J.; Brown, A.; Bresnahan, J.C.; et al. Large animal and primate models of spinal cord injury for the testing of novel therapies. *Exp. Neurol.* **2015**, *269*, 154–168. [CrossRef]
77. Fukuda, S.; Nakamura, T.; Kishigami, Y.; Endo, K.; Azuma, T.; Fujikawa, T.; Tsutsumi, S.; Shimizu, Y. New canine spinal cord injury model free from laminectomy. *Brain Res. Protoc.* **2005**, *14*, 171–180. [CrossRef]
78. Tator, C.H.; Fehlings, M.G.; Yon, D.K.; Ahn, T.K.; Shin, D.E.; Kim, G.I.; Kim, M.K.; Kaye, I.D.; Vaccaro, A.R.; Pallottie, A.; et al. Review of the secondary injury theory of acute spinal cord trauma with emphasis on vascular mechanisms. *J. Neurosurg.* **1991**, *75*, 15–26. [CrossRef]
79. Hawryluk, G.; Whetstone, W.; Saigal, R.; Ferguson, A.; Talbott, J.; Bresnahan, J.; Dhall, S.; Pan, J.; Beattie, M.; Manley, G. Mean Arterial Blood Pressure Correlates with Neurological Recovery after Human Spinal Cord Injury: Analysis of High Frequency Physiologic Data. *J. Neurotrauma* **2015**, *32*, 1958–1967. [CrossRef]

Disclaimer/Publisher's Note: The statements, opinions and data contained in all publications are solely those of the individual author(s) and contributor(s) and not of MDPI and/or the editor(s). MDPI and/or the editor(s) disclaim responsibility for any injury to people or property resulting from any ideas, methods, instructions or products referred to in the content.

MDPI
St. Alban-Anlage 66
4052 Basel
Switzerland
www.mdpi.com

Biomedicines Editorial Office
E-mail: biomedicines@mdpi.com
www.mdpi.com/journal/biomedicines

Disclaimer/Publisher's Note: The statements, opinions and data contained in all publications are solely those of the individual author(s) and contributor(s) and not of MDPI and/or the editor(s). MDPI and/or the editor(s) disclaim responsibility for any injury to people or property resulting from any ideas, methods, instructions or products referred to in the content.